ANIMAL SOCIAL PSYCHOLOGY

ANIMAL SOCIAL PSYCHOLOGY

A READER OF EXPERIMENTAL STUDIES

ROBERT B. ZAJONC

JOHN WILEY & SONS, INC.

New York · London · Sydney · Toronto

Copyright © 1969 by John Wiley & Sons, Inc.

10 9 8 7 6 5 4 3 2 1

Library of Congress Catalog Card Number: 68-55338
SBN 471 98105 2
Printed in the United States of America

PREFACE

A FEW years ago I was asked by one of my colleagues to write a short text in experimental social psychology. Since I would have found it rather unexciting to deal with material that was widely and very adequately represented in existing textbooks, I decided to give attention to literature which, in my opinion, was relevant for social psychologists, but which was generally missing in standard texts. One such area was animal experimentation on social behavior.

Before undertaking a systematic search of the literature on animal social psychology I thought—probably like most social psychologists—that it consisted of a few descriptive studies, some scattered experiments poorly conceived and carelessly executed, and of a number of vague theoretical articles, all having rather tangential relevance for experimental social psychologists. My search of the literature, however, revealed a volume of material entirely surprising in scope, quality, and significance.

This vast material served as basis for a few seminars on animal social psychology held at the University of Michigan. It was largely the enthusiasms of the participants of our seminars that prompted me to put this volume together. I am grateful to Professors J. David Birch, Eugene Burnstein, Nickolas B. Cottrell, and Robyn M. Dawes, and to Philip Brickman, John DeLamater, Donald A. Dewsbury, Louise Hauenstein, Alex Heingartner, Ed Herman, Ann D. Kahalas, Stuart Karabenick, David Olton, Esther Goodman Sales, Stephen M. Sales, Stephen J. Sherman, Carol Tavris, James J. Taylor, John H. Todd, Dik VanKreveld, Bruce R. Weinert, and Robert J. Wolosin. They contributed many insights during our weekly discussion.

I also acknowledge the kind permission of both the authors and the publishers who have allowed me to reproduce the studies in this volume.

University of Michigan   *Robert B. Zajonc*

CONTENTS

PART TWO: BEHAVIORAL INTERDEPENDENCE

INTRODUCTION

THERE are today no texts of social psychology that make serious and systematic use of experimental findings in the area of animal social behavior. There are no such texts today because there are hardly any social psychologists who are engaged in animal research. The social psychology of the 1960's, and in particular experimental social psychology, studies primarily those species that can earn academic credit for participating in an experiment. The social psychologist may thus take pleasure in regarding himself as the only scientist whose laboratory work is of instant benefit to his subjects. What is even more unique about his work is that this instant (albeit modest) benefit is totally independent of the social psychologist's experimental results and of his theoretical conclusions. One is prompted to ask, however, whether these benefits compensate for the dangers of parochialism which such a practice obviously holds.

The reasons for social psychologists' neglect of animal behavior are not entirely clear, but they are to be found mainly in the scientific climate prevailing when social psychology made its great leap forward. The major developments in social psychology took place shortly after World War Two. The war and the post-war years brought about pronounced social unrest and social change throughout the world. A strong emphasis on the scientific approach to problem solving emerged. Man turned for answers to the social sciences, and among them to social psychology. And social psychology eagerly responded to the call. The studies of the authoritarian personality, of ethnocentrism, intergroup and interracial conflict, aggression, prejudice, persuasion, juvenile delinquency, political behavior, rumor, and a host of other such issues, were the expression of a mounting interest in socially significant problems. "Action research" became a key phrase. Naturally, the enormous emphasis on problems generated by social ills and formulated not by the scientist but by the military, the politician, the church, the advertising industry, the activist or the social worker, overshadowed problems generated by whatever theory there was. And, of course, it also overshadowed whatever interest there was in the social behavior of animals. Post-war social psychology took upon itself the responsibility of becoming the science of man, and preferably, of the suffering and of the underprivileged man.

1

The interest in man, and in socially significant problems, was dominated by a firm dedication to the cognitive approach. While there were some brave attempts to extend it to animal research (e.g. Krechevsky, 1932; Lawrence & Festinger, 1962; Tolman, 1932), this approach is particularly clumsy when applied to lower animals. With a commitment to social ills and with the cognitive approach as his major theoretical tool, the post-war social psychologist could not easily become involved in work with animals. But, should he become involved?

What benefits can be derived from animal research? Some of these were described by Hebb and Thompson (1954). Animal research, they said, needn't guarantee an explanation of human problems. "Proving something true of an animal does not prove it for man" (p. 532), and no one should expect it to do so. Animal research will have made an important contribution if it merely clarified a human problem, and if it merely helped to formulate it without "proving" anything at all. As an exercise, I sometimes ask students to design an experiment on some particular research question. They most often come up with one involving human subjects. I then have them redesign the experiment for animal subjects. After having gone through this process they often develop a considerably more elegant and more effective design than their previous one.

But, there is a host of other advantages. First, a given psychological effect is quite often difficult or even impossible to isolate using human subjects. This same effect, in a fairly pure form, may be quite easy to observe when animal subjects are used. Take the effects of hunger, for example. If you deprive a rat of food for 48 hours, you get a hungry rat. If you deprive a sophomore of food for 48 hours, you get a hungry sophomore, but often also one who is trying to impress the experimenter with how well he can endure suffering and who suppresses all behavior that may be construed as deriving from his inability to tolerate hunger. If the sophomore believes that the proper thing to do is to show signs of hunger, you will observe the opposite forms of behavior.

Second, results obtained on animal subjects do not tempt us, as do results obtained on human subjects, to begin looking for their explanation in our own experience—more often the source of deception than of conception. As human beings, we know *too much* about man—too much to be able to pull out from this tangle the fundamental and elementary threads that form the basis of a systematic theory. Because our personal information about man tends to fuse itself with scientific information, the essentials escape us. We have less personal knowledge about animals, but not less scientific knowledge. We have less involvement with them and can look at them dispassionately. We are less likely to confuse the scientific significance of an observed phenomenon with its personal or moral significance; we needn't be preoccupied with the moral implications of our findings. Solving the scientific questions about behavior is hard enough.

Third, in assessing psychological states of an animal, we are forced to use non-verbal measures. We cannot ask him how he feels, or what he thinks, or what he intends to do. It is, of course, possible, by means of verbal responses, to obtain some reliable measures of psychological states of human subjects. But these measures leave much to be desired, and one fears that they may soon outlive their usefulness altogether. Human subjects are becoming more sophisticated. Results of psychological research and the techniques used in this research are beginning to be widely known. Psychologists will soon have to look for techniques that are less transparent to the human subject, and invent more subtle behavioral indices. Moreover, verbal responses are not easily comparable across cultures. Findings obtained by means of noncomparable measures do not cumulate to form safe empirical generalizations. As long as the lion's share of social psychological research is done in English-speaking countries we have little to worry about, provided we earmark our empirical generalizations as applying exclusively to English-speaking populations. But there are strong indications that social psychology will soon be confronted with a large volume of research on nonEnglish-speaking subjects—some of it consistent and some probably inconsistent with what is now known. We must invent measures that are standardized and that are reliable across cultures and across languages. In all likelihood, the role of verbal behavior in psychological measurement will sharply decline. Animal studies show us how to be inventive in the construction of nonverbal measures of psychological states and variables.

Fourth, and most important, there is a host of problems which at present cannot be dealt with at all if we insist on human subjects. The study of behavior genetics, hormonal influences upon social behavior, brain and social behavior, the physiological consequences of social conditions and ecology, special environmental conditions which we would not dare to impose on human subjects, and a great many others, could not be carried out at all if they confined themselves to the college sophomore. These fields have recently made enormous strides and they have significantly advanced our understanding of behavior in general, and of social behavior in particular.

It is interesting in this respect that, in 1948, Hilgard warned experimental psychologists working in the area of learning against their overdependence on animal experimentation.

> A price is always paid for the convenience of a given approach to a problem. The price to be paid for over much experimentation with animals is to neglect the fact that human subjects are brighter, are able to use language—*and probably learn differently because of these advances over lower animals.* (Hilgard, 1948, p. 329.)

Hilgard's position raises an important question of scientific strategy for the study of learning that applies equally to social psychology. If there exist fundamental discontinuities between man and lower animals along *all* dimensions of behavior, of its antecedents, and of its consequences, then we will surely be misled in our attempts to understand human behavior, by the results of animal experimentation. Even if these discontinuities exist only for the critical basis of behavior, the results of animal experimentation will have little to offer to a learning psychologist or a social psychologist.

But the existence of such discontinuities is today more a matter of opinion and conjecture than a matter of established fact. What is then the optimal strategy to follow? To begin with, we can assume that man is thoroughly unique and divide psychology into human and animal subdisciplines, each with a unique theoretical orientation and each with unique methods, units of measurements and descriptive concepts. Because of its parochialism such a strategy has an excellent chance of failing to discover and to formulate general *behavioral* laws should they in fact be there to be found and formulated.

However, general laws are not really "found" in nature. They are *imposed* on it.

They are the products of ingenious abstractions, of rearrangements of observables and concepts, arrived at by a process of successive approximations. Perhaps Hull's $sE_R = f(D \times sH_R)$ is one in the series of such approximations which will eventually lead to a law that applies equally to all species, or better yet, to all *behavior*, independent of species.

There is another strategy which the social psychologist (and psychologists in other areas) will find more promising: for each empirical generalization and for each theoretical formulation, unless we know *explicitly* otherwise, let us assume that it applies generally to all species. Let us restrict the applicability of the empirical generalization or the theoretical formulation only if it is confronted with solid contradictory evidence. If it is possible to formulate general laws of behavior, then this strategy is more likely to promote their formulation. While the learning theorist's parochialism consisted of an over-reliance on animal studies, the social psychologist's parochialism is not to be regretted less because it consisted of an over-reliance on human studies. *General* laws of social behavior are more likely to emerge when research on the social behavior of man is carried out side-by-side with research on the social behavior of other species. A social psychology confined to man is as parochial as a chemistry confined to gold.

ORGANIZATION OF THIS BOOK

It is the purpose of this volume to widen the experimental interests of social psychology, to introduce social psychologists to the vast and exciting literature on animal social behavior, and to show that a concern with animal social behavior must become an integral part of social psychology. This book brings together a series of *experimental* studies on animal social behavior which are likely to bear upon popular theoretical formulations in social psychology.

By social psychology we shall understand the study of behavioral dependence and interdependence among individuals. The study of behavioral dependence requires that the behavior of one individual be explained and analyzed in terms of influences upon it which derive from the behavior of other individuals. The study of behavioral interdependence considers at least two individuals whose behavior is mutually and reciprocally influenced. The organization of this volume of readings follows from these requirements of the definition of social psychology. We begin with the study of behavioral dependence, and we begin in Chapter One with a classical paradigm of social psychology, namely, *coaction*. In Chapter One experiments are presented to demonstrate the effects of the mere presence of other individuals on performance and learning. These results are explained in terms of motivation principles. Chapter Two shows that the presence of other individuals may act not only as a motivational force, but that it may also provide cues as to the appropriate responses in a situation. Here several examples of studies in the area of *imitation* and *observational learning* are included. Chapter Three considers the general effects upon the behavioral makeup of the animal which derive from a continued and prolonged presence of others—namely, the effects of *socialization*. Also considered are the effects of *social deprivation*. The tendencies of animals to seek out the presence of others—*affiliative* tendencies—are studied in Chapter Four.

Chapter Five deals with the negative side of this coin—namely, with *aggression*. This concludes our considerations of behavioral dependence. The remainder of the book deals with *behavioral interdependence*. The first chapter of this part (Chapter Six) considers *communication* among animls, i.e. reciprocal influence by means of signals. It consists of experimental studies on various forms of transmission of information among animals, and on various conditions which affect it. The classical problems of *competition* and *cooperation*, i.e. reciprocal influence by means of control of reinforcement, are then examined in Chapters Seven and Eight. We conclude with the *effects* of competition. cooperation, and conflict upon the social structure through studies of *dominance* and *dominance-structure*. The organization of this volume follows the organization of a short book on experimental social psychology which I wrote three years ago (Zajonc, 1966), and which relies most heavily on evidence collected on human subjects. By comparing the experimental findings in the two volumes, the reader will discover that many empirical phenomena hold true for humans as well as for lower species.

We have confined the readings to experimental studies alone, and we have done so not because we think that they are more interesting or because they constitute more definitive evidence. The purpose of this volume is to bring evidence from work with animals to bear on research and theory in social psychology. The significance and implications for social psychology of findings obtained with animals will be seen more readily in experimental work than in observational and field studies, because the critical variables emerge with a greater clarity, and because their similarity to social psychological research with humans is greater. Observational and field studies of animal behavior have fewer parallels in social psychology and are less likely, therefore, to have an impact on it.

REFERENCES

HEBB, D. O. and THOMPSON, W. R. The social significance of animal studies. In G. Lindzey (Ed.), *Handbook of Social Psychology*. Cambridge: Addison-Wesley, 1954.

HILGARD, E. R. *Theories of Learning*. New York: Appleton–Century–Crofts, 1948.

KRECHEVSKY, I. Hypotheses in rats. *Psychol. Rev.*, 1932, **6**, 516–532.

LAWRENCE, D. H. and FESTINGER, L. *Deterrents and Reinforcement: The Psychology of Insufficient Reward*. Stanford, Calif.: Stanford University Press, 1962.

TOLMAN, E. C. *Purposive Behavior in Animals and Men*. New York: Century, 1932.

ZAJONC, R. B. *Social Psychology: An Experimental Approach*. Belmont, Calif.: Wadsworth, 1966.

PART ONE

BEHAVIORAL

DEPENDENCE

CHAPTER ONE

COACTION

THE first experiment in social psychology was carried out at the University of Indiana in 1897. It was performed by Norman Triplett and it dealt with the effects of pacing and of competition on skilled performance. Triplett was impressed with the fact that cyclists consistently obtain better times when racing in competition, or when paced by others, than when racing alone against time. The record for a 25-mile distance for an unpaced rider racing alone against time was then 1 hr. 25 min. 27.5 sec. The record for a cyclist also racing against time, but having one pacer, was 48 min. 7.5 sec. However, these times were made by different men, and Triplett correctly observed that the difference could have been due to individual factors rather than the conditions of the race. He therefore searched through the official records of the League of American Wheelmen for further evidence. Not only was his previous finding fully confirmed for other distances and for other seasons, but he also found evidence free of possible confusion with individual difference factors. According to these records, Arthur Gardiner, a prominent racer of the late 1890's, rode his fastest unpaced mile in 2 min. 3.8 sec. Paced, Gardiner's fastest mile was 1 min. 39.6 sec. Another famous cyclist of that time, Earle Kiser, also achieved a considerably faster speed when paced. His personal record for the unpaced mile was 2 min. 10 sec., and for the paced mile 1 min. 42 sec. These differences could not have been due to individual factors. But Triplett wasn't fully satisfied. The personal records of Gardiner and Kiser were obtained during different stages of these men's careers, and factors other than the very conditions of the race could have been operating to produce differences between unpaced and paced records. He turned for the answer to the best of all sources: the laboratory. His subjects, children and adults of both sexes, were required to execute 150 winds on a fishing reel. Times to complete the task were measured for subjects working alone and working pairs. Triplett found to his satisfaction that this sort of performance also improves in the presence of partners.

By definition, Triplett's problem represents one of the basic experimental concerns of social psychology: the dependence of the behavior of one individual on the behavior of another. His experiment examined the most elementary forms of this dependence. Triplett's experimental paradigm, which Allport (1924) denoted as

9

"coaction," involved the observation of performance of several individuals engaged in the same task in the presence of one another.

It is instructive to recall how Triplett explained performance enhancement obtained under conditions of coaction:

> The bodily presence of another rider is a stimulus to the racer in arousing the competitive instinct; that another can thus be the means of releasing or freeing nervous energy for him that he cannot of himself release; and, further, that the sight of movement in that other by perhaps suggesting a higher rate of speed, is also an inspiration to greater effort (p. 516).

The modern account of coaction effects differs in language, in emphasis, and in that it does not assume the existence of a "competitive instinct." It doesn't differ from Triplett's, however, in assuming that the presence of others is a source of arousal that energizes behavior (Zajonc, 1965). It goes somewhat further than Triplett's account because it assumes that the arousal brought about by the presence of others energizes *all* responses that are made salient by the stimulus situation confronting the individual at the moment. Among those, the dominant responses (i.e. those most likely to be emitted) are assumed to derive the greatest benefit from the presence of others. Thus, the presence of others is assumed to act as a source of arousal having properties and consequences of a generalized drive state, D, specified by Hull (1952) and Spence (1956).

One property of drive D, according to Hull and Spence, is that it affects behavior by combining multiplicatively with underlying habits. Thus, reaction potential, sE_R, is said to be a function of both the underlying habit strength and the generalized drive state present at the moment; specifically, $sE_R = f(D \times sH_R)$. According to the classical formulation, when more than one response is evoked by the stimulus situation, the probability that a given response will occur depends on the extent to which its underlying reaction potential, sE_R, exceeds all others. Given two responses, R_1 and R_2, and $sH_{R_1} > sH_{R_2}$, we have

$$sE_{R_1} - sE_{R_2} = f[(D \times sH_{R_1}) - (D \times sH_{R_2})] = f[D(sH_{R_1} - sH_{R_2})].$$

An increase in drive must result in an increased difference between sE_{R_1} and sE_{R_2}, and, therefore, in an increased likelihood of the emission of the dominant response.

Now, if sE_{R_1} happens to be the response required by the situation, the presence of others will enhance performance. In observing professional cyclists, Triplett dealt with *correct* dominant responses. The responses that dominated their behavior at the time of the race—riding the bicycle—were based on well-established habits and were entirely appropriate to the situation. The presence of other racers, of pacers or of spectators, by increasing drive, could only make these responses more vigorous. The race times, therefore, improved. But suppose that sE_{R_1} is a wrong response, while sE_{R_2} is the correct one. Such is often the case during the early stages of learning a new skill. For instance, the experimenter defines the correct response as turning left in a T-maze. If the rat has some preference for right turns, the wrong response will be dominant at the early stages of learning. If he happens to be blessed by the presence of another rat at that time, his learning of the T-maze will be delayed by the increased arousal. Briefly, then, the presence of coactors will have beneficial effects if the stimulus situation happens to bring out primarily correct responses and detrimental effects if it brings out incorrect ones.

The studies brought together in this chapter illustrate the above principle. When we deal with cases where the dominant responses are appropriate at the outset, enhancement effects are observed. This is true of the feeding studies by Harlow and by Tolman and Wilson, of the results on copulation among rats reported by Larsen, and of the findings on the nest-building behavior of the ant. Where occasions exist for incorrect response tendencies to be dominant, interference is found when others are present. The maze learning performance of the cockroach reported by Gates and Allee shows the effects of interference due to the presence of coactors. Rasmussen's study on the learning of avoidance responses also shows interference due to the presence of others. It should be noted that the rat's task in his study was to inhibit approaching the water dish and drinking. If the drinking response, like the eating response, is enhanced by the presence of others (and our own experience as well as an experiment by R. H. Bruce (1941) both contribute convincing evidence on the subject), then the learning of the avoidance response in Rasmussen's experimental situation should indeed suffer by the presence of other animals. From his data in Figure 1 one may well conclude that the group situation delayed learning, rather than that the solitary situation enhanced it.

Is there any independent evidence that the presence of others is drive-producing? I have reviewed elsewhere (Zajonc, 1965) data indicating, perhaps somewhat indirectly, that the presence of others is indeed a significant source of arousal. The activity of the endocrine system, which is a fair index of arousal, appears to be increased by the continued presence of others. Under these conditions, monkeys as well as humans (Mason & Brady, 1964), and mice (Thiessen, 1964a; 1964b) show heightened endocrine activity (evidenced by increased adrenal weights, elevated hydrocortisone levels, or intensified susceptibility to excitation by emphetamine).

It must be observed, finally, that the presence of others may have effects considerably more elaborate than simply increasing arousal. The individual may gain from others cues as to appropriate behavior—he can benefit from their presence by copying or imitating their responses. By observing their reactions he can reassure himself, when in stress, that the danger is not as great as he thought, or he can convince himself that the danger is indeed serious. However, the experimental paradigm of coaction in its pure form considers only the effects of *mere* presence of others: when their presence is not a source of appropriate cues, when it does not control the organism's reinforcement, when it is, in short, behaviorally nondirective. New factors come into play when there is an opportunity to imitate or to seek reinforcement from others. But these factors are meant to be held constant in the experimental paradigm of coaction. They are, of course, the subject of other studies and will be taken up in future chapters.

REFERENCES

BRUCE, R. H. An experimental analysis of social factors affecting the performance of white rats. I. Performance in learning in a simple field situation. *J. comp. Psychol.*, 1941, **31**, 363–377.

HULL, C. L. *A Behavior System*. New Haven: Yale Univ. Press, 1952.

MASON, J. W., and BRADY, J. V. The sensitivity of psychoendocrine systems to social and physical environment. In Leiderman, P., and Shapiro, D. (Eds.), *Psychobiological Approaches to Social Behavior*. Stanford, Calif.: Stanford Univ. Press, 1964.

SPENCE, K. W. *Behavior Theory and Conditioning.* New Haven: Yale Univ. Press, 1956.

THIESSEN, D. D. Population density, mouse genotype, and endocrine function in behavior. *J. comp. physiol. Psychol.*, 1964, **57**, 412–416. (a)

THIESSEN, D. D. Amphetamine toxicity, population density, and behavior: A review. *Psychol. Bull.*, 1964, **62**, 401–410. (b)

TRIPLETT, N. The dynamogenic factors in pacemaking and competition. *Amer. J. Psychol.*, 1897, **9**, 507–533.

ZAJONC, R. B. Social facilitation. *Science*, 1965, **149**, 269–274.

SOCIAL FACILITATION OF FEEDING IN THE ALBINO RAT*

H. F. Harlow

THE effect of social influences on the drives of the normal human being is quite obvious. We inhibit certain tendencies and reinforce others in the presence of individuals like ourselves. Eating is influenced, probably not so much as to quantity as to appreciation. A good meal tastes better if we eat it in the company of friends. Likewise, it tastes much worse if we eat it in disagreeable company.

In the study of animals we are reduced to a more quantitative attack on the problem. It is impossible to determine the rat's increased appreciation of food (if any such exists) but we can measure the amount ingested. It has seemed desirable to study the effect of a social situation on this elementary type of response by comparing the amounts of food taken in solitary feeding as compared with the amount eaten by the same animal when feeding in a group.

PERTINENT LITERATURE

Fischel (1927) was able to show that hens, though exhibiting no signs of hunger, were stimulated to begin eating again by the sight of another hen feeding. Bayer (1929) completed a series of researches designed to check and extend this observation. In these investigations,

The research was carried out as follows: a hungry V-animal (experimental animal) was placed before a large heap of wheat (a food well-liked by the hens) and allowed to eat until satiated. Then a hungry A-animal (exciting animal) was introduced.

Two measures of the effect of the social facilitation of the feeding responses which were subsequently obtained were made by Bayer. The first of these was the changed behavior of the hen elicited by the above situation, and the second was the amount of food that the V-animal ate after the A-animal was admitted to the cage.

Since one of the two hens almost invariably dominated the other, Bayer arranged the situation so that the dominant animal was alternately the V-animal and the A-animal.

The behavior of the hens is described by Bayer as follows:

As soon as the A-animal was placed in front of the food, it began to eat with great zeal. Now how did the V-animal behave? If it were the dominant animal it began to attack the A-animal immediately. To one who is not a behaviorist this obviously indicates that the V-animal attempts to prevent the A-animal from eating as a result of envy over the food (*Futterneid*). As soon as the V-animal noticed, however, that it was having little success, and that the A-animal continued to eat in spite of everything, it began once more to eat the food, intermittently striking at the other animal. This behavior was facilitated by the fact that the V-animal no longer ate with its original zeal. If the V-animal was the subordinate animal it made no effort to hinder the

---

* From the *Journal of Genetic Psychology*, 1932, **43**, 211–221. Copyright 1932 by The Journal Press, Provincetown, Mass.

feeding of the other upon its entrance. Instead, it began to eat again in spite of the fact that the dominant animal would strike at it from time to time. The sight of the dominant animal eating was apparently a strong enough stimulus to inhibit the fear which the subordinate animal normally felt. One could note that the subordinate animal behaved cautiously and ate, so to speak, "behind the back" of the other.

The social increment as measured by the increased amount of food eaten depended somewhat upon the particular situation. Using one V-animal and one A-animal, Bayer experimented on eight hens. Social increments (amount of food eaten by the V-animal after the A-animal was admitted to the cage) were obtained in every case and averaged 34 per cent, the lowest being 25 per cent, and the highest 43 per cent of the amount eaten to induce satiation.

Experiments on the same animals in which the situation was altered so that there was always one V-animal and three A-animals gave even greater social increments, averaging 53 per cent, and ranging from 33 to 67 per cent.

Where the situation was reversed so that there was only one A-animal and three V-animals, the effect was very much less. One out of four hens was entirely passive, and no animal showed a social increment in excess of 33 per cent. With the three V-animals eating in the same compartment, results obtained on four animals gave a social increment averaging 10 per cent, and, with the three V-animals eating in separate compartments, results obtained on four animals gave a social increment of 21 per cent.

A fourth experimental situation was so arranged that four hens would eat separately one day and together the next. Two experiments were completed and social increments (increased consumption of food for day 1 over day 2) were obtained in every instance. These ranged in the individual animals from 33 to 200 per cent, and averaged 96 per cent.

Seven experiments have been conducted.[1] In the following study, all have the same purpose but vary slightly in technique. For purposes of convenient understanding, the technique and results will be given for each separately before any attempt is made to compare the results or draw general conclusions.

TECHNIQUE AND RESULTS

Experiment 1. In this experiment the animal was allowed to feed to satiation, an hour period being assumed to meet this requirement. At the end of this time a hungry rat (24-hour food deprivation) was introduced and allowed to feed for an hour. The object was to find if the satiated rat would be stimulated to feed further in the second hour by the presence of the hungry rat. Corn was used as food which made it easy to observe the amount eaten by the satiated animal as the grains seized could be counted. The results were tabulated for eight animals on each of three days, and in no case were they distinctly favorable to the existence of a social facilitation process, as the satiated animal ate no more in the second hour in the presence of the hungry rat than when alone.

Experiment 2. In this experiment young rats without previous experience with solid food were used. Twenty rats were weaned at the age of 18 days, and, as the mothers had been fed outside the cage for the three preceding days, they could have had no experience with solid food. They were fed a liquid diet for a certain number of days (2, 7, 12 and 17, respectively, for the four groups of five rats each) so that, on the day of starting

TABLE 1. *Comparison of Amount of Corn Eaten and Spilled by Albino Rats Feeding Separately and in Groups of Five*

Group No.	Age (Days)	Individual Feeding		Group Feeding	
		Grains Eaten	Grains Spilled	Grains Eaten	Grains Spilled
1	24–25	14	67	21	204
2	29–30	18	59	30	221
3	34–35	21	84	41	242
4	39–40	24	78	49	237
Total		87	288	141	904

[1] Experiments 1–6 were conducted at Stanford University under Dr. C. P. Stone. The work was financed by the Thomas Welton Stanford Fund. Experiment 7 was done at the University of Wisconsin.

the experiment, the groups were 20, 25, 30 and 35 days old. They were then placed in individual cages and fed corn during two one-hour periods per day for five days. On the sixth day, all five animals were placed in one cage and allowed to feed.

Objective comparison was made between the fifth day (last individual feeding) and sixth day (first group feeding). The criteria used were the number of grains eaten and the number of grains spilled on the floor. The results are summarized in Table 1.

The number of grains eaten by the 20 rats on the last day of individual feeding was 87. On the first day of group feeding the total number of grains eaten jumped to 141. This amounts to an increase of about 70 per cent. The amount of corn spilled was even more strikingly increased. The 20 rats spilled 288 grains when feeding individually, but when feeding in groups they spilled 904 grains, an increase of about 200 per cent.

The behavior of the rats was even more markedly different. The rat eating alone characteristically fed leisurely and with occasional wandering about the cage. In the group situation, the performance was hurried and apparently competitive. Each rat would hasten to a corner, or two rats would eat back to back. The process of getting grains from the pan occasioned much scrambling and pushing, which was the chief cause of the large amount spilled. The group performance appeared much more highly motivated than the process of individual feeding.

Experiment 3. In this experiment 20 animals, all males, were reared in isolation from the age of 20 to 40 days. From the fortieth to the forty-fourth day, inclusive, the amount of standard diet eaten by each animal in his two feeding periods was recorded, and on the forty-fourth day they were grouped by weight into ten pairs. From the forty-fifth to the forty-ninth day they were fed in pairs and the total amount eaten by the pair compared with the sum of the amounts eaten by the two individually. Some of these pairs showed a social increment, some no effect, and some a social decrement. An observation of the behavior of these animals indicated that in many cases the two would spend the feeding hour in play, even though a steady decline in weight indicated that they must be in a state of hunger.

Experiment 4. In this experiment 20 males, 40 to 45 days old (with two exceptions, 61 days), were used. The situation was the same as in the third experiment, with the chief exception that previous to the beginning of the individual feeding, they had been housed and fed in groups of

four or five, so that they were accustomed to a social feeding situation.

The experiment lasted 25 days, including five individual days, five social, five individual, five social, and a final five individual days. These will be designated control 1, experimental 1, control 2, etc.

TABLE 2. *Amount of Standard Diet Eaten by Young Albino Rats Feeding Separately and in Groups of Two*

Period	Total Grams Eaten	Average Grams Eaten	
Control 1	917	45.85	
Experimental 1	1184	59.20	
Control 2	794	39.72	
Experimental 2	1220	61.10	
Control 3	797	39.85	
Mean differences:	Diff.	S.D. diff.	C.R.
E1—C1	13.35	1.69	7.90
E1—C2	19.50	1.67	11.67
E2—C2	21.30	1.46	14.58
E2—C3	21.15	1.42	14.89

The results on amount of food eaten are summarized in Table 2. The differences here are strikingly in favour of the concept of social facilitation. The mean difference between the first control and the first experimental period is 13.35 grams and the S.D. of this difference is only 1.69, giving a critical ratio of 7.90. The other differences are even greater, and the critical ratios correspondingly higher.

The amount of gain in weight made by each rat was also computed. These results are given in Table 3. The differences here are all large and reliable except for the comparison of experimental 1 and control 1. The conclusion which is obvious from these two criteria is that social facilitation of feeding responses unquestionably does occur, and that it is important.

Experiment 5. This experiment differs from the previous one in that only nine animals were used, and during the two experimental periods they were fed in groups of three. Since the number of cases is so small, the results are not given, but the critical ratios were significant for both food eaten and weight gained, in every comparison of the group and individual situations. The results did not indicate, however, that there was greater facilitation when three animals were feeding together than in Experiment 4 in which they fed in pairs.

TABLE 3. *Amount of Weight Gained by Young Albino Rats Feeding Separately and in Groups of Two*

Period	Total Grams Gained	Average Grams Gained		
Control 1	294	14.70		
Experimental 1	333	16.65		
Control 2	50	2.50		
Experimental 2	459	22.95		
Control 3	75	3.75		
Mean differences:	Diff.	S.D. diff.	C.R.	
E1—C1	1.95	1.41	1.38	
E1—C2	14.15	1.54	9.18	
E2—C2	20.45	1.93	10.59	
E2—C3	19.20	1.48	12.97	

Experiment 6. Since Bayer (1929) claimed that "envy" was one of the factors operating in the social facilitation process, an attempt was made to eliminate the competitive factor. A cage, 7 in. long, 3 in. high, and $1\frac{1}{2}$ in. wide, was used to confine a large adult rat. An opening in the front end allowed him to reach out his head, but he could not escape or turn around. When, thus confined, he was placed in the cage with a feeding rat, the other rat was not stimulated to eat more, whether or not the confined rat was placed near enough to the food to eat from his position. It does not seem likely, then, that "envy" or personal animosity is a factor operative in the social facilitation situation.

Experiment 7. The first six experiments were considered as more or less exploratory, and an attempt was made to duplicate Experiment 4 under more carefully controlled conditions. Since we consider this the crucial experiment of the series, the technique is detailed fully.

The rats were weaned at 18 days and reared in individual cages until 30 days old. Their two daily feedings were one-half hour in length, at 9 A.M. and 5 P.M. A double pan was used for feeding with food only in the inner pan, so that most of the spilled food was collected in the outer. Food falling on the floor of the cage was caught by a sheet of oilcloth placed underneath.

Only the amount of food eaten was taken as a measure of facilitation. Exactly 20 grams of food were furnished the animal daily, and the amount eaten was computed by collecting carefully all spillage and reweighing. The computations were then made to 0.1 gram on scales which are accurate to 0.05 gram.

At the beginning of the experimental period,

34 animals were divided into 17 pairs on the basis of weight. On the 31st day of life, they were given their first group feeding, in pairs, and from then on to the 49th day, every odd-numbered day was an experimental period and every even-numbered day (32 to 50) a control day. To avoid any effect of familiarity with surroundings, the animals were alternated between the home cage of one member of the pair and the other.

TABLE 4. *Average Amount Eaten by Pairs of Albino Rats Feeding Together and Separately on Alternate Days*

Day of experiment	Group feeding	Individual feeding
1	10.35	
2		9.22
3	11.67	
4		10.43
5	12.57	
6		11.20
7	12.74	
8		12.09
9	13.24	
10		12.24
11	14.08	
12		13.32
13	13.66	
14		13.56
15	15.42	
16		13.83
17	15.84	
18		14.69
19	16.33	
20		14.75
Average	13.59	12.53
difference	1.057	
S.D. diff.	0.294	
C.R.	3.59	

Table 4 summarizes the results of this experiment. In the social situation the average amount of food eaten was 13.59 grams, while in the individual situation only 12.53 grams were eaten. The difference is 1.058, and is 3.59 times its standard error. It is therefore a statistically "true" difference.

It seems important not only that the group means are significantly differentiated, but also that the difference from day to day is consistent. The differences and critical ratios for each two-day sequence have been computed, but in general (as a result of the small number of cases, no doubt) these ratios are less than three. A definite tendency is shown, however, for group feeding to manifest a significant facilitating influence upon the amount of food ingested.

The behavior of the animals in this experiment (alternating daily between group and solitary feeding) manifested even more excitement than in the previous experiments in which five-day periods had been used.

DISCUSSION

The results of the several experiments argue clearly for a process of social facilitation of the feeding response in the rat. If we rely upon differences in food ingested, in food spilled or in weight gained, we see that there is in practically every case a significant superiority of the group situation.

The observations of behavior also indicate a process of facilitation. The animals in the group situation were characteristically highly motivated, displaying an excess of activity in getting food, struggling, crowding other rats from the food dish, etc. There was a distinct difference in the leisurely feeding of the individual rat, which was frequently interrupted by exploration of the cage, etc.

These generalizations hold true only when the various rats are free and competing. The experiments in which this condition did not prevail (Experiments 1 and 6) do not show social facilitation.

The observation of the behavior of these animals in the individual and social situations leads to another conclusion which is of considerable importance in the understanding of hunger motivation. It has often been believed that the immediate stimulus for ingestion of food was the stimulation derived from contractions of the stomach. Yet it is well known that struggling and excessive striped muscle activity tend to inhibit the contractions of the smooth musculature. We have, then, in this experiment, the anomalous condition of animals in which the stomach contractions have undoubtedly been inhibited or diminished eating greater quantities of food than they did when under comparable conditions without the external disturbance. We wish to suggest, then, that ingestion of food is determined more by the external situation than by the actual interoceptive stimulation. This conclusion is substantiated by much common knowledge of human behavior, such as the influence of music, pleasant surroundings, the holiday atmosphere, etc.

BASIC MECHANISMS OF SOCIAL FACILITATION

1. *Social facilitation does not depend upon learning.* Rats without previous experience with solid food (Experiment 2) and rats without previous experience with food in the presence of other rats (Experiment 7) manifest the effects of facilitation on the first day of the experiment.

2. *The process is not subject to change and adaptation.* Five periods of five days each (Experiment 4) and 20 days of alternate feeding (Experiment 7) showed differences at the end of the experiment equal to or greater than those at the beginning.

3. *It does not depend upon imitation.* A satiated rat watching a hungry rat eating (Experiment 1) is not thereby motivated to eat more.

4. *It does not depend upon "envy."* The presence of a restrained, non-competing rat (Experiment 6) does not motivate the free rat to ingest larger amounts of food.

5. *It is not a function of the size of the group.* Experiment 2 (five rats in group), Experiment 4 (two rats), and Experiment 5 (three rats) show differences of similar size.

6. *It may be related to age.* Table 1 (Experiment 2) indicates that the older rats increase their food intake proportionately more than the younger rats. Evidence from the individual records of some of the other experiments, however, does not substantiate this finding.

7. *The essential condition for the occurrence of social facilitation is the presence of rats unrestrained and actively competing with each other for food.*

SUMMARY

1. The effect of social facilitation upon the feeding response in the albino rat has been demonstrated.

2. This social facilitation takes place only between rats that are unrestrained and freely competing.

3. Facilitation is independent of previous experience.

4. It does not depend upon imitation or envy.

REFERENCES

BAYER, E. Beiträge zur Zweikomponententheorie des Hungers. *Zsch. f. Psychol.*, 1929, **112**, 1–53.

FISCHEL, W. Beiträge zur Sociologie des Haushuhns. *Biol. Centbl.*, 1927, **47**, 678–696.

SOCIAL FEEDING IN DOMESTIC CHICKS*

C. W. Tolman and G. F. Wilson

CONSIDERING the apparent gregariousness of the domestic chicken, it is not surprising to find that conspecific companions exert considerable influence upon the individual in a wide range of activities. One activity for which such influence has been demonstrated and studied is feeding. In 1929, Bayer demonstrated that adult hens, satiated in isolation, resumed eating when placed with a hungry hen. Furthermore, the amount eaten was greater when the satiated hen was placed with three hungry companions. Hungry hens were also found to eat more in groups than in social isolation. Tolman (1964), using young chicks, confirmed this last finding and found this effect to be little influenced by lack of prior social experience but dependent upon unrestricted interaction between companions while feeding.

The present research constitutes a further examination of both social and non-social variables affecting the feeding of individuals in groups.

EXPERIMENT 1

Several variables were taken under investigation in this experiment. Chicks were tested for amount eaten under a wide range of social conditions selected to yield information on the effects of number of companions, degree of social contact with companions, and the differential food deprivation of companions. At the same time the effect of food deprivation was investigated by testing under each of the conditions at 0, 6, 12 and 24 hours of food deprivation.

SUBJECTS. The chicks were White Leghorn cockerels brought into the laboratory on the day of hatching from a commercial hatchery. They were maintained in groups of from 60 to 100 in brooder batteries. The food used in maintenance and testing was Purina Chick Startena. Except for the food deprivation specified in the procedure section below, all chicks were given free and continuous access to plentiful amounts of food.

APPARATUS. Testing for this experiment was

carried out in a battery of small enclosures. The basic enclosure was 40 in. long, 7 in. wide, and 7 in. high. Fixtures on the walls of this enclosure allowed the insertion of up to three partitions making four small enclosures, each 10 in. long.

The walls and floors of the enclosures were smooth and painted grey. The openings at the top were covered by quarter-inch wire mesh.

PROCEDURE. The following conditions were used.

1. Isolates. Individual chicks were tested in the small compartments. Isolation was principally tactual and visual. No attempt at auditory isolation was made.

2. Pairs. Two chicks were tested at the same time in a small compartment.

3. Groups of four. Four chicks were tested simultaneously in a double compartment (20 in. by 7 in. by 7 in.).

4. Groups of 16. Sixteen chicks were tested simultaneously in the larger, basic enclosure.

5. Single chicks separated by Plexiglass from a single companion (S/S). This is similar to the Isolate condition except that the partition separating two chicks was Plexiglass thus allowing visual, as well as auditory contact.

6. Paired chicks separated by Plexiglass from another pair (P/P). This is similar to S/S except that pairs were used in place of single chicks.

7. A deprived individual separated by Plexiglass from a non-deprived individual (D/N). This is similar to the S/S condition except that one of the chicks was never deprived of food.

8. Non-deprived individual separated by Plexiglass from a deprived individual (N/D). These chicks were the companions to the D/N chicks.

9. Non-deprived individual separated by Plexiglass from another non-deprived individual (N/N).

Each of these conditions was represented by an independent group at four tests (see Table 1). The first test occurred at 8 A.M. on the fifth day from hatching, the second at 2 P.M. of that day, the third at 8 P.M. of the same day, and the fourth at 8 A.M. on the following day. Except

---

* From *Animal Behavior*, 1965, **13**, 134–142. Copyright 1965 by Ballière, Tindall & Cassell, London.

where otherwise called for by the testing condition, food was removed for all chicks at 8 A.M. on the first test day, i.e. the fifth day from hatching. Thus, chicks were tested in all conditions except N/D and N/N at 0, 6, 12 and 24 hours of food deprivation. The chicks in conditions N/N were tested at the same times of day to provide a control for possible contamination of the deprivation data by other variables associated with the time of day and to provide a baseline for comparison with certain other groups.

TABLE 1. *Summary of Number of Chicks Used in Each Condition*

Time	8 A.M.	2 P.M.	8 P.M.	8 A.M.	Totals
Isol.	16	16	16	16	64
Pairs	16	16	16	16	64
Fours	8	8	8	8	32
Sixteens	16	16	16	16	64
S/S	16	16	16	16	64
P/P	8	8	8	8	32
D/N	8	8	8	8	32
N/D	8	8	8	8	32
N/N	8	8	8	8	32
Totals	104	104	104	104	416

Prior to any particular test, food was weighed

out in the amount of approximately 6 g. per chick and placed in each test compartment. At the designated time chicks were placed into the compartments according to their respective conditions, allowed to eat for one hour, then removed. Droppings were removed from the remaining food. The food was then weighed and the difference between this weighing and the earlier one served as the estimate for amount eaten. Where more than one chick was in the compartment, the estimate was divided by the number to give the estimated amount eaten per individual.

Weighing was done on an analytic balance with an accuracy of approximately ± 0.003 g. for the range of weights being measured here. Readings were taken to the nearest hundredth.

There was no water in the compartment during the one-hour test period.

RESULTS. Statistical comparison of the various groups is made somewhat less straightforward than is desirable by the fact that the means of only five of the nine groups were obtained from individual data. The remaining four means were obtained from the joint performance of from 2 to 16 individuals. In order to overcome this apparent obstacle the assumption was made that while means may differ, the variances of the groups for which variance could not be computed do not differ significantly from those of the groups for

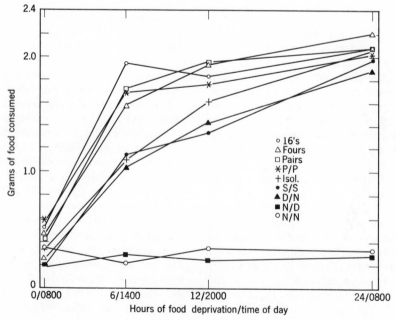

FIGURE 1. The amount of food consumed as a function of hours of deprivation and social condition.

which they could be computed. This assumption allowed the computation of a standard error from the available individual data for use in testing all mean differences. A simple analysis of variance was carried out for the five groups of individual data at each of the four test times. The resulting within-group variance estimates were then used as a basis for common standard errors according to the procedure outlined by McNemar (1949, p. 245). Table 2 summarizes the resulting statistical evaluation of the mean differences.

As can be seen by inspection of Fig. 1, the general effect of increasing food deprivation is a negatively accelerated increase in the amount of

TABLE 2. *Summary of Group Mean Differences at 0, 6, 12 and 24 Hours' Food Deprivation*

	Condition	n	N/D 8	N/N 8	D/N 8	P/P 8	S/S 16	16/s 16	Fours 8	Pairs 16
0 Hour Deprivation	Isol.	16	0.16	0.01	0.11	0.23*	0.14	0.18	0.13	0.08
	Pairs	16	0.24*	0.07	0.19	0.15	0.22*	0.10	0.05	
	Fours	8	0.29*	0.12	0.24	0.10	0.27*	0.05		
	16's	16	0.34*	0.17	0.29*	0.05	0.32*			
	S/S	16	0.02	0.15	0.03	0.37*				
	P/P	8	0.39*	0.22	0.34*		n	S_e	$\bar{X}-\bar{X}$ for t 0.05	
	D/N	8	0.05	0.12			16+16	0.0911	0.183	
	N/N	8	0.17				16+ 8	0.1114	0.224	
							8+ 8	0.1288	0.259	
6 Hour Deprivation	Isol.	16	.81*	.88*	0.06	0.59*	0.04	0.84*	0.48*	0.62*
	Pairs	16	1.43*	1.50*	0.68*	0.03	0.58*	0.22	0.14	
	Fours	8	1.29*	1.36*	0.54*	0.11	0.44*	0.36		
	16's	16	1.65*	1.72*	0.90*	0.25	0.80*			
	S/S	16	0.85	0.92*	0.10	0.55*				
	P/P	8	1.40*	1.47*	0.65*		n	S_e	$\bar{X}-\bar{X}$ for t 0.05	
	D/N	8	0.75*	0.82*			16+16	0.1594	0.320	
	N/N	8	0.07				16+ 8	0.1952	0.392	
							8+ 8	0.2254	0.453	
12 Hour Deprivation	Isol.	16	1.37*	1.27*	0.19	0.14	0.30	0.21	0.31	0.34
	Pairs	16	1.71*	1.61*	0.53	0.20	0.64*	0.13	0.03	
	Fours	8	1.68*	1.58*	0.50	0.17	0.61*	0.10		
	16's	16	1.58*	1.48*	0.40	0.07	0.50			
	S/S	16	1.07*	0.97*	0.11	0.44				
	P/P	8	1.51*	1.41*	0.33		n	S_e	$\bar{X}-\bar{X}$ for t 0.05	
	D/N	8	1.18*	1.08*			16+16	0.2435	0.489	
	N/N	8	0.10				16+ 8	0.2983	0.600	
							8+ 8	0.3444	0.692	
24 Hour Deprivation	Isol.	16	1.78*	1.73*	0.18	0.04	0.08	0.02	0.15	0.02
	Pairs	16	1.80*	1.75*	0.20	0.06	0.10	0.00	0.13	
	Fours	8	1.93*	1.88*	0.33	0.19	0.23	0.13		
	16's	16	1.80*	1.75*	0.20	0.06	0.10			
	S/S	16	1.70*	1.65*	0.10	0.04				
	P/P	8	1.74*	1.69*	0.14		n	S_e	$\bar{X}-\bar{X}$ for t 0.05	
	D/N	8	1.60*	1.55*			16+16	0.1962	0.394	
	N/N	8	0.05				16+ 8	0.2400	0.482	
							8+ 8	0.2773	0.557	

*Significant at the 0.05 level.

food consumed. Examining the differences between Isolates and Pairs as the basic evidence for
social facilitation, a rather interesting interaction
between food deprivation and the companion as
variables controlling food consumption can be
seen. At 0 hours there is no apparent difference
between Isolates and Pairs. At 6 hours, however,
a rather substantial difference appears. By 12
hours the difference has again diminished to an
insignificant level.

Although there is a substantial increase in the
amount eaten at 6 hours deprivation among Pairs
as compared with Isolates, these data give no
evidence for a continued increase with larger
groups. At no point on the graph do Pairs, Fours
and Sixteens significantly differ from one another.

The introduction of the S/S and P/P conditions
was prompted by the question whether S/S would
behave as Isolates or Pairs and, in anticipation
of a possible difference between Pairs and Fours,
whether P/P would behave as Pairs or Fours. The
anticipated difference between Pairs and Fours
did not occur and P/P behaved as both. At no
point did P/P differ significantly from either. S/S,
however, clearly behaved like Isolates, being at no
point significantly different from them. With
respect to the general question of social contact,
the results indicate only one variable among those
tested as necessary and sufficient for maximal
social facilitation, and that is unrestricted physical
contact. At what appears to be the crucial point
of food deprivation those groups in which unrestricted physical contact is present do not differ
significantly among themselves but differ as a
whole from those groups in which there is either
no contact or where contact is limited. Like the
unrestricted contact groups, the restricted contact
groups do not differ among themselves.

Although there was no difference between Pairs
and Isolates at the 0-hour point, some differences
between restricted and unrestricted groups did
occur. There are more such differences than
would be expected by chance suggesting either
that some social facilitation is possible at this
point or that not all chicks were actually 0 hours
deprived. The latter is possible since the chicks
had been on *ad lib.* feeding and there is no
assurance that each chick had satiated himself
prior to the beginning of deprivation.

Differential deprivation, as tested here under
restricted conditions (N/D group) did not seem to
have any apparent effect. Non-deprived chicks,
although viewing a deprived chick did not eat
more than the N/N group. Similarly, the view of
a non-deprived chick did not, in any apparent
way, alter the behavior of deprived chicks. The

D/N chicks at no point differed significantly in
their performance from Isolates or S/S.

The performance of the N/N and N/D groups
over the four points in time do not appear to
differ appreciably from a straight line, suggesting
that contamination from extraneous time-of-day
variables was no more than minor.

EXPERIMENT 2

In an earlier paper (Tolman, 1964) it was
suggested that social facilitation was caused by
physical interaction in the feeding behavior of
companions. Another possible interpretation, and
in many ways a simpler one, is that with restricted
or no contact with a companion, a chick is more
emotional. It would be presumed, of course, that
emotionality interferes with eating. Emotionality,
or relief from it by the presence of a companion,
would then constitute the mechanism of social
facilitation.

The present experiment approaches this emotionality hypothesis via the correlational method.
Emotionality, as judged by the rate of defecation,
is measured in five of the conditions used in
Experiment 1. If the hypothesis is correct a
negative correlation between the emotionality
data and feeding data should be expected.

SUBJECTS. This experiment used 780 White
Leghorn cockerels obtained from the same source
and maintained in the same manner as in
Experiment 1.

APPARATUS. The apparatus used here was the
same as that used in Experiment 1.

PROCEDURE. The chicks were tested at two ages:
480 chicks at three days of age and 300 chicks at
seven days of age. These age groups were divided
evenly among five test conditions. Isolates, Pairs,
Fours, S/S, and P/P.

The testing proceeded as follows. Four chicks
per condition were removed from the brooder
and placed into the test compartments for a
period of three minutes, then placed into another
brooder. The number of defecations in the test
compartments for each group of four was then
counted, the compartments cleaned, and the procedure repeated until all the chicks were tested.

It will be noted that the score obtained by this
procedure for purposes of statistical treatment
was number of defecations per four chicks. Thus
there were 24 scores per group for the 3-day-olds
and 15 scores per group for the 7-day-olds.

RESULTS. The results are presented graphically
in Fig. 2. Analysis of variance showed the differences among the 3-day-olds to be significant

($F = 11.22$, 115 & 4 df). The differences among the 7-day-olds, although the means maintained approximately the same rank order as those for the 3-day-olds ($r_S = 0.80$), were not significant ($F = 2.25$, 70 & 4 df). Further analysis of the 3-day-old data showed the Isolates to be significantly different from each of the other groups ($p < 0.01$ for each comparison). Taking number of chicks and presence or absence of the restricting Plexiglass as independent variables, the remaining four groups of 3-day-olds were analysed in a 2×2 analysis of variance. This indicated that the only manipulated variable contributing significantly to the over-all variance was the presence or absence of Plexiglass ($F = 7.00$, 92 & 1 df), i.e. S/S and P/P together were significantly different from Pairs and Fours together. Comparison of these results with the feeding data for 6 hours' deprivation reveals a negative, although far from significant correlation ($r_S = -0.30$ for each day).

As can be seen by inspection of the data the results from the Isolates, Pairs and Fours meet the requirements of the hypothesis, i.e. Isolates are low on feeding and low on defecation. Difficulty for the hypothesis arises primarily from the S/S and P/P groups in their failure to differentiate on either of the test days. It will be recalled that these groups were clearly differentiated in feeding.

Comparison of the two ages shows that only the emotionality in the Isolate condition had significantly decreased ($t = 3.19$, 37 df).

EXPERIMENT 3

The feeding of one chick appears to depend on what his companion does, not on his mere presence. The results of Experiment 1 confirm the earlier findings (Tolman, 1964) upon which this conclusion was reached. The results of Experiment 2 similarly, although indirectly, do not support a "mere presence" explanation.

If the subject's feeding behavior depends upon the behavior of his companion, and this presumably would also be feeding behavior, then it should be possible to manipulate the subject's feeding behavior in a quantitative fashion by varying the companion's feeding behavior. The most obvious way of doing the latter is to vary the state of deprivation of the companion. In Experiment 1 this was attempted in the N/D condition but probably due to restricted contact with the companion, no positive results were obtained.

In the present experiment 6-hour deprived chicks were placed with no companion, a like

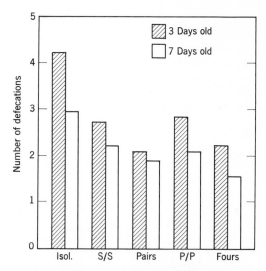

FIGURE 2. Emotionality indicated by amount of defecation as a function of social condition and age.

deprived companion, and with differentially deprived companions, one non-deprived and one 24-hour deprived.

SUBJECTS. Eighty chicks of the same variety and source as in experiment 1 were used. Maintenance was also the same as Experiment 1.

APPARATUS. The apparatus consisted of two observation boxes 12 in. by 12 in. by 8 in. which were solid enclosures except for the fronts which were Plexiglass and the tops which were 0.25 in. wire-mesh screen.

PROCEDURE. The chicks were divided into four groups of 20 each. Beginning at 7 A.M. on the seventh day after hatching, four from each group were placed in a retaining cage with water but no food available. At 8 A.M. a similar number of chicks was placed in another retaining cage. This was repeated each hour until all 80 chicks were deprived of food. Testing began 6 hours after the first groups were deprived, i.e. at 1 P.M. Testing of the first groups was completed by 2 P.M. whereupon testing of the next groups began. This continued until all chicks were tested. The subjects, then were all between 6 and 7 hours food deprived when tested. For convenience they will be referred to as 6-hours deprived. Another group had been deprived the previous day at 1 P.M. and consequently 24 to 29 hours deprived during the testing period. For convenience they will be referred to as 24-hours deprived.

The four testing conditions were 6-hour deprived chicks with no companion (designated as 6-n), with a non-deprived companion (6-0), with a like deprived companion (6-6), and with a

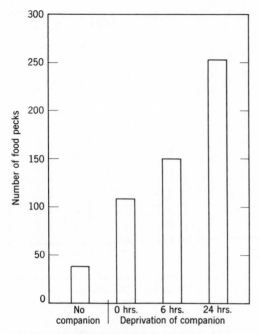

FIGURE 3. Eating behavior indicated by number of food pecks by 6-hour deprived chicks as a function of the deprivation state of the comparison.

APPARATUS. The observation boxes described for Experiment 3 were used.

PROCEDURE. Forty-eight chicks were tested when non-deprived and 7 days old. Half were tested with one 24-hour deprived companion and half were tested with two 24-hour deprived companions. The testing procedure was similar to that used in Experiment 3. Tests were 5 minutes long and number of food pecks were counted.

Thirty-two chicks were tested when 6-hours' food deprived and 4 days old. Half were tested with one 24-hour deprived companion and half with three 24-hour deprived companions. The testing was conducted in the same manner as for the 7-day-old chicks. The deprivation procedure was the same as for the 6-hour deprived chicks in experiment 3.

RESULTS. A group comparison for the 7-day-old, non-deprived, chicks showed no significant difference due to the number of companions. Similarly, no significant differences were found for the 4-day-old, 6-hour deprived chicks. The results are presented graphically in Fig. 4.

24-hour deprived companion (6-24). The feeding behaviour of only the 6-hour deprived subjects was measured.

The measure used was number of food pecks during a 5-minute test period. Testing was carried out in the observation boxes, the floors of which were covered with Purina Chick Startena. Food pecks were counted by the observers and recorded on Veeder-Root hand counters.

RESULTS. An analysis of variance of the results presented graphically in Fig. 3 yielded an F of 15.79 which is significant at the 0.001 level. Comparison of individual groups showed all differences to be significant except the one between 6-0 and 6-6. The difference between 6-0 and 6-n was significant at the 0.05 level: all of the others were significant at the 0.01 level.

EXPERIMENT 4

The results of Experiment 1 indicated no effect upon individual eating by increasing the number of companions. The present experiment is an attempt to confirm this finding using procedures similar to those employed in Experiment 3.

SUBJECTS. Eighty White Leghorn cockerels from the same source and maintained in the same way as in the previous experiments were used.

DISCUSSION

At least four relatively distinct categories of variables affecting social feeding have been investigated in these experiments: social contact, emotionality, number of companions, and food deprivation. In order to facilitate discussion, each of these categories will be handled separately.

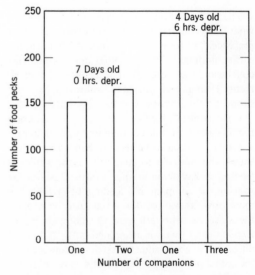

FIGURE 4. Eating behavior indicated by number of food pecks as a function of number of companions.

SOCIAL CONTACT. The information regarding social contact gained particularly from Experiment 1 is not new but rather is confirmatory. Data reported in an earlier paper (Tolman, 1964) yielded substantially the same conclusion, and that is that in order for social facilitation of feeding to occur among young chicks, unrestricted social contact between the interacting companions must prevail.

EMOTIONALITY. As pointed out in an earlier section, the conclusion regarding social contact can lead to at least two hypotheses regarding the mechanism of social facilitation. First, the mechanism is one of behavioral interaction such as bill pecking as was proposed in an earlier paper by Tolman (1964), or local enhancement of cues as proposed by Thorpe (1956, p. 121). Second, it is possible that isolated or otherwise socially restricted animals are emotional and this emotionality interferes with feeding. The results of Experiment 2 showed that under the conditions of that experiment a chick placed on the other side of a Plexiglass barrier from a companion displays less emotion than an isolate. The feeding data from Experiment 1, particularly at 6 hours of deprivation, indicate, however, that a chick feeding in that condition eats no more than an isolate. That is, although less emotional than an isolate, he nonetheless eats no more.

It could well be argued that the data of Experiment 2, although seemingly upsetting to the emotionality hypothesis, do not really invalidate it since we are depending on a single measure of emotion and it is well known that emotion is no unitary process. Another measure may have given results more in line with the hypotheses.

In response to this objection, it might be mentioned that, although we did not have the equipment to measure it, we found that the only condition that could be reliably differentiated on the basis of intensity of calling was isolation. This observation corresponds with the findings of Bermant (1963) in which significantly more distress calling occurred in isolation than in a condition comparable to our S/S condition. The companions used for Bermant's chicks were hens but it would not be unreasonable to expect similar results with a chick of similar age.

While none of this rules out emotionality as playing a role in social facilitation it does seem that the behavioral interaction hypothesis handles more of the present data. Not only is this hypothesis more compatible with the data of Experiment 2, but it also fits the data of Experiment 3. If the interaction operates through feeding behavior then an increase in that behavior of the companion should lead to increased facilitation. The emotionality hypothesis in this situation would have to be considerably less straight-forward.

NUMBER OF COMPANIONS. Bayer (1929) reported that social facilitation was greater with three companions than with one. We fully expected such a result to emerge from Experiment 1, but as can readily be seen from the data it did not. In our Experiment 1 the data were obtained from equally deprived chicks, whereas Bayer's results were obtained from birds which were differentially deprived, i.e. his subjects were satiated while the companions were starved. Our Experiment 4 which uses differential deprivation, however, again did not confirm Bayer.

One possible explanation for this discrepancy lies in the fact that Bayer used adult hens whereas we used chicks. Another, and perhaps more likely, explanation is methodological. Bayer's conclusion is based on the observation of eight birds tested at one time with a single companion and, after the passage of some unspecified amount of time, tested again with three companions. This rather unbalanced experimental design clearly leaves the results subject to other variables which may have been different at each testing. Bayer did not analyse his data statistically. We performed such an analysis, however, and found them to be significant. But in spite of this significance, it is interesting to note that his mean difference was 6 gm. while his measuring technique was apparently accurate only to the nearest 5 gm. Such a mean difference could easily be accounted for by measurement error.

FOOD DEPRIVATION. The general slope of the curve shown in Figure 1 shows that chickens are similar in this regard to nearly all other vertebrates that have been tested. The most interesting function revealed by these data is a secondary one obtained by taking the difference between the upper curves on the graph (Pairs, Fours, etc.) and the middle curves (Isolates, S/S, etc.). This would seem to represent social facilitation among like-deprived chicks as a function of hours of deprivation. The general shape of this function is a rise from near zero at 0-hours' deprivation to a high level at 6-hours and a return to near zero through 12 and 24 hours. This again seems to be understandable from the behavioral interaction hypothesis. At 0-hours the rate of eating by chicks is low enough to be increased but too low to be stimulating to a companion. At around 6-hours the rate is low enough to be increased and high enough to be stimulating. At the later hours, the rate is high enough to be stimulating but too high to show any increase.

These results of Experiment 1, taken together with the results of Experiment 3, in which the deprivation of the companion was varied while that of the subject was held constant, suggest the following hypothesis concerning the two variables, deprivation of the subject and deprivation of the companion. The amount of food consumed by the subject member of a pair is an increasing function of the number of hours the subject has been deprived of food. The amount of food consumed by the subject is also an increasing function of the number of hours his companion has been deprived, although the extent of this influence exerted by the companion decreases with increased deprivation of the subject.

SUMMARY

1. A series of experiments related to the feeding of individual chicks in groups is reported.

2. In Experiment 1 chicks were tested for amount of food consumed under a variety of social conditions at varying lengths of food deprivation. The data obtained (*a*) confirmed an earlier finding that social facilitation depends for its occurrence on unrestricted social interaction among chicks; (*b*) showed that social facilitation is optimal in the area of 6 hours of food deprivation and only minimal, if existent, at 0, 12 and 24 hours of food deprivation; and (*c*) showed that increased number of like-deprived companions does not increase the amount of social facilitation.

3. Experiment 2 was designed to produce data on emotionality in some of the conditions used in Experiment 1 for the purpose of testing the hypothesis that the differences in Experiment 1, i.e. social facilitation, could be accounted for by emotion produced by restricted contact with a companion. The hypothesis was not confirmed by the data.

4. In Experiment 3 deprivation of the companion was varied while that of the subject was held constant. It was found that the amount of feeding behavior of the subject varied directly with the deprivation of the companion.

5. Experiment 4 was concerned with the effect of number of companions. The companions in this experiment were more deprived than the subjects. The results confirmed those of Experiment 1 in that no difference could be demonstrated due to number of companions.

6. It was concluded that social facilitation is the product of some specific kind of behavioral interaction and that the amount of food consumed by the subject member of a pair is positively related to the hours of food deprivation of the subject and companion member, however, the size of the effect of increasing the companion's deprivation is inversely related to the hours of food deprivation of the subject.

ACKNOWLEDGEMENTS

This investigation was supported by a U.S. PHS research grant (MH 10034-01) from the Institute of Mental Health, Public Health Service, and carried out at Idaho State University. Much of the pilot work directly leading to this investigation was supported by the Graduate Council of the University of South Dakota.

REFERENCES

BAYER, E. (1929). Beiträge zur Zweikomponenten-theorie des Hungers. *Z. Psychol.*, **112**, 1–54.

BERMANT, G. (1963). Intensity and rate of distress calling in chicks as a function of social contact. *Anim. Behav.*, **11**, 514–517.

McNEMAR, QUINN (1949). *Psychological Statistics*. New York: John Wiley & Sons, Inc.

THORPE, W. H. (1956). *Learning and Instinct in Animals*. Cambridge, Mass. Harvard University Press.

TOLMAN, C. W. (1964): Social facilitation of feeding behaviour in domestic chicks. *Anim. Behav.*, **12**, 245–252.

THE EFFECT OF GROUP ACTIVITY ON THE SEXUAL BEHAVIOR OF THE INDIVIDUAL ANIMAL*

Knutt Larsson

1. INTRODUCTION

OCCASIONAL observations have suggested that the presence of other copulating animals stimulates the sexual activity of the individual rat. Soulairac has reported one experiment on this theme. He introduced active males into a cage with inactive ones and found that the inactive animals were stimulated to copulate normally (Soulairac, 1950).

Grunt and Young made a similar observation on the male guinea pig. Normally the male does not copulate after the first ejaculation. In one observation, however, it was noted that the introduction of a new female released a second ejaculation. In several others intromission appeared (Grunt and Young, 1952 b).

In the investigation presented in the next section it was found that the intensity of sexual activity measured in ejaculations per hour was increased by allowing the animals to copulate in groups of three in the same observation cage. It was discovered furthermore that this increase in the intensity of the sexual reflexes did not modify the length of the refractory periods but was only related to the series of copulations.

In another experiment it was found that the efficiency of senescent rats was considerably heightened by collective copulating.

2. THE EFFECT OF GROUP IN ANIMALS AGED 8–10 MONTHS

EXPERIMENTAL CONDITIONS. The animals were observed under three different conditions, denoted A, B and C.

A. Three males and three females were allowed to copulate in the same cage.

B. One pair was allowed to copulate in each of three cages placed beside each other. This was the normal method of observation in our laboratory.

C. One observation of one pair of copulating animals was made. No other animals were in the same room.

Half of the animals were first subjected to A and then to C test; the other half to test C and then A. Series B was not recorded until the end of the two other series.

The observations were timed to a hundredth of a minute.

Before the first observation the rats had not been sexually inactive for three weeks. The different observations made on the same animals were separated by an interval of 14–21 days.

Sixty-two animals took part in the experiment. Three animals were withdrawn because of diseases before the series of observation B. The rats had a mean weight of 333 ± 4 gm. and were 8–12 months of age.

The investigation was performed in the autumn of 1955. Miss Iris Jönsson, M.A., acted as technical aid.

RESULTS. Table 2 presents the number of ejaculations, intromissions and attempts per hour under the different experimental conditions.

In the A group the number of ejaculations attained in one hour was 5.5. During copulation in isolation the frequency was 4.2. This difference is statistically assured. Studying in Table 1 the number of different ejaculations achieved, we find that group copulation diminishes the number of non-ejaculating animals from 5 to 1, and the average frequency of copulations per hour increases throughout.

The rise in the frequency of ejaculations was accompanied by a slight decrease in the frequency of complete and incomplete copulations. This decrease, however, was not statistically significant.

It was then revealed that group copulations increase the excitatory value of each individual copulation and that with increased nervous excitation a greater number of ejaculations is made.

---

* From *Conditioning and Sexual Behavior in the Male Albino Rat*, Stockholm: Almqvist and Wiksell, 1956. Pp. 145–156. Copyright 1956 by Almqvist and Wiksell, Stockholm.

In B the animals achieved an average of 4.9 ejaculations. While no statistically assured difference between A and B was to be found, a slightly significant difference between B and C does exist. We therefore conclude that the copulating pairs of animals in the other observation cages produce sensory stimuli which have in principle the same effect, but less intense, as when the animals are in the same box.

Tables 3–6 reveal the change in the pattern of the sexual behavior accompanying copulation in isolation.

Examining the number of intromissions we find a slight overall increase in the number of copulations preceding ejaculation when the rats are isolated. This tendency, however, is slight and only shows statistical significance for the first ejaculations. The number of intromissions is 12.3 in A and 17.1 in C.

TABLE 1. *The Distribution of the Frequency of Ejaculations*

| | | | | | | No. of Ejaculations | | | | | |
	N	0	1	2	3	4	5	6	7	8	9
A	62	1	61	60	60	58	52	33	11	3	1
B	59	1	58	56	52	45	37	27	14	2	0
C	62	5	57	53	47	39	35	24	7	1	0

TABLE 2. *Effect of Collective Copulation on Sexual Activity: The Number of Ejaculations, Intromissions and Attempts in Three Experimental Conditions*

	A			B			C		
	N	M	SD	N	M	SD	N	M	SD
Ejac.	62	5.5	1.45	59	4.9	1.90	62	4.2	2.20
Intr.	62	41.6	10.28	59	42.9	13.35	62	44.1	14.18
Att.	62	26.7	19.64	59	26.0	18.87	62	29.6	21.30

	A–B			B–C			A–C		
	df	t	p	df	t	p	df	t	p
Ejac.	58	0.60	—	58	2.06	<0.05	61	4.24	<0.001
Intr.	58	0.65	—	58	0.44	—	61	1.31	—
Att.	58	0.04	—	58	0.96	—	61	1.15	—

TABLE 3. *Effect of Collective Copulation: The Frequency of Intromissions*

Ejac.	A			B			C		
No.	N	M	SD	N	M	SD	N	M	SD
I	61	12.3	8.09	58	14.4	7.14	57	17.1	11.11
II	60	5.4	2.44	56	5.6	3.30	53	6.2	3.82
III	60	5.5	3.61	52	5.5	2.85	47	6.1	4.14
IV	58	5.4	3.28	45	6.9	6.11	39	5.8	2.66
V	52	7.6	6.89	37	7.2	4.71	35	8.5	5.40

	A–B			B–C			A–C		
	df	t	p	df	t	p	df	t	p
I	56	1.50	—	52	1.79	—	56	2.80	<0.01
II	53	0.26	—	46	1.13	—	51	1.98	—
III	49	0.26	—	38	0.42	—	45	1.52	—
IV	41	1.83	—	29	1.44	—	37	1.46	—
V	31	0.15	—	20	0.17	—	32	0.40	—

TABLE 4. *Effect of Collective Copulation: The Number of Minutes to Attain Ejaculation*

Ejac.	A		B		C	
No.	M	SD	M	SD	M	SD
I	4.9	5.21	6.4	4.94	7.9	7.66
II	1.9	1.45	2.5	2.56	3.1	4.62
III	1.9	1.82	2.3	2.15	2.8	3.64
IV	2.3	2.65	2.8	3.54	2.5	2.35
V	4.0	4.59	3.7	4.19	4.1	3.95

	A–B			B–C			A–C		
	df	t	p	df	t	p	df	t	p
I	56	1.58	—	52	1.58	—	56	2.41	< 0.05
II	53	1.47	—	46	1.26	—	51	2.16	< 0.05
III	49	0.74	—	38	0.77	—	45	2.48	< 0.02
IV	41	1.27	—	29	0.83	—	36	1.78	—
V	31	0.20	—	20	0.14	—	32	0.15	—

TABLE 5. *Effect of Collective Copulation: The Total Number of Intromissions Per Minute*

Ejac.	A		B		C	
No.	M	SD	M	SD	M	SD
I	2.9	1.00	2.8	1.11	2.9	1.38
II	3.7	2.00	3.2	1.75	3.6	2.05
III	3.9	2.01	3.5	1.95	3.7	2.69
IV	3.3	1.60	3.2	1.69	3.6	2.24
V	3.1	1.79	3.3	1.82	2.7	1.46

	A–B			B–C			A–C		
	df	t	p	df	t	p	df	t	p
I	56	0.42	—	52	0.04	—	56	0.03	—
II	53	1.41	—	46	0.46	—	51	0.07	—
III	49	1.00	—	38	0.44	—	45	0.94	—
IV	41	0.79	—	29	1.13	—	36	1.18	—
V	31	0.56	—	20	1.56	—	32	1.18	—

TABLE 6. *Effect of Collective Copulation: The Length of the Refractory Periods*

Ejac.	A			B			C		
No.	N	M	SD	N	M	SD	N	M	SD
I	61	4.6	0.72	58	4.5	0.75	57	4.5	1.33
II	60	5.5	0.90	56	5.6	1.29	52	5.2	1.55
III	60	6.8	1.33	52	6.8	1.86	46	6.6	1.78
IV	58	8.5	1.87	43	8.0	2.60	39	7.8	2.34
V	47	10.9	2.65	35	9.5	2.58	35	10.1	4.51

	A–B			B–C			A–C		
	df	t	p	df	t	p	df	t	p
I	56	0.58	—	51	0.72	—	55	0.11	—
II	53	1.33	—	44	1.32	—	50	1.18	—
III	49	0.15	—	37	0.68	—	44	0.50	—
IV	39	0.67	—	27	0.51	—	36	0.79	—
V	25	1.39	—	18	0.75	—	29	0.86	—

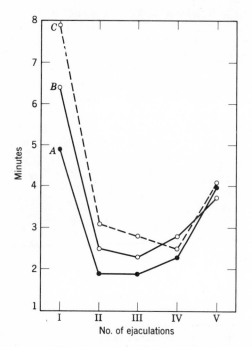

FIGURE 1. The effect of collective copulation on the time to achieve ejaculation by three different groups. A, three couples of the same age; B, in three cages placed side by side; C, one isolated couple in one cage.

The means of series B lay between those of series A and C.

Here we find a slight tendency to a greater frequency of intromissions preceding the different ejaculations by animals copulating in isolation.

Table 4 presents the time in minutes preceding ejaculation. While the first ejaculation is reached in A after 4.9 minutes, in C it took 7.9 minutes. The second ejaculation is reached after 1.9 and 3.1 minutes respectively. These differences are statistically assured in the first three copulation series.

Thus we find that copulation in isolation prolongs the duration of the series of ejaculations. This fact is illustrated in Fig. 1.

The slight increase in the frequency of intromissions, however, counteracts this time extension, and we get the same number of copulations per minute regardless of the conditions of experiment (see Table 5).

The duration of ejaculation is slightly prolonged in B in comparison with A and decreased in comparison with C.

The refractory periods presented in Table 6 do not show any changes. After the first ejaculation the refractory periods are 4.6, 4.5 and 4.5 minutes in series A–C. After the second ejaculation the values are 5.5, 5.6 and 5.2. There is a shortening of the refractory periods when the animals are isolated. This tendency may be due to the fact that there is a selection of highly potent animals with shorter refractory periods. This selection becomes more marked in C and in A, where the external sensory stimulation is stronger.

The figures indicating the very stable refractory periods, which remain totally uninfluenced by the sensory stimulation or collective copulation, emphasize the findings above. The increase in duration of ejaculation may be still more pronounced than can be seen in the figures, which are influenced by the selection of high potency rats in C.

When group copulation is compared to isolated copulation the effects are:

(1) The number of ejaculations per hour is increased.

(2) The number of intromissions is slightly decreased.

(3) The duration of the series of copulation is shortened and the frequency of intromissions diminishes, while the number of intromissions per minute remains stable.

(4) The refractory period is uninfluenced.

TABLE 7. *Effect of Collective Copulation on Sexual Activity in Senescent Rats: The Number of Ejaculations, Intromissions and Attempts*

	A			B			C		
	N	M	SD	N	M	SD	N	M	SD
Ejac.	15	3.8	2.57	12	2.9	1.83	15	2.1	1.62
Intr.	15	19.3	12.95	12	22.3	11.51	15	23.6	12.98
Att.	15	17.7	12.65	12	22.7	15.16	15	37.3	15.24

	A–B			B–C			A–C		
	df	t	p	df	t	p	df	t	p
Ejac.	11	2.78	<0.02	11	0.56	—	14	2.88	<0.02
Intr.	11	0.31	—	11	1.54	—	14	1.12	—
Att.	11	0.57	—	11	2.92	<0.02	14	5.16	<0.001

The ordinary experimental method of placing three observation cages side by side is shown to increase the number of ejaculations achieved compared to a situation where only one pair was observed in one observation.

3. THE EFFECT OF COLLECTIVE COPULATION IN SENESCENT ANIMALS

EXPERIMENTAL CONDITIONS. In this experiment we tried to discover whether the decreasing sexual activity of senescent rats could be stimulated to higher achievements by copulating in group.

Fifteen rats aged 25 months were observed under the same conditions as in the previous study:

A. Three pairs of animals copulated in the same observation box.

B. Three observation cages were placed side by side. One box held an old male, the two others contained young males.

C. Only one pair was allowed to copulate at the same time in the laboratory.

The animals weighed 390 ± 10 gm. Three animals taking part in A and C had to be killed before series B was started. The study was made in December 1955.

RESULTS. The number of ejaculations, intromissions and attempts during one hour's observation is presented in Table 7. With isolated copulation the number of ejaculations is 2.1, with collective copulation 3.8.

TABLE 8. *Effect of Collective Copulation on Sexual Activity in Senescent Rats: The Number of Intromissions, Minutes and Intromissions Per Minute to Achieve Ejaculation*

	A			B			C		
	N	M	SD	N	M	SD	N	M	SD
Intr.									
I	11	7.7	3.87	10	12.2	6.43	12	13.0	5.54
II	11	4.0	2.05	9	3.6	3.57	8	4.8	2.66
III	11	4.2	2.05	8	4.1	1.80	7	6.6	3.41
IV	10	3.4	1.65		—			—	
Min.									
I		7.0	4.24		17.3	8.39		20.0	14.81
II		2.7	2.19		4.3	7.80		2.9	1.72
III		2.4	2.24		3.4	1.89		6.2	4.98
IV		2.5	3.82		—			—	
Intr./min.									
I		1.2	0.52		0.8	0.38		0.9	0.48
II		1.9	0.99		1.7	1.14		1.8	0.49
III		2.3	1.48		1.3	0.34		1.5	0.83
IV		2.1	1.08		—			—	

	A–B			B–C			A–C		
	df	t	p	df	t	p	df	t	p
Intr.									
I	9	2.17	—	9	0.64	—	9	3.11	<0.02
II	8	0.17	—		—		6	0.03	—
III		—			—			—	
IV		—			—			—	
Min.									
I	9	2.31	<0.05	9	0.10	—	9	2.22	—
II	8	0.75	—		—		6	0.86	—
III		—			—			—	
IV		—			—			—	—
Intr./min.									
I	9	2.00	—	9	1.00	—	9	0.17	—
II	8	1.04	—		—		6	0.04	—
III		—			—			—	
IV		—			—			—	

The number of complete copulations in isolation increases from 19.3 to 23.6 and the number of incomplete copulations from 17.7 to 37.3.

In B the number of ejaculations is 2.9. This figure significantly differs from the corresponding number in A but not from that in C. Thus the effect of the sensory stimulation from copulating couples in other cages is not as great in senescent rats as in young animals.

However, it should be noted here that the B observation was made 20 days later than the A and C observations, and a decrease in sexual potency cannot be totally excluded, the changes at this age being rather rapid.

Table 8 reveals changes in the isolated components of behavior. Unfortunately, the small number of animals makes any accurate statement impossible.

TABLE 9. *Effect of Collective Copulation on Sexual Activity in Senescent Rats: The Length of the Refractory Period*

Ejac.	A			B			C		
No.	N	M	SD	N	M	SD	N	M	SD
I	11	5.7	1.66	9	5.3	1.27	11	5.5	1.06
II	11	6.5	1.67	9	6.2	1.01	8	6.1	1.07
III	11	8.1	2.58	8	7.6	0.93	7	8.3	1.67
IV	10	8.3	2.18	—			—		

	A–B			B–C			A–C		
	df	t	p	df	t	p	df	t	p
I	8	0.41	—	8	0.02	—	9	0.48	—
II	8	0.50	—	—			6	1.58	—
III	—			—			—		
IV	—			—			—		

While the first ejaculation is achieved after 7.7 intromissions, the corresponding figure in isolated copulation is 13.0. This difference is statistically assured. As in the previous experiment this difference is much more marked in the first series of copulations than in those following after.

The duration of ejaculation is prolonged from 7.0 to 20.0 in C. The difference, however great, is not statistically assured. Between A and B, however, where the means are 7.0 and 17.3, the difference is significant.

The number of intromissions per minute does not change considerably, nor do the lengths of the refractory periods (Table 9).

Group copulation has an even greater influence on the sexual activity of the senescent animals, the number of ejaculations when isolated decreasing to 55.3 per cent in comparison with 76.4 per cent for the younger rats (Table 10).

TABLE 10. *Comparison Between the Effects of Collective Copulation on the Number of Ejaculations in Senescent and Non-senescent Animals*

Exp. Cond.	Non-Sen.	% of A	Sen.	% of A
A	5.5	100.0	3.8	100.0
B	4.9	89.1	2.9	76.3
C	4.2	76.4	2.1	55.3

This fact indicates that the maximal level of activity in the senescent rat cannot be determined by the usual method of observation.

4. DISCUSSION

It is an experimentally well-founded fact that the rat eats more in the presence of other animals than when alone (Harlow 1932; Rasmussen 1939; Bruce 1941; Ross and Ross 1949; James 1953; Soulairac and Soulairac 1954; James and Gilbert 1955). The strengthening effect of sexual excitation when animals, especially birds, are copulating in groups is a common experience to biologists.

In this investigation we have not only observed an increase in sexual activity. The augmented number of ejaculations achieved was seen to be dependent on a heightening of the nervous stimulation eliciting ejaculation. The refractory periods remained stable while the copulatory reflex grew more intense. The components directly controlled by the cerebral cortex were affected by the sensory stimulation, while the metabolic processes remained unaffected.

It was further seen that the sensory stimulation not only referred to the immediate visual environment of the animal but also to distant non-visual stimuli originating from copulating animals in the same room.

Similar observations on female minks have been reported by Enders (Beach, 1952). He placed the females partly within the living quarters of other animals, partly without the immediate physical contact with the herd. In the females placed in the quarters he noted greater follicular development than in animals placed without the contact of the other minks. Distant sensory stimuli from other animals hastened the follicular development. The minks ordinarily ovulate only after copulation and even when no ovulation occurred here, the sensory stimulation was seen to influence the gonadal development.

To what extent can non-sexual stimuli contribute to the activation of sexual behavior? We have seen that the intervention of the experimenter enforcing intervals increases the activity. Certainly other stimuli can have a similar effect.

The increased dependence of the senescent animal on sensory stimulation reveals a heightened threshold for nervous stimulation in old age.

SUMMARY

1. Animals copulating in groups were seen to reach a higher number of ejaculations per hour and to achieve ejaculation in a shorter time than when isolated. The length of the refractory periods remained unaffected.

2. The presence of other copulating animals in the room had a similar but less intense effect.

3. The sexual activity in senescent rats was considerably increased when they were allowed to copulate in groups.

REFERENCES

BEACH F. A. (1952). "Psychosomatic" phenomena in animals. *Psychosom. Med.*, **14**, 261.

BRUCE, R. H. (1941). An experimental analysis of social factors affecting the performance of white rats. I. Performance in learning a simple field situation. *J. Comp. Psychol.*, **31**, 363.

GRUNT, J. A. and YOUNG, W. C. (1952). Psychological modification of fatigue following orgasm (ejaculation) in the male guinea pig. *J. Comp. Physiol. Psychol.*, **45**, 508.

HARLOW, H. F. (1932). Social facilitation of feeding in the albino rat. *J. Genet. Psychol.*, **41**, 211.

JAMES, W. T. (1953). Social facilitation of eating behavior in puppies after satiation. *J. Comp. Physiol. Psychol.*, **46**, 427.

JAMES, W. T. and GILBERT, T. F. ('955). The effect of social facilitation on food intake in puppies fed separately and together for the first 90 days of life. *Brit. J. Animal Behav.*, **3**, 131.

RASMUSSEN, E. W. (1939). Social facilitation in albino rats. *Acta Psychol.*, **4**, 275.

ROSS, S. and ROSS, J. G. (1949). Social facilitation of feeding behaviour in dogs: I. Group and solitary feed. *J. Genet. Psychol.*, **74**, 97.

SOULAIRAC, A. (1950). L'effet de groupe dans le comportement sexuel du rat mâle. *Coll. Internat. de C.N.R.S. Paris*, **34**, 91.

SOULAIRAC, A. and SOULAIRAC, M.-L. (1954). Effets du groupement sur le comportement alimentaire du rat. *C.R.S. Soc. Biol.*, **148**, 304.

SOCIAL MODIFICATION OF THE ACTIVITY OF ANTS IN NEST-BUILDING*

Shisan C. Chen

SOCIAL behavior of various groups of animals has been studied by many naturalists, and important contributions have already been made to our knowledge about social behavior, especially of the social insects. Recently, some biologists have attempted to study the general social phenomena in the animal world and have started various lines of study. One of these is the study of animal aggregations, started by Allee (1931). This line of work is concerned with the physiological effects

* From *Physiological Zoology*, 1937, **10**, 420–436. Copyright 1937 by University of Chicago.

of crowding upon the individuals composing the crowd. Through the work of Allee and his followers many interesting facts concerning the effects of crowding have already been discovered.

Another line of the study of general social phenomena is that of population growth started by Pearl (1925). This line of work is concerned with the experimental studies of the growth of animal population in comparison with that of human population.

The third line of study of general social phenomena of animals has been independently carried on by Deegener (1918), Alverdes (1925), Picard (1933), and others. These authors have done a great deal in collecting the isolated facts, scattered in numerous special publications, about the social life of various groups of animals and in organizing them into comprehensive synthetic treatises.

The work which I am going to report in a series of papers, of which the present paper is the first, is concerned with experimental studies of some general social phenomena in ants. The special social activities of ants, or activities which are peculiar to ants, such as division of labor, care of young, common activities in defense and attack, slavery, relation with aphids and other insects, etc., have already been extensively and intensively studied by numerous entomologists, especially Wheeler (1910, 1923), Forel (1928), Wasmann (1915), Escherich (1917), and Brun (1924). In spite of the richness of the contributions which the myrmecologists and entomologists have made to our knowledge about the special social activities of ants, there has been little experimental work from the standpoint of general sociology.

My experimental study of general social phenomena in ants was started in the spring of 1936 by a series of analytic and synthetic studies of their nest-building activities. The results obtained are reported in two separate papers, of which the present one is concerned with the modification by social environment of the nest-building activities of individual ants.

MATERIAL AND METHODS

The common black ant, *Camponotus japonicus* var. *aterrimus* Emery, was used for all experiments reported in the present paper. A colony of this variety of ants is usually composed of a few wingless females, a few or many winged females and males, and thousands of workers. The workers vary greatly in size and may be classified, according to size, into at least three classes: the

small, medium-sized, and large workers. The average body length of the small workers is about 9 mm; that of the medium-sized workers and large workers, about 11.5 mm and 15 mm.

Camponotus japonicus var. *aterrimus* Emery is widely distributed in China, and its occurrence has been recorded in the following provinces: Hopei, Shantung, Hunan, Kiangsu, Chekiang, and Fukien (Wheeler, 1930). This variety of ants was selected for the following reasons: (1) The large size allows their activities to be observed easily and clearly. (2) Polymorphism in the size of workers allows individual ants to be easily distinguished or recognized without marking them with coloring matter or other materials, which often produce injurious effects. (3) On account of the wide distribution and abundance of this variety of ants, they are easily collected.

Workers of various size were used for experimental studies, and sexual forms were included because (1) since the primary object of the present study is to find out the general relations between the individuals in a society, it is better to eliminate the disturbing factor of sex by using a group of animals without sexual difference; (2) the sexual forms of ants ordinarily do not take part in the nest-building activities of the colony.

The ants were collected on the campus or in the vicinity of the National Tsing Hua University. Collections were made frequently, at least once a week, in order to insure a constant supply of fresh material for study. They were generally collected a few days before the beginning of an experiment. Only in exceptional cases, when experiments were continued for several weeks, were ants used after a month or more in the laboratory.

In order to avoid injurious effects of environmental conditions in the ordinary laboratory rooms, the experiments were carried out in a building situated in the Goldfish Garden of the University, where the conditions are nearly similar to the natural environmental conditions. In this building there is a spacious room which is well ventilated, evenly lighted, and yet not exposed to direct sunlight. The temperature in this room varied from 25°C to 32°C during the period, in which these experiments were carried on.

The ants collected from the field were kept in large glass containers, supplied with a piece of wet sponge or a little moist earth, covered with cheesecloth, and put in a dark cool place. Ants belonging to different colonies were kept in separate containers. They were occasionally fed with honey or with sugar water on a small piece of sponge. At the beginning of each experiment the ants to be used were carefully examined in order to avoid use of individuals which were accidentally injured in collecting and handling them.

NEST-BUILDING ACTIVITIES OF ANTS IN ISOLATION

The nest-building activities of ants have long been well known to naturalists, and the different types of ant nests have already been studied and described in detail (Wheeler, 1910; Forel, 1928). But this activity as a general phenomenon of co-operative social work has not yet been investigated. My own studies along this line have been first analytic, then synthetic.

The analytical part of the work consists of a study of the nest-building activity of individual ants in isolation. Normally, nest-building is a social activity and individual ants work as members of a group. But we cannot conclude from the ability of nest-building by a group that all ants can also build a nest when in isolation. Experiment is needed to settle this question.

The experiment planned for studying nest-building in isolation was started on May 19; 70 workers were used. Each worker was introduced into a separate working bottle, shown in Figure 1, which was made of a half-pint milk bottle covered by a piece of cheesecloth fastened to the mouth of the bottle by a rubber band. As nest-building material, about 130 cc of air-dried sandy soil, which had been finely pulverized and passed through a sieve, was put into each bottle. This was followed by 35 cc of water, which soaked into the soil and made it moist enough for nest-building. The surface of the moist earth was made smooth by sprinkling over it at first a little very finely pulverized dry soil and afterward a fine spray of water from a washing bottle such as used in the chemical laboratories. The 70 working bottles thus made were not only similar in size and form but also in the volume and the moisture content of the contained earth.

The 70 workers used for this experiment were not selected with reference to size; they included large, medium-sized and small workers. They were introduced into the 70 working bottles, one ant into each bottle. Each ant was then in a new world entirely isolated from the others. After a few minutes in this isolated condition, some ants started the work of nest-building. Four hours after introduction into the bottle, 47 of the 70 workers had already started excavation of nests. Twenty-three hours after being introduced into the bottle, 62 of the 70 ants had started the work

FIGURE 1. Working bottle.

of nest-building. Of the remaining 8 workers which had not yet started, 6 had begun work $21\frac{1}{2}$ hours later, and the other two had begun work 70 hours after introduction. Hence, within 70 hours all the 70 ants had started nest-building.

This experiment shows that all workers are capable of nest-building in isolation but that this ability is not identical among the individual ants. Some ants, in response to the nest-building stimuli (absence of old nests, presence of moist earth, etc.) in the new environment of isolation, began building very early; others began later or very late.

The isolated ants differ not only in the time of beginning the building but also in many other ways as regards details of construction. Most ants selected a spot near to the source of the light, i.e. the southern window of the room, and against the glass wall of the bottle for beginning work. Others started building in other places in the bottle. Most ants worked consistently in one spot; but a few dug two or more shallow cavities in different places in the bottle instead of concentrating their energy in constructing one deep nest. Some ants worked continuously and persistently until the nests or holes which they made were one or more inches deep and a considerable amount of earth which they dug out of the nests had accumulated around the openings. Others worked intermittently and dug out only a little earth to deposit around the openings of the shallow cavities.

TABLE 1. *Reaction Time of Ants Working in Isolation and in Groups**

Ants	In Isolation (June 4–6)		In Groups of 2 (June 8–10)		In Groups of 3 (June 11–13)		In Isolation (June 14–16)	
	M.R.T.	Dev.	M.R.T.	Dev.	M.R.T.	Dev.	M.R.T.	Dev.
L1	182+	8	22	3	25	1	360+	159
L2	130+	60	10	15	12	12	6	195
L3	360+	170	30	5	13	11	360+	159
L4	142+	48	10	15	27	3	142+	59
L5	49	141	35	10	2	22	360+	159
L6	360+	170	4	21	4	20	2	199
L7	247+	57	13	12	11	13	240+	39
L8	244+	54	4	21	3	21	124+	77
L9	141+	49	45	20	20	4	265+	64
L10	147+	43	8	17	131	107	241+	40
L11	156+	34	8	17	19	5	74	127
L12	124+	66	108	83	24	0	242+	41
Av.	190+	75	25	20	24	18	201+	110
M1	212+	9	17	1	12	30	271+	144
M2	230+	27	2	14	3	39	298+	171
M3	207+	4	57	41	125+	83	62	65
M4	225	22	8	8	20	22	126+	1
M5	360+	157	30	14	2	40	121+	6
M6	5	198	14	2	5	37	2	125
M7	259+	56	8	8	3	39	6	121
M8	248+	45	12	4	257+	215	360+	233
M9	26	177	7	9	4	38	127+	0
M11	298	95	1	15	15	27	6	121
M12	158+	45	20	4	21	21	20	107
Av.	203+	76	16	11	42+	54	127+	99
S1	7	175	4	40	14	18	19	134
S2	96	86	5	39	19	13	242+	89
S4	57	125	13	31	20	12	243+	90
S5	360+	178	22	22	20	12	244+	91
S6	338+	156	24	20	122+	90	13	140
S7	175+	7	1	43	2	30	136+	17
S8	63	119	127+	83	3	29	8	145
S9	244+	62	157+	113	12	20	173	20
S10	228	46	8	36	98	66	74	79
S11	300+	118	10	34	17	15	282+	129
S12	133+	49	108	64	21	11	248+	95
Av.	182+	102	44+	48	32+	29	153+	94
A.A.	192+	84	28+	26	33+	34	160+	101

*M.R.T., mean reaction time; Dev., deviation; Av., average; A.A., average of three averages.

34

NEST-BUILDING ACTIVITIES OF
ANTS IN ASSOCIATION

The synthetic part of this study, i.e. the study of the nest-building activities of ants in association, will be reported in the following: On June 3, I selected from a stock of *Camponotus* 36 sound and healthy workers—12 large, 12 medium-sized and 12 small workers. The large workers varied in body length from 14.1 mm to 16.6 mm, averaging 15.03 mm; the medium-sized, from 10.5 mm. to 12.2 mm, averaging 11.23 mm; the small, from 8.2 mm to 9.8 mm, averaging 9.15 mm. The body lengths of these ants were measured at the end of this experiment by means of a caliper, after the ants were etherized.

Each of these 36 ants was distinguished by a letter, "L," "M," or "S," and a number, from 1 to 12. The letter "L" meant large; "M," medium-sized; "S," small worker. The numbers from 1 to 12 were the numbers of the individuals belonging to each size-class.

These 36 ants were kept in isolation or in association in two kinds of bottles: the working bottles and the resting bottles. The working bottles with moist earth in them were prepared according to the method described in the previous section. In these bottles the ants carried on their work of nest-building. In order to prevent the ants from excessive nest-building work, no moist earth was supplied; rather, there was in each resting bottle a piece of wet sponge for supplying moisture. Occasionally, the ants were fed with dilute sugar water while they were in the resting bottles; the frequency of feeding and the concentration of sugar water was identical for all the 36 ants.

During the period of experimentation the ants were transferred from the resting bottles into the working bottles for a period of 6 hours every day, after which they were transferred back into the resting bottles.

The experiments with the 36 ants lasted 12 days. During the first 3 days, the ants were kept isolated in 36 resting bottles and were transferred into 36 working bottles, one in each bottle, to work in isolation for 6 hours daily. In the second 3 days, the 36 ants were combined into 18 pairs. Each of these 18 pairs was kept in a separate resting bottle during the resting hours and was introduced into a separate working bottle, where the ants were allowed to work in association for 6 hours every day. During the third 3 days, they were recombined 3 by 3 into 12 groups, and these 3-ant groups were kept in 12 resting bottles and allowed to work for 6 hours daily in 12 working bottles.

In the last 3 days, they were separated again and kept isolated both during the working and during the resting hours.

Concerning the daily work of each ant or group of ants, the following records were taken: (1) the reaction time and (2) the weight of earth excavated. The reaction time is the time elapsed between the entrance of ants into the working bottle and the beginning of their building activity. This interval of time was recorded in number of minutes. In case the ant did not start to work in the 6-hour working period, it was recorded as "360 + ." The weight of earth excavated is the amount of the earth which the ants brought up in the form of pellets from the nest and deposited around the entrance of the nest. While preparing the working bottles, special attention was paid to making the surface of the earth smooth and fairly compact so that the pellets deposited on it could be brushed away. At the end of the 6-hour working period, all the pellets accumulated on the surface of the earth were carefully brushed together and transferred into vials. After having been air-dried in the vials, the collected masses of pellets were weighed. The weight of these masses of pellets was taken to represent the amount of work done by the ant, or the ants, within the 6 working hours.

The data obtained from the experiments described above are recorded in Tables 1–4. M10 and S3 died within the period of experimentation; hence their data are not included in the tables. From the data in these tables the following conclusions were derived.

1. *The reaction time was shorter when the ants worked in association than when they worked in isolation.*—Table 1 shows that when the ants worked in 2-ant or 3-ant groups the mean reaction time was, on the average, 28 + minutes or 33 + minutes; when they worked in isolation the mean reaction time was 192 + minutes or 160 + minutes.

2. *The amount of accomplished work was more when the ants worked in association than when they worked in isolation.*—When isolated, each ant excavated, on the average, 0.207 gm of earth in 6 hours (Table 2). When associated, the amount of excavated earth was greatly increased; it was 3.649 times more when the ants worked in 2-ant groups (Table 3) and 3.456 times more when they worked in 3-ant groups (Table 4) than when they worked in isolation.

3. *The associated ants worked with greater uniformity than the isolated ants.*—Based upon the data of the mean reaction time, average deviations had been calculated and recorded in Table 1.

TABLE 2. *Weight of Earth Excavated by 1 Ant Working in Isolation*

Ants	Weight of Earth Excavated in Isolation (in Milligrams per 6 Hours of Work)		
	Averages of three trials of first period of isolation (June 4–6)	Averages of three trials of second period of isolation (June 14–16)	Averages of both periods of isolation
L1	627	0	314
L2	356	553	455
L3	0	0	0
L4	61	243	152
L5	676	0	338
L6	0	146	73
L7	43	58	51
L8	195	90	143
L9	32	75	54
L10	85	95	90
L11	162	655	409
L12	278	23	151
Av.	210	162	186
M1	90	11	51
M2	446	11	229
M3	165	168	167
M4	191	239	215
M5	0	21	11
M6	1,023	657	840
M7	86	957	522
M8	147	0	74
M9	434	104	269
M11	67	496	282
M12	166	197	182
Av.	256	260	258
S1	671	663	667
S2	521	18	270
S4	111	18	65
S5	0	148	74
S6	30	61	46
S7	592	57	325
S8	39	52	46
S9	45	56	51
S10	87	177	132
S11	115	23	69
S12	335	98	217
Av.	231	125	178
A.A.	232	182	207

*Av., average; A.A., average of three averages.

TABLE 3. *Weight of Earth Excavated by 2 Ants Working in Isolation or in Groups**

Ants	Weight of Earth Excavated by 2 Ants (in Grams per 6 Hours of Work)		
	When Working in Isolation (1)	When Working in Groups (2)	Ratio (2):(1)
L1, M1	0.365	1.555	4.260:1
L2, M2	0.684	1.920	2.807:1
L3, M3	0.167	0.893	5.347:1
L4, M4	0.367	2.646	7.210:1
L5, M5	0.349	2.529	7.246:1
L6, M6	0.013	1.682	1.842:1
L7, S7	0.376	1.329	3.535:1
L8, S8	0.189	0.829	4.386:1
L9, S9	0.105	0.335	3.190:1
L10, S10	0.222	1.694	7.631:1
L11, S11	0.478	1.200	2.510:1
L12, S12	0.368	1.380	3.750:1
M7, S1	1.189	2.492	2.096:1
M8, S2	0.344	1.652	4.802:1
M11, S5	0.356	1.331	3.739:1
M12, S6	0.228	1.002	4.395:1
Av.	0.419	1.529	3.649:1

*Data in column (1) were obtained by adding the data of individual ants in Table 2.

TABLE 4. *Weight of Earth Excavated by 3 Ants Working in Isolation or in Groups**

Ants	Weight of Earth Excavated by 3 Ants (in Grams per 6 Hours of Work)		
	When working in isolation (1)	When working in groups (2)	Ratio (2):(1)
L1, M1, S1	1.032	2.134	2.068:1
L2, M2, S2	0.954	3.375	3.538:1
L4, M4, S4	0.432	2.715	6.285:1
L5, M5, S5	0.423	3.810	9.007:1
L6, M6, S6	0.959	1.123	1.171:1
L7, M7, S7	0.898	3.499	3.896:1
L8, M8, S8	0.263	0.708	2.692:1
L9, M9, S9	0.374	1.700	4.545:1
L11, M11, S11	0.760	2.085	2.743:1
L12, M12, S12	0.550	1.826	3.320:1
Av.	0.665	2.298	3.456:1

*Data in column (1) were obtained by adding the data of individual ants in Table 2.

TABLE 5. *Data showing that the Accelerating Effect of Association is Greater for Slow Workers than for Rapid Workers*

Ants	Mean Reaction Time in Minutes						Lengthening (+) or Shortening (−) of Reaction Time by Working in Groups
	When working in isolation			When working in groups			
	June 4–6	June 14–16	Aver- age	In 2-Ant Groups June 8–10	In 3-Ant Groups June 11–13	Aver- age	
Rapid workers:							
M6	5	2	3.5	14	5	9.5	(+) 6
S1	7	19	13	4	14	9	(−) 4
S8	63	8	35.5	127 +	3	65 +	(+) 29.5 +
Slow workers:							
L3	360 +	360 +	360 +	30	13	21.5	(−) 338.5 +
M8	248 +	360 +	304 +	12	257 +	134.5 +	(−) 169.5 ±
S5	360 +	244 +	302 +	22	20	21	(−) 281 +

These deviations are much smaller (26 and 34) in ants working in groups, and larger (84 and 101) in ants working in isolation, showing greater uniformity in the former than in the latter.

4. *The accelerating effect of association was greater for the slow workers than for the rapid workers.*—Table 5 shows the mean reaction times of the three best workers (M6, S1, S8) and three poorest workers (L3, M8, S5) of the 36 ants investigated. It is clear from these data that there may be some increase in the mean reaction time by working in association with other ants in the case of the best workers, but there is a great reduction in the case of the poor workers. Thus, the mean reaction time of the best worker (M6) is 3.5 minutes in isolation, but 14 minutes in the 2-ant group and 5 minutes in the 3-ant group. In this case the reaction time is increased. The poorest worker (L3) did not work at all during the 6 hours in isolation. But, when associated with other ants, it started to work within 30 or 13 minutes after introduction into the working bottle.

REVIEW OF LITERATURE AND DISCUSSION

The modification of the activity of individuals by social environment has long been studied in the field of human social psychology. Triplett, Mayer, Moede, Allport, and others (Allport, 1924; Murphy and Murphy, 1931) found that, in both physical and mental work, children or adults accomplish more work, and with greater uniformity or less deviation, when working in asso-

ciation with other co-workers than when working in isolation. They found that the work of an individual within a group is accelerated by a more rapid co-worker but retarded by a slower one, and that the accelerating effect of association is greatest for the slowest and least for the most rapid members of the group because the average speed of activity of the group is more rapid than the speed of the slow members of the group but less rapid than that of the rapid members of the group. According to Allport (1924), two social factors may be recognized in the social influence on activity of individuals. "The first is social facilitation, which consists of an increase of response merely from the sight or sound of others making the same movements. The second is rivalry, an emotional reinforcement of movement accompanied by the consciousness of a desire to win."

While the social influence on the activity of individuals has been much studied in man, it is very little explored in other animals. Bayer (1929) found that hens eat more when in groups than when isolated. In 4-hen groups the social increment ranged from 33 per cent to 200 per cent, and averaged 96 per cent.

Harlow (1932) investigated the social facilitation of feeding in the albino rat and found that the amount of food eaten by 20 rats was 70 per cent more when these were fed in groups of 5 than when fed individually. In many cases, however, the difference between the experimental and the control animals were just statistically significant. Harlow also observed that social facilitation is not a function of the size of the group, the results of the experiments on groups of 5 rats, groups of 2 rats, and groups of 3 rats being similar. Here, too, the numerical data which the author reported seem too few to substantiate his conclusion.

In fishes, Welty (1934) found that goldfishes become conditioned to run a simple maze more rapidly when grouped than when isolated; the larger the group the quicker the conditioning. He also found that a trained goldfish in a maze with an untrained fish will speed up the conditioning of the latter. In some experiments the fish which gave the most rapid reaction was found among the isolated fishes and not in the groups. Likewise the slowest fish to respond was found among the isolated fishes. This shows a greater variation in reaction time among the isolated fishes. Concerning social facilitation in feeding, Welty found that fishes eat more food per fish in groups than when isolated. He also observed that under certain experimental conditions mud minnows in groups exhibit an antagonistic behavior that inhibits learning.

Although the social life of ants has long been studied in detail by numerous entomologists, no one has yet investigated the problem of social modification of individual activities in ants as Triplett, Mayer, Allport, Bayer, Harlow, Welty, and others studied this problem in man, hens, rats, and fishes. The experimental results reported in the present paper agree essentially with the principles established by earlier workers for man and other animals. They agree with the work on men, hens, rats and fishes in that individuals react more rapidly and accomplish more when in groups than when isolated. However, social increment in ants is much greater than in man or other animals. When the ants worked in 2-ant or 3-ant groups, the reaction time was, on the average, 28 + minutes or 33 + minutes; when they worked in isolation, the reaction time was 192 + minutes or 160 + minutes. The isolated ants excavated, on the average, 0.207 gm of earth per ant in 6 hours. When grouped, the amount of excavated earth was greatly increased; it was 3.649 times more when the ants worked in 2-ant groups and 3.456 times more when they worked in 3-ant groups than when they worked in isolation· The social increment in man and other animals is never so great as that of ants, in some cases the difference between the data of the grouped individuals and those of the isolated individuals being only just statistically significant. The very large social increment in ants is probably due to the fact that ants are much more highly socialized than men, hens, rats, and goldfishes; hence they are much more affected by being isolated or grouped.

In ants, I observed the fact that the grouped individuals worked with greater uniformity than the isolated individuals, that the accelerating effect of association was greater for the slower workers than for the more rapid workers, and that the more rapid co-worker has an accelerating effect, and the less rapid co-worker has a retarding effect, upon the work of an individual within a group. Mayer and others (Allport, 1924; Murphy and Murphy, 1931) observed similar facts in their studies of human social behavior; and Welty (1934), in his experiments on goldfish.

In regard to the effect of the size of the group on the activity of an individual, the experimental data which I obtained show that there is no difference between the effect of a group of 2 ants and that of a group of many more ants on the rate of work of an individual ant within the group. This problem has not yet been extensively investigated in man and other animals. Incidentally, contradictory results have been obtained by Harlow (1932) and Welty (1934) in their studies on rats and goldfishes. Harlow obtained results, which agree with my observations, showing that the social facilitation of feeding in the albino rat was not a function of the size of the group, but Welty found that the larger the group the quicker the goldfishes become conditioned to run a simple maze.

SUMMARY

The author reports in this paper the results of studies on the social modification of the activity of individuals in a variety of common Asiatic ants, *Camponotus japonicus* var. *aterrimus* Emery. From the experimental data the following conclusions were derived:

1. All the workers of *Camponotus* are capable of nest-building in isolation, but building work varies with different individuals.

2. The reaction time is shorter when the ants work in association than when they work in isolation.

3. The amount of work accomplished is more when the ants work in association than when they work in isolation.

4. The associated ants work with greater uniformity than the isolated ants.

5. The accelerating effect of association is greater for the slow workers than for the rapid workers.

6. The rapid co-worker has an accelerating effect, and the slow co-worker has a retarding effect, upon the work of an individual.

7. There is no difference between the effect of a group of 2 ants and that of a group of many more ants on the rate of work of the individual ant within the group.

8. The results of the present study in ants, as reported above, agree in principle with those of similar studies in men, hens, rats, and goldfishes.

REFERENCES

ALLEE, W. C., 1931. *Animal aggregations.* Chicago: University of Chicago Press.

ALLPORT, F. H., 1924. *Social psychology.* Boston: Houghton Mifflin Co.

ALVERDES, F., 1925. *Tiersoziologie.* Leipzig: Hirschfeld.

BAYER, E., 1929. Beiträge zur Zweikomponenten-theorie des Hungers. *Zeit. f. Psychologie*, **112**, 1–54.

BRUN, R., 1924. *Das Leben der Ameisen.* Leipzig: B. G. Teubner.

DEEGENER, P., 1918. *Die Formen der Vergesellschaftung im Tierreiche.* Leipzig: Veit.

ESCHERICH, K., 1917. *Die Ameise.* Braunschweig: Verweg. u. Sohn.

FOREL, A., 1928. *The social world of the ants compared with that of man.* London and New York: G. P. Putnam's Sons.

HARLOW, H. F., 1932. Social facilitation of feeding in the albino rat. *Jour. Genetic Psychol.*, **41**, 211–21.

MURPHY, G. M. and MURPHY, L. B., 1931. *Experimental social psychology.* New York and London: Harper & Bros.

PEARL, R., 1925. *The biology of the growth of population.* New York: Knopf.

PICARD, F., 1933. *Les phénomènes sociaux chez les animaux.* Paris: Librairie Armand Colin.

WASMANN, E., 1915. *Das Gesellschaftsleben der Ameisen.* Münster: Aschendorffsche Verglagsbuchhandlung.

WELTY, J. C., 1934. Experiments in group behavior of fishes. *Physiol. Zoöl.*, **7**, 85–128.

WHEELER, W. M., 1910. *Ants, their structure, development, and behavior.* New York: Columbia University Press.

——, 1923. *Social life among insects.* New York: Harcourt Brace.

——, 1930. A list of the known Chinese ants. *Peking Nat. Hist. Bull.*, **5**, 53–81.

CONDITIONED BEHAVIOR OF ISOLATED AND GROUPED COCKROACHES ON A SIMPLE MAZE*

Mary Frances Gates and W. C. Allee

THERE is nothing new in the demonstration that an insect, even such a generalized insect as the cockroach, can show conditioned as well as innate behavior responses. Roaches are strongly negative in their reactions to light (Graber, 1884). Modification in this tropistic behavior was demonstrated by Szymanski (1912) in experiments which showed that the direction of movement of these insects could be changed by punishment with electric shocks so that they would remain in light instead of entering an otherwise available dark chamber.

Turner (1912) confirmed the results of these experiments and (1913) also conditioned cockroaches to run a maze. He used the oriental cockroach, *Periplaneta orientalis L.* The maze was a series of runways with blind alleys; it was made from a sheet of copper which was supported over a tray containing water. An incline at the end of one of the runways led to the glass container to which each individual was accustomed. All of the experiments were conducted by daylight, and the roaches were conditioned after successive trials with short intermissions between tests. Eldering

* From the *Journal of Comparative Psychology*, 1933, **15**, 331–358. The present investigation was aided in part by a grant to The University of Chicago from The Rockefeller Foundation.

(1919) also found that cockroaches (*P. americana*) showed improvement from day to day in time spent in running a simple maze.

Also there is nothing new in the idea that grouped animals may behave differently from those that were isolated. Tarde (1903) wrote of this effect under the heading of inter-psychology and many others have recorded observations showing that numbers of arthropods together may give different reactions than do isolated individuals. Much of this literature has recently been reviewed by Allee (1931). More recently Welty (1933) has shown that four fishes will become conditioned to run a simple maze more rapidly than will a similar number of isolated fishes similarly treated. Stimulated by Welty's experience and by the work of Szymanski, Turner and Eldering, it has been the object of these experiments to compare the maze behavior of a single roach when isolated, paired, and a member of a group of three. Certain effects of group living have long been known for ants, bees and wasps, representatives of highly specialized insects; termites nearer relatives of the roaches, also have highly developed social life; but the extent to which social attributes exist in a primitive sub-social type, such as the cockroach, is another issue.

METHODS

The roaches used were all females of *Periplaneta americana*; only females 3 cm in length were selected. The individuals selected for testing were isolated in milk bottles plugged with cotton; they were given an abundance of bananas for food and were kept in the dark-room in which the maze tests were made.

Two types of mazes were tried; both were constructed of galvanized iron. Both were similar to the one used by Turner in regard to having water as a punishment for leaving the maze platform, but the design was less complex. The first was a platform with four runways set in a tray containing about an inch of water. Food, the reward, was placed at the end of one of the middle runways. This maze was used in experiments of series I. The maze used in all the later work had a platform with only three runways, as shown in Figure 1. The reward was furnished by the dark shelter given by a small bottle covered with black paper; it was placed at the end of the center runway and corresponded to the dark places where roaches usually live. Maze dimensions as used for experiments of series II are given in Figure 3.

FIGURE 1. A flashlight photograph of the maze as used in series 2 with three roaches on the runways.

In the last group of experiments (series III) the runways were all shortened 2 inches in an attempt to secure a still simpler maze.

For convenience in noting errors on the maze, the runways were *A*, *B*, and *C*, from left to right, and the corners of the platform *A* corner and *C* corner. All experiments were conducted in a well-ventilated dark-room under a red light of intensity as shown in Table 1. The light was placed 15 inches above the *B* runway and was so placed that the ends of the runways were less intensely illuminated than were the corners of the platform.

The roaches, when on the maze, could not touch the sides of the tray, but could reach the water below with their antennae. When first placed on the platform, there was a tendency for the roach either to fall or to run off into the water, after which some time was spent in cleaning itself; this cleaning time was deducted from the learning time. A pair of forceps and some cotton, onto which the roach would crawl, was useful in removing it from the water.

All the trials were timed with a stop watch; 2 minutes were allowed between trials. The average number of trials per day per individual was from 15 to 25 (see Fig. 2 for the first-day trials of one roach).

The roach was allowed to crawl from the bottle, onto the platform opposite the *B* runway in such a way that it was directed away from the rewarding shelter. It moved about on the maze until it reached the dark bottle; to do this it had to pass through the area receiving the most light.

TABLE 1. *Relative Intensity of the Light at Different Points on the Maze in Terms of Percentage of the Greatest Intensity*

	A	B	C
A. Series II			
End of runways	51.6	52.7	51.3
Corners	54.5		
Beginning of runway		77.3	
Directly under light		100.0	
B. Series III			
End of runways	46.1		50.0
Corners	77.8		74.4
Beginning of runway		90.6	
Directly under light		100.0*	

*Absolute value on a Weston Illuminometer, 1.8 foot candles

After the roach had entered the dark shelter, this bottle was removed and replaced by another the same size; the maze was wiped with alcohol to destroy all chemical traces that might remain, and after a 2-min. rest period during which the alcohol evaporated, the roach was allowed to crawl out of the bottle and onto the maze for the next trial. None of the roaches were handled directly; by tipping the bottle they were induced to crawl up the sides and out onto the experimental platform. This method is an improvement over the use of forceps or other devices for handling insects when using them in maze experiments.

A record was kept of each movement, change in direction, pause and all behavior reactions. Movements to the *A* or *C* corners, to the end of *A* or *C*, were counted as errors.

The procedure for the roaches when paired and grouped was the same as that used when they were isolated; when it was difficult to tell the insects apart, a marking system was used, i.e. a speck of oil paint was placed on the tegmina. When groups were tested, the 2-min. intermission between trials started when the last roach had entered the dark bottle.

RESULTS
Series I

In this series food (banana) was used as a reward on a four-runway maze. The banana was placed at the end of one of the center runways.

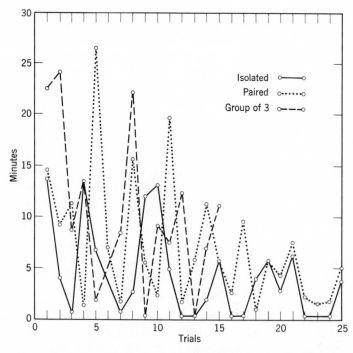

FIGURE 2. Average time per trial of roach 5, for the first day on maze isolated, first day paired with roach 6, and first day when grouped with roaches 4 and 6.

TABLE 2. *Experiments Using Dark Shelter as a Reward*

Date	Result	How Tested	Roaches	Number of Days for Each Roach
		Series II		
Spring, 1931	Conditioned	Isolated	A	2
			B. C. D. E.	1
			I	7
			II	5
		Paired	I, II	4
	No conditioning	Isolated	6	1–4
	Died		1	
Autumn, 1931	Conditioned	Isolated	III, IV, V, VI	5
			VII	1
			VIII	3
		Paired	III, IV	5
			V, VI	5
		Group of three	IV, V, VI	5
	No conditioning		2	3–5
	Died		1	
		Series III		
Spring, 1932	Conditioned	Isolated	IX, X, XI	5
		Paired	IX, X	5
			X, XI	5
		Long rest periods	XI, XII, XIII	1–5
	No conditioning		8	1–5
	Died		1	

After tests continuing as long as 25 days with 25 trials per day, only 1 of 10 roaches had shown evidence of improvement in time taken to run the maze and the improvement in this one case was not marked. The failure under these conditions was probably due to the fact that roaches must be given frequent trials frequently repeated and food did not afford a constant reward, but tended to produce variable results when it was necessary for 25 trips to be made to it in one day. The conclusion that these roaches did not show improvement in running the maze under the conditions just described is not entirely correct for it does not consider the fact that about two-thirds of the roaches tested did become maze-conditioned to the extent that they did not continue to fall or to run off the maze into the water below. When time spent in cleaning themselves after such an emersion is eliminated, as was the invariable custom in all these experiments, only one of the insects tested showed even slight improvement.

Series II and III

There were roaches which could not be conditioned to show improvement on the simplified maze with three runways even with a dark receptacle as a reward. Those which gave the best results in preliminary trials were selected and conditioned when isolated. They were then paired with one another to find what effects one had on the other, and finally they were tried in a group of three to see how the presence of these small numbers on the maze at one time affected the behavior of each individual in running the maze. Table 2 gives a summary of the experiments upon which the remainder of this report is based.

In all experiments an individual roach, when first placed on the maze, followed the outer edges of the platform and runways. Each pause resulted in a turn to either right or left. If at the end of the *C* runway (Fig. 1), for example, and at the right side, a right turn was made and the roach followed the outer edge back to the corner on this

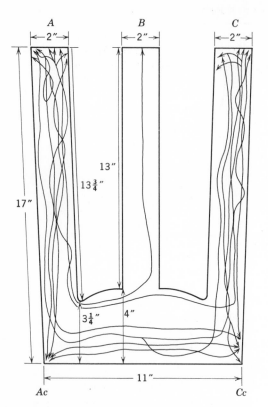

FIGURE 3. Activity on the maze as exhibited by roach 5 on its first trial (isolated). Numbers give dimensions of the maze for series 2. In series 3, each runway was shortened 2 in.

side of the platform. If on the left side, it followed back on the inside to the platform, and to the corner on the opposite side or directly down the *A* runway. Finally, this tendency to go to the outside was overcome, perhaps accidentally, and the roach went down on the inside of the runways, and so found the *B* runway.

After several trials in succession with 2-min. intervals, improvement was noted by decreases in the amount of time required per trial, and in the number of errors made before successful entrance to the *B* runway. Upon once entering *B*, even unconditioned roaches never turned back, but went straight ahead into the sheltering bottle at its end. An unaccustomed observer might have concluded from their accelerated movement that the roaches detected the black bottle some centimeters before reaching its shelter but the fact that they were moving from the area of bright light to the end of the center runway where there was less intensity was probably responsible for the observed behavior. This interpretation is strengthened by the

fact that the same tendency to go ahead and to accelerate were evident after the roaches had progressed some distance toward the end of *A* and *C* runways.

A typical case of the time and errors per trial for one roach when isolated, paired, and in a group of three, is shown in Figure 2; the course of a roach in one trial on the maze is given in Figure 3. All of the records are available in similar form, as well as in detailed tables showing trial by trial the time taken and the errors made.

When isolated, the roach almost always used more time for the initial trial than for any succeeding trial. As the tendency to remain near the outside of the maze was overcome, the time and errors for each successive trial decreased. The roach tended to go to either side of the maze

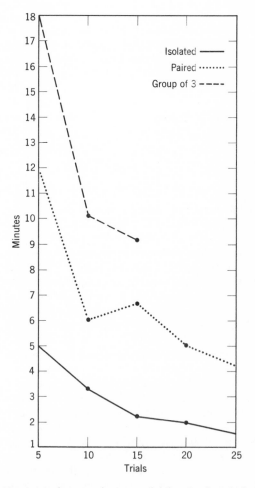

FIGURE 4. Average time per trial for the first day's tests under the respective conditions for all roaches tested in series 2.

a like number of times. The presence of another roach acted as an interference. Conditioning removed this in part after the preliminary trials, but it was always present when the roaches were paired, and still more evident in a group of three as shown by increased time per trial.

The presence of additional roaches did not affect the proportion of errors on different parts of the maze. Considering all the data at hand the errors appear to be entirely symmetrical and random.

There are two ways of studying the data obtained. We may examine the average performance of a given roach on all the days on which it was tested, or we may escape from individual deviations by showing the mean reactions of all the roaches on successive days. In both it appears best to present the data in sets of five successive trials rather than trial by trial. This permits a cross comparison of sets of five trials from day to day.

TABLE 3. *Average Time and Errors per Trial of All the Roaches for Successive Sets of Five Trials for Successive Days When Isolated*

Times Per Trial

25 Trials Sets of 5	Series II					
	First day	Second day	Third day	Fourth day	Fifth day	Mean average
1	4.864	4.152	6.968	6.650	9.377	6.402
2	3.249	3.152	3.777	3.863	4.065	3.621
3	2.181	3.019	2.441	2.845	3.644	2.826
4	1.918	2.179	1.625	2.799	2.000	2.104
5	1.579	1.728	2.014	1.722	1.665	1.742
Mean average:						
15 trials	3.431	3.441	4.395	4.453	5.695	4.285
25 trials	2.758	2.846	3.365	3.576	4.150	3.339
Series III. Mean averages						
15 trials	7.454	8.271	10.774	8.356	11.379	9.247
Series II and III. Mean Averages						
15 trials	5.04	5.406	6.947	6.014	7.969	6.275

Errors Per Trial

	Series II					
1	7.9	6.5	8.966	8.1	10.3	8.353
2	6.333	4.866	5.333	4.5	4.5	5.106
3	3.533	3.866	3.066	3.666	4.1	3.046
4	3.233	4.166	2.32	3.966	1.733	3.084
5	3.033	3.633	3.2	1.7	2.533	2.870
Mean average:						
15 trials	5.922	5.077	5.788	5.422	6.3	5.702
25 trials	4.806	4.606	4.577	4.386	4.633	4.602
Series III						
15 trials	6.667	7.350	6.383	4.267	4.667	5.867
Mean Averages for Series II and Series III						
15 trials	6.22	5.987	6.027	4.96	5.647	5.767

TABLE 4. *Average Time and Errors per Trial of All the Roaches for Successive Sets of Five Trials for Successive Days When Paired*

Times Per Trial

25 Trial sets of 5	Series II					
	First day	Second day	Third day	Fourth day	Fifth day	Mean average
1	11.897	12.159	12.809	9.671	12.855	11.878
2	5.972	5.609	5.466	6.202	4.581	5.566
3	6.596	4.891	3.450	3.770	5.414	4.824
4	4.993	3.113	3.632	3.461		3.800
5	4.234	2.152	2.824	2.983		˙3.048
Mean average:						
15 trials	8.155	7.553	7.242	6.548	7.617	7.423
25 trials	6.738	5.585	5.636	5.217		5.794
	Series III. Mean averages					
15 trials	15.602	16.162	14.125	18.742	23.853	17.697
	Mean averages for Series II and Series III					
15 trials	11.133	10.997	9.995	11.425	14.111	11.532

Errors Per Trial

	Series II					
1	9.966	9.933	8.3	6.866	9.9	8.993
2	6.5	6.033	3.533	6.133	3.75	5.190
3	6.633	10.15	4.7	3.866	5.4	6.150
4	4.95	6.05	6.1	4.45		5.387
5	7.3	4.5	4.75	2.55		4.775
Mean average:						
15 trials	7.700	8.705	5.511	5.622	6.35	6.777
25 trials	7.070	7.333	5.477	4.773		6.162
	Series III					
15 trials	6.367	7.967	8.633	6.400	5.633	7.000
	Mean averages for Series II and Series III					
	7.133	8.39	6.76	5.933	6.063	6.856

A reservation must be made for the values given for the fifth day in Tables 3, 4 and 5 since it does not include records for roaches I and II which were only conditioned for four days. The group of three was made up of roaches IV, V and VI. These three tables show the data on the basis of successive days; Figures 4 to 7 give graphic representations of all the data. Figure 4 gives graphically the mean time per trial for the first day's tests whether isolated, paired, or member of a group of three of all the roaches tested in series II. Figure 5 shows the same sort of data for errors for the same tests. Graphs at hand for the successive days that these tests ran are impressive only by their similarity to those given. Figure 6 summarizes all the data available from series II so far as time per trial is concerned and Figure 7 does the same for errors per trial. The graphs and the tables on which they are based are smoothed somewhat by considering averages for successive graphs of five tests rather than listing or plotting the tests singly.

As the figures indicate the roaches showed the results of conditioning on any given day as indicated by the fact that time per trial and errors per trial decreased with training. A comparison of the time per trial or errors per trial in the sets of five trials on successive days shows that there is no marked difference or trend in either direction from day to day in the isolated, paired and grouped cases. With the isolated roaches in series II there is some evidence of more rapid and more complete elimination of errors per trial on the fourth and fifth than on the first three days.

In a comparison of the mean time per trial for the roaches when paired and isolated the rate and amount of decline of the time curve, Figure 6, is greater for the roaches in the former case between the initial five trials and those first succeeding after which, the rate of decrease is about the same for the roaches whether paired or isolated. The ultimate trend of the curve for the roaches when in a group of three is not apparent from the data at hand, but a rapid rate of decrease persists at least to the fifteenth trial.

In the mean average time per trial there is a suggestion that the isolated roaches ran the maze less rapidly from day to day; the paired roaches tend to show some reduction in time per trial required from day to day, but the results in the group of three show no tendency in either direction. The total time change, whether an increase or decrease from day to day, is not significant from the data presented; there is no real evidence of conditioning being carried over from one day to the next. It must be realized, however, that such experiments lasted only five days. It is true that fish in a simple maze show decided conditioning even in this short period when given only one trial per day.

Evidence was obtained in series III that the effects of partial conditioning were not carried with certainty over an interval of one hour.

The number of errors tend to fall off with successive trials on any given day whether the roach is isolated, paired, or a member of a group of three. In general, error curves show the same sort of decrease as do those based on time per trial (see Fig. 7). In the errors per trial in the final sets of five trials on successive days (Tables 3, 4 and 5), as in the case of time per trial, there was no trend in either direction from day to day in the isolated cases. When the roaches were paired there was somewhat of a decrease; in the grouped cases the same was true with the exception of the third day. In considering the mean of the average errors per trial on successive days, isolated roaches showed no consistent improvement from day to day whether isolated, paired or in groups of three.

TABLE 5. *Average Time and Errors per Trial of All the Roaches in Series II for Successive Sets of Five Trials for Successive Days—Group of Three*

15 Trials sets of 5	First day	Second day	Third day	Fourth day	Fifth day	Mean Average
			Time per trial			
1	17.823	14.551	15.986	18.013	16.788	16.632
2	10.118	8.990	9.568	11.119	12.108	10.381
3	9.171	6.875	12.058	7.711	6.716	8.506
Mean average: 15 trials	12.371	10.139	12.537	12.281	11.871	11.840 (Mean)
			Errors Per Trial			
1	13.533	10.2	11.933	10.466	10.533	11.333
2	7.4	6.733	8.4	7.533	7.666	7.546
3	7.533	7.533	10.733	6.466	5.533	7.560
Mean average: 15 trials	9.489	8.155	10.355	8.156	7.911	8.813 (Mean)

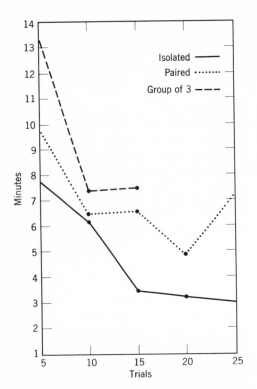

FIGURE 5. Average errors per trial for the first day's tests under the respective conditions for all roaches tested in series 2.

found that these insects of a primitive type show learning curves both as regards time to run the maze and the elimination of errors on successive trials on a given day, but unlike the experience of Eldering there is little or no evidence that this experience was carried over from one day to the next. However, such retention could not be determined definitely until the insects were tested over a longer period of days. We have found that paired roaches interfere with the initial speed of running the maze, but that later speed increases with successive trials, in fact the improvement but not the actual rate may be more rapid than if the roaches are isolated. The roaches when paired and in a group of three require more time per trial to run the maze than when isolated, and the difference in this is significant. Similar differences in errors are not statistically significant, but the average mean errors of the roaches when paired and in a group is larger than when isolated. For a proper comparison, errors and time must both be considered.

Before attempting a discussion of the problems presented, let us consider the activity of an isolated roach when first placed on the maze. As shown in part in Figure 3, it tends to follow the outside of the runways. It acts as though it possessed a certain amount of energy and when this is spent a pause results. The pause is followed by a movement to the right or left as the roach again continues. Equal motions toward and away from the light are characteristic of all specimens.

As mentioned before, after several trials, the tendency to stay near the outside of the runways is overcome, and finally, entrance to the B runway is made from the inside of A or C. After passing through the area receiving the most light the roaches get to the dark receptacle at B; once they enter this they remain within.

This tendency to remain near the outside of the runways can be explained by the fact that the roaches are photonegative (the light was over the center of the maze). Schneirla (1928) working with ants on a maze has suggested that centrifugal force tends to force these animals to the outside. If this is a factor with cockroaches then for conditioning to take place both tendencies must be overcome so that the roach no longer responds to the force or the combination of forces that originally determined its behavior.

The roaches used their antennae in exploring the edges of the maze, especially, when moving along the inside of the runways and making the turn down the center which would indicate that light is not the only stimulus acting.

The errors per minute made by the roaches in successive sets of five trials for successive days indicate that, when isolated, there is an increase in activity from the initial five trials to the final set. The mean average errors per minute from day to day show an apparent decrease. The same is true of those roaches when paired, and in a group of three, as concerns the increase in errors per minute with successive trials, but there is not a consistent decrease of mean average errors per minute from day to day. The significant differences between the isolated, paired and grouped cases, in regard to errors per minute, indicate that these roaches show less activity per unit of time in the two latter cases, especially during the initial trials. The marked individuality shown by cockroaches in conditioning experiments has been emphasized by all previous workers.

DISCUSSION

In a consideration of the behavior of roaches on a maze we have confirmed earlier workers who

The light was of great importance as a stimulus, and as an orienting factor. Roaches are negatively phototropic and the light, no doubt, has a disturbing effect on their visual sensory receptors, the nature of which has not been explained.

As shown in Table 1 the relative intensity of the light conditions on the two sides of the maze is about the same; so one would expect the errors to the two sides to be of the random type under such conditions. The two corners of the platform were illuminated slightly more than the corresponding ends of the runways. This may account for the proportionate increase of errors to the ends as compared to the corners, since there would be a tendency for these photonegative insects to move to the areas of less intensity of illumination. Distribution of errors to the ends and corners of the maze is a constant factor, and is not markedly affected when the roaches are paired or grouped.

Although these insects are photonegative their initial activity on the maze indicated that after any given pause there was an equal chance that the next move would be toward or away from the light. Conditioning removed in part this tendency to turn from the light, but the roaches did not become photopositive or the dark shelter would not have acted as a reward, and the light would not have continued effective as an orienting factor.

The importance of the light in a central position over the maze is emphasized by observations which showed that when it was slightly moved the behavior of the roaches was affected. Placing it at the end of the *A* runway for example, after training had been completed, upset the previous conditioning so much that the roach would not turn into the *B* runway; it also tended to make more movements toward the light in the new position at the end of *A*.

It is evident from Figure 1 and Table 1 that the place on the platform with the greatest light intensity was the entrance to the *B* runway. This area would be the last to be entered by the roach though it would tend to go to both sides of the maze a like number of times because of the similarity of conditions on either side. Once having entered the middle runway and the dark bottle the insect was sheltered from the light and other conditions on the maze. Repetition of the entrance of the runway leading to the shelter resulting at first in random movements, produced, after only one or two experiments, a transitory change in the time response to the light which lasted but a short period. From the reactions such as are recorded in Figure 2 it appears

that there was increased sensitivity to conditions on the maze with successive trials so less time was required with each trial until a certain amount of exposure had been experienced, then a lengthening of the reaction time occurred which again was rapidly shortened. The learning curve shows a number of these apparent reversal giving a sawtooth type of curve with steadily diminishing amplitude of variations.

The presence of more than one roach was an opposing factor to the orienting effects of the light in that it prolonged the time for successful entrances to the dark shelter and made the light less effective as a sign of the route to shelter, especially in the initial trials. If the light produced certain photochemical effects in the receptors, longer rest periods in the paired and grouped cases meant greater loss of such effects from trial to trial for the first roach to enter the dark chamber. Under these conditions with 15 trials per day for 6 roaches, a total of 90 trials, it

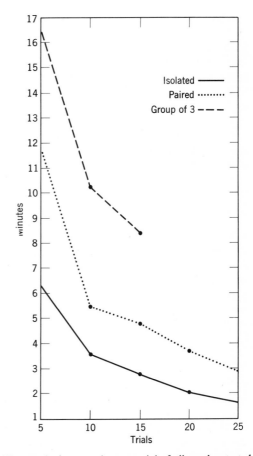

FIGURE 6. Average time per trial of all roaches tested in series 2.

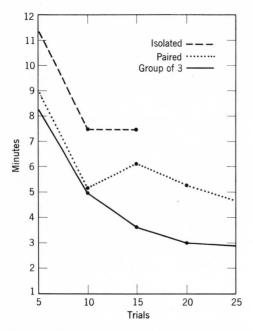

FIGURE 7. Average errors per trial of all roaches tested in series 2.

was found that when isolated the animals spent on the average 4.076 minutes per trial on the lighted maze with mean intervals in the dark bottle of 2.0 minutes. When paired the corresponding amounts of time are 7.19 and 3.879 minutes respectively. One would expect that with a greater amount of time spent in darkness the paired roaches would require more time in the light to obtain the same orientation and the expectation is realized. The reactions as presented are capable of partial explanation by a relatively simple hypothesis. Assuming that the conditioning as shown in Figure 2 in the first trial is carried over to the next in spite of the 2-min. rest period we see that the effects of preceding experiences remain and become accentuated in the third trial and then almost vanish. The conditioning effects of the fourth trial start a new series of decreasing reaction times which in this case last through the seventh trial. This type of response is characteristic of all the curves for roaches. As one might expect, the peaks and depressions in the graphs for the paired and grouped cases are more accentuated; the reason perhaps is to be found in the greater average length of the rest periods. Toward the end of the 25 trials the rest periods for the roaches were shorter, and the loss of conditioning less in proportion, so that the time curves of the roaches when isolated and paired tend to be more similar. Turner's and Schneirla's

curves show similar peaks in the learning trends.

An explanation which may apply to the type of learning curves presented, may be given in the following general terms for cockroaches, under the conditions of these experiments.

A roach being photonegative, when placed on the maze, was stimulated by light. During the first, usually extended, period of random movement on the maze, the light stimulus would have an extended time of action. This may have served to produce certain photochemical changes in the sensitive parts of the eyes while it probably induced pigment migration. For both of these reasons the eyes were probably physiologically different after a long period of exposure to the light. During the long periods of exposure on the maze the animals are fairly active and hence would be accumulating within their bodies certain of the products of metabolism which might collect in sufficient quantity to affect their behavior. After such a period of extended exposure some or all of the accumulated physiological effects would carry over past the 2-min. rest period and would leave the animal more sensitive for the following trial. As these trials grow shorter and the intervening rest periods take a larger proportion of the time, the physiological effects which result from the initial long exposures would cease to be effective and the roach would again remain a longer period on the maze. The difference in the length of this longer period from the time spent in the initial trial is an expression of the amount of conditioning that has taken place. Again the longer time on the maze would set up the physiological conditions following the original long exposure, and again these would gradually lose their effectiveness until again recalled by another period of more than usual time exposed to maze conditions. Some such explanation was proposed by Shelford and Allee (1914) for similar periodic behavior in fishes.

In testing this working hypothesis to explain the observed cycles of behavior changes while being conditioned on the maze, roaches were exposed to a light of greater intensity than that over the maze for varying lengths of time. When placed on the maze platform after such an exposure, they oriented more rapidly than before but were still negative to light and attempted to move under the platform away from it. This tendency was more marked after exposure to bright light than when roaches were transferred to the maze from the usual period in dark bottles. An uncovered bottle was used as a reward in these tests to avoid a reversal of physiological changes which had accompanied the exposure of the roaches to light. Resting for 2 minutes in a dark bottle or five exposures to the dim light of the maze gave time for the loss of effects produced by such exposures to higher illumination.

Conditioning of cockroaches as Szymanski (1912), Turner (1912) and Eldering (1919) practiced, using electrical shocks for punishment would also be expected to yield physiological results which would pass fairly rapidly and call for repeated stimulation as was observed at least by the first two. Such a relatively simple physiological explanation of conditioning does not necessarily explain all the facts observed in our experiments although it does fit with the rapid loss of conditioning on standing as well as with the saw-tooth character of the learning curve. It will not, however, completely account for the maze training reported by Turner (1913) where the roaches were trained on a maze exposed to north daylight in summer and were apparently rewarded by attaining the supposedly clear, jelly glasses in which they ordinarily lived. Turner does not mention evidences of fatigue after a long day of training with 30-min. rest periods between tests; evidently even in his experiments the physiological effects of conditioning behavior upon the organism as a whole, as well as upon its nervous system must be considered.

With our own experiments, our suggestion is that the so-called learning is due to a general conditioning of the organism as a whole as well as to the possible or even probable conditioning of the nervous system.

In the paired and grouped cases, longer rest periods for the first of a pair or for the first and second of a group of three roaches to enter the dark bottle, would mean a greater loss of these physiological effects and of conditioning in general, hence a greater time should be taken to acquire fairly consistent speed in running the maze.

The fact that there were fewer errors per minute in the paired and grouped cases suggests that there must have been factors which were not present or effective in the isolated cases. The in-creased time required at first where more than a single roach was present on the maze indicates that there were forces opposing the smooth operation of the physiological adjustments just postulated to help explain the maze behavior of isolated roaches or rapidly repeated exposures to maze conditions. The nature of these inhibiting forces is speculative, but the fact of some sort of group interference is obvious.

The presence of other roaches did not operate to change greatly the movements to different parts of the maze, but did result in increased time per trial. The roaches tended to go to the corner or end of a runway and remain there a longer time when another roach was present than when alone; the other roach was a distracting stimulus. Chemical traces introduced by one of the other roaches simultaneously occupying the maze may have acted to interfere with orientation. Social stimuli which in goldfishes tend to increase the rate of reacting to a maze by members of a group, here have the opposite effect.

SUMMARY

1. Cockroaches can be conditioned to run a simple maze when isolated or paired, or present in groups of three.

2. A roach when isolated can be conditioned with less time and errors per trial than when paired or when a member of a group of three.

3. In the two latter cases activity and accordingly the errors per minute are less.

4. Conditioning was carried over from trial to trial in isolated, paired, and grouped cases when the interval between periods of exposure on the maze is brief but there is little or no evidence that there was retention of the effects of maze conditioning from day to day.

5. The evidence presented indicates that the maze behavior of these animals is affected by the conditioning of the organism as a whole as well as by the possible or even probable conditioning of the nervous system.

REFERENCES

ALLEE, W. C., 1931. *Animal Aggregations*. University of Chicago Press, 431 pp.

ELDERING, F. J., 1917 (1921). Sur les habitudes acquises chez les insectes (d'après des expériences sur Periplaneta americana). *Troisième Reunion Ann. d. Physiol. Arch. Ne'er. d. Physiol.*, v, 129.

ELDERING, F. J., 1919. Acquisition d'habitudes par les insectes. *Arch. Ne'er. d. Physiol.*, iii, 469–490.

GRABER, V., 1884. Grundlinien zur Erforschung des Helligskeits- und Farbensinnes des Tiere. *Prag. s.* 147–157.

SCHNEIRLA, T. C., 1928–1929. Learning and orientation in ants. *Comp. Psych. Monographs*, v–vi.

SHELFORD, V. E. and ALLEE, W. C., 1914. Rapid modification of the behavior of fishes by contact with modified water. *Jour. Animal Behavior*, iv, 1-30.

STUDENT, 1925. New tables for testing the significance of observations, v, No. 3, 1925, p. 26.

SZYMANSKI, J. S., 1912. Modification of innate behavior of cockroaches. *Jour. Animal Behavior*, ii, 81-95.

TURNER, C. H., 1912. An experimental investigation of an apparent reversal of responses to light of the roach, Periplaneta orientalis L. *Biol. Bull.*, xxiii, 371-386.

TURNER, C. H., 1913. Behavior of the roach, Periplaneta orientalis L., on an open maze. *Biol. Bull.*, xxv, 348-361.

WELTY, JOEL CARL, 1933. Experimental explorations into group behavior of fishes: A study of the influence of the group on individual behavior. *Physiol. Zool.*, vi.

SOCIAL FACILITATION:
AN EXPERIMENTAL INVESTIGATION WITH ALBINO RATS*

E. Wulff Rasmussen[1]

SOCIAL psychological observations show that human beings in groups react differently than when alone. The same kind of observations can also be made in regard to animals, but only a few experiments have dealt with the problem. Katz has shown experimentally that hens will eat more when they are together with other hens than when eating alone. "Thus, in each case the first hen, which has already been fed, begins to eat again under the influence of the example set by the second, although she has already eaten to full satisfaction. An additional 60 per cent or more may be eaten under the social influence of the second hen."[2]

The present investigation deals with a similar problem: Will rats receiving electric shocks react differently when in groups as when alone? Will they act as if more "courageous", more "self-reliant" when together than when alone?

APPARATUS

The shock-apparatus is placed on a 50 cm tall box. It rests on four 10 cm long feet, measures $50 \times 50 \times 25$ cm, and is painted black inside. The floor consists of 1 mm thick brass-wires, 67 in number, with a distance from each other of 7.3 mm from centre to centre of the wire. The wires being fastened to screws on both ends every other one of them are connected with one of two metal bands on opposite sides. Thus, there is one end of each wire which is not closed (ends in the screw). The wires as well as the bands are fastened to a wooden frame, 4×3 cm thick and richly applied with voltwax.[3] Between the frame and the lower edge of the walls are inserted four pieces of rubber, leaving an opening 4mm high. The two bands may be connected up with opposite poles. Then a rat, sitting on the grill, will get a shock when the current is closed. In our experiment both bands are connected up with the same pole. The other pole is led to a small water-container. Then a rat, standing on the grill and beginning to drink, closes the current and gets a shock.

As roof of the experimental box is used a 5 mm thick glass plate. The box is illuminated by a 60-watt bulb, placed in a reflector made of 4 mm

* From *Acta Psychologica*, 1939, **4**, 275–294. Copyright 1939 by North Holland Publishing Co., Amsterdam.

[1] The author wishes to thank professor Harald K. Schjelderup for having supported this experiment, the expenses of which have been partly covered by a grant from the Psychological Institute. We are also deeply grateful to professor, dr. med. Klaus Hansen, director of the Pharmacological Institute, who has supplied room for the animals, and been helpful in many other ways.

[2] Katz, D. *Animals and men*, p. 163.

[3] The bands are fastened underneath the frame, so that the wires run from the upper side of the frame and down the sides, where they are fastened to the screws, and end in the bands.

scaleboard. The reflector widens from 13×13 cm to 21×21 cm. Outside and low down on the box is a shelf for recording the date, 35×22 cm. It is illuminated by a small desk-lamp. The experiment is carried on in a dark-room. Thus, the laboratory as well as E cannot be seen by the rats. When the latter have received a shock and again approaches the water, they are very watchful. The least sound is sufficient to make them jump back. So it is important that *no* sounds are allowed to disturb the animals.

In the main experiment of our investigation we have used a shock of 500 V. (alternating current, 50 cycles), a resistance of 2,702,700 ohms, and a current of 0.185 milliamperes. In order that individual differences in the resistance of the rats shall play a minimal role, it is necessary to use high voltage and great resistance. Warden[1] assumes that the resistance in rats may vary from 100–5000 ohms. Muenzinger[2] has found, however, that the resistance is much greater. "Next we determined the skin resistance of 20 rats, 1 to 3 months old, with the Wheatstone bridge and found an average of 300,000 ohms with a range of 75,000 to 1,000,000 ohms."[3] Still greater resistances were found on our rats by assistant lecturer Aars at the Physical Institute, when using the Wheatstone bridge. They ran up in several million ohms when using direct current (2 V.). With alternating current (2–4 V.), the average was 510,000 ohms with a range from 200,000–1,500,000 ohms. (The ten animals measured had an average weight of 362 gm with a range of 316–405 gm.) If the rat remained on the grill for a while, the resistance would change continually, and it is arbitrary, of course, which reading shall be chosen as representative for the rat in question.

The present writer has tried himself to determine the resistance in the following way: The rat is placed in the experimental box and gets a shock of 200 ms.[4] The amperemeter will then give constant reading, but the reading will be somewhat higher than when the current is closed with a switch. We used a voltage of 475 V. and a resistance of 317,000 ohms, giving a reading of 1.5 milliamps with the switch on the grill. If, instead, the current is closed by means of the shock-apparatus, giving a shock of 200 ms., the amperemeter will read 1.77 milliamps. If, now, the rat is placed on the grill and the shock is given, the reading may be e.g. 1.4 milliamps. Using the formula $1.77 : 1.5 = 1.4 : x$, we can find what the

reading would have been if the current had been made in the usual way. The correctness of this formula has been controlled by making the current in the usual way, and with the shock-giving instrument with various resistances connected up with the grill. Then we get $x = 1.13$; The resistance of the rat will be as follows: Voltage : (resistance already present + resistance of the rat) $= 1.13$. Or: $475 : (317,000 + y) = 1.13$; $y = 103,000$. In other words, the resistance of the rat will be 103,000 ohms.

By this method the resistance of 12 male rats (average weight 362 gm, range 316–405) has been found to be on the average 158.000 ohms with a variation ranging from 43,000–193,000 ohms. Eight females of the same age and five young rats (mean weight 28 gm, variation 24–34) gave similar results. We must reckon with an observation-error of from 0.05–0.1 milliamps. Finally, the above mentioned male rats gave corresponding results when measured by the same method in an apparatus where the current went through the rat from forelegs to hindlegs.

We cannot find any objections to the method. When resistances corresponding in size to those of rats, are connected up with the grill, the ampere-meter will give readings comparable to those found with rats. We may explain the differences between our measurements and those made by Muenzinger as well as those performed at the Physical Institute in the following way: Since rats weighing 30 gm have resistances similar to those weighing 360 gm, it seems that we do not deal with an inner resistance of the rat, but with a transitional resistance, which differs according to whether direct or alternating current is used. Possibly, the voltage applied may be of some influence.

Assumed that our resistance-measurements are correct, we must take of a variation in the shock-effect of maximally 5–6 per cent when the external resistance is 2,702,600 ohms. Even if the resistance for the different rats should vary more, it would not be of any consequence for our experiment because we are dealing with group results.

The electrical arrangements are placed in a wooden box. From the transformer, which may be regulated by means of a variable resistance from 400–600 V., the current passes a series of resistances. Warden reports that he has tried "some of the better types of standard radio resistance,"[5] but that they were too unstable. It

[1] Warden, C. J. *Animal motivation*, p. 28.

[2] Muenzinger, Karl F. and Mize, Robert H. The sensitivity of the white rat to electric shock. Threshold and skin resistance. *Jour. Comp. Psychol.* 1933, **15,** 139–148.

[3] *Op. cit.,* p. 145.

[4] Ms. = millisecond = 1/1000 sec.

[5] *Op. cit.,* p. 27.

was in 1930, and radio resistances are now greatly improved. We made use of "Always" radio resistances, and they have caused no difficulties during ten months of experimentation. One (for 2 watts) could stand 200,000 ohms. One series of resistances (one for 0.5 mg ohms and nine for 0.2 mg ohms, all ten for 0.5 watt) and another series of nine for 0.025 mg ohms (0.5 watt) had regulating switches, so that practically any current between 0.185 and 2 milliamps could be obtained.

The city line (230 V.) varies between 5–6 V., so the voltmeter was controlled before each test. The amperemeter, type KNW, Neuberger, has a voltage drop of 1 V.

is 150 days average $209 + 2.03$ S.D.: 17.15 Range 171–239. Among these 32 animals only 8 weighed less than 200 gm. As can be seen, the weight of the animals is comparable with, to say the least, weight-reports from all other laboratories. The weight of five other litters (12 females which have been used) is 180 days average $207 + 3.46$ S.D. = 20.74 range 177–149. That the weight of these animals is somewhat less than of those weighted at the age of 150 days, is supposedly a matter of chance. Unfortunately, the weight of litter number 9 cannot be reported, but Table 3 shows that the weights of females numbered 95, 96, 98 and 99 do not differ from the average.

THE ANIMALS AND THEIR CARE

The rats have all been born in the laboratory and descend from albino rats received from various institutes in Norway. (On account of an investigation on heredity in rats, we could not use near related animals and had to get them from different places, a procedure which preferably should be avoided.) The food has been Mc-Collums diet:

	per cent
Whole wheat, ground fine	67.5
Casein	15.0
Whole milk powder	10.0
Sodium chloride	1.0
Calcium carbonate	1.5
Butter fat	5.0

We used whole milk powder of 28 per cent. To this diet was added 5 per cent dry yeast on recommendation from the State Vitamin Institute. In addition, the animals got plenty of vegetables, 15–20 gm per animal per week, and 10 gm maize (unground) per animal per week. The feeding took place every evening from 7–9 P.M., and it was controlled that the animals had food and water in the cage all the time. After weaning at the age of 30 days, the animals in this experiment were placed together with males in a big cage, 170×140 cm. Three and a half months old males and females were separated, the latter bringing forth and feeding children before they were used in the experiment. Although the weight of each animal is given in Tables 2 and 3, we shall list here the weight at 5 and 6 months of those litters which have been tested. These figures may serve as basis for comparison with weights reported from other laboratories. The weight of 11 of the litters used

METHOD AND PROCEDURE

Forty-eight hours before the tests, the animals were weighed and removed to another cage, where they did not get water. McCollums diet, usually mixed with water, was given dry. On account of the great percentage of water in vegetables, this part of the diet was withheld, too, during the thirsting period.[1] The need of water reaches maximum after 1 day of thirst measured by the "Obstruction method."[2] In some experiments with hunger, using our shock-method, the maximal need of food fell somewhat later than for the obstruction method. We had reason to believe, therefore, that the same would be the case in regard to thirst. So 2 days of thirst were chosen.

Two parallel experiments were performed; in the one case the rat was alone in the test-box, in the other case three rats were tested at a time. The animals were given 3 min. for exploring the box. Then they were taken out and a glazed pottery-dish (10 cm diameter, 2 cm deep) with 35 ccm water was placed in one corner of the box. A stiff isolated copper-wire was put up between the wires of the grill, bent around the edge of the dish and down into the water. Then the rat (or rats) were put back into the box. In the course of a few minutes, most often a few seconds, they get in touch with the dish, and start drinking. After 5 sec. of drinking, the current was closed. The rat would get a shock and run away from the dish. The current remains on with one pole in the grill and one in the water. Thus, every time the rat tries to drink, she will get a shock. The resistance of ordinary running water varies a little. When the poles are placed farthest away from each other in the water of the dish, the resistance will be about 15,000 ohms. On account

---

[1] Cfr. Warden, *Op. cit.*, p. 103.

[2] See Warden, *Op. cit.*, p. 109, Table 3.

of the high external resistance, the variations in the water-resistance should not be of any importance. Perhaps it would have been most correct to add e.g. some calcium carbonate to the water. But then it must have been done previously in the cage, too.

We recorded the number of times the rat tried to drink and got a shock, and the number of approaches-withdrawals without drinking. The latter behavior is very characteristic: The rat *approaches slowly* and cautiously, but when it comes near the dish, it *withdraws* or jumps back *quickly*. In some cases it may remain motionless for several seconds, leaning over the dish, but then withdraws suddenly. Some rats will stand still, moving only the head back and forth several times. These reactions were not recorded as approaches-withdrawals. As criterion for withdrawal was used, that the animal at least moved the forelegs.

A third measure of the behavior of the rats has also been employed. In a series of preliminary experiments with thirst as well as hunger, we found that some rats would eat and drink in other apparatus and cages than those associated with

the experiments, while other rats were unwilling to do so—seemingly dependent upon how strongly the negative conditioning to the dish with water and food was. So after the rat had been observed in the shock-apparatus for 50 min., it was moved over to the transporting cage (bottom of zinc, the rest netting, size $30 \times 30 \times 12$ cm).

As the rats were allowed to drink while in this cage, we have called it the "allowance cage". The drinking dish was disconnected and moved over into the allowance cage, and the time was recorded which would pass until the rat drank, if at all during a period of 3 min. Finally, the animal was weighed and moved over to a male rat to see if she was in the sexual phase.

RESULTS

After some trial experiments, the first preliminary ones were carried out August 21–23, 1938, from 10 P.M. till 2 A.M. Two parallel series for males and females at the age of 2–3 months gave the following results:

TABLE 1. *6 Males and 6 Females, 2–3 Months of Age Shock: 475 V. 0.3 Milliampere. One Main Wire Electrified*

	In Groups of Three							Alone						
						in allow							in allow	
	w. I	w. II	d.	sh.	ap.	dr.	ap.		w. I	w. II	d.	sh.	ap.	dr.	ap.
♂ 36 lit. a	196	172	24					♂ 34 lit. a	168	146	22	3	17	0	3
„ 37 „ c	112	90	22					„ 35 „ c	144	128	16	10	21	1	0
„ 38 „ e	140	118	22					„ 39 „ e	125	105	20	3	32	0	2
Sum	448	380	68	37	38	3	0	Sum	437	379	58	16	70	1	5
Average	149	126	23	12	13	0.15	0	Average	145	126	19	5	23	0.4	1.7
♀ 62 lit. a	121	106	15					♀ 61 lit. a	131	115	16	4	9	0	8
„ 63 „ c	133	117	16					„ 65 „ c	156	137	29	2	0	0	0
„ 64 „ d	203	180	23					„ 66 „ d	190	167	23	4	51	0	1
Sum	457	403	54	19	101	3	0	Sum	477	419	58	10	60	0	9
Average	152	134	18	6	34	0.09	0	Average	159	139	19	3	20	0	3
Sum ♂ + ♀				56	139	6	0	Sum ♂ + ♀				26	130	1	14
Average				9.3	23	0.12	0	Average				4.3	22	0.4	2.3

lit. = litter. w. I = weight before thirst. w. II = weight after 49 hours thirst. d = difference, weight loss. sh. = number of shocks. ap. = number of approaches-withdraws. in allow. = in "allowance cage". dr. = drinking in allowance cage. ap. = number of approaches-with-draws in allowance cage.

The results for animals in groups are given in the left half of the table. In the first column we find the number of the rat and the litter to which it belongs. It will be noticed that the corresponding animal, tested alone, always is of the same litter, and the animals are placed so that the

average weight is as similar as possible for each corresponding couple. Then we get the weight before and after thirsting and the loss of weight. The fifth column lists the number of shocks which the rats have received, i.e. the number of times they have tried to drink, closed the current and gotten a shock. The sixth column gives the number of approaches-withdrawals. These, as well as the shocks, were so frequent for the animals in groups, that is was difficult to keep count of each separate rat. Therefore, the results were recorded for all three rats taken together (first row) and then divided by three to get the average (second row). The reaction in the allowance cage is indicated in columns 7 and 8, the former giving the time before the rat drank (if at all), the latter the number of back-and-forth-movements. The sum in column 7 gives the number of the three animals which drank in the allowance cage, the row average giving the average time elapsing until they drank. The indi-

vidual drinking time is later given for all animals in Table 4. The number of approaches-withdrawals in the allowance cage represents the total for all three "grouped" animals, but for the animals alone, individual results are also given.

As can be seen from the table, the number of shocks was about twice as many for rats in groups as for rats alone, males 37 to 16, females 19 to 10. The number of back-and-forth-movements is greater for males alone, but together with females alone, it is slightly less than for the grouped animals. All group-rats drank immediately afterwards in the allowance cage, while this was done only by 1 of the 6 animals tested alone.

TABLE 2. *12 Females, 4–5 Months of Age*
Shock: 500 V. 0.3 Milliampere. Both Main Wires Electrified

	In Groups of Three								Alone						
						in allow.								in allow.	
	w. I	w. II	d.	sh.	ap.	dr.	ap.		w. I	w. II	d.	sh.	ap.	dr.	ap.
♀ 87 lit. 7	215	193	22					♀ 86 lit. 7	217	188	29	2	7	1	1
,, 89 ,, 7	205	178	27					,, 88 ,, 7	195	172	23	2	4	0	3
,, 92 ,, 8	183	157	26					,, 91 ,, 8	190	168	22	4	30	1	3
Sum	603	528	75	13	24	3	1	Sum	602	528	74	8	41	2	7
Average	201	176	25	4.3	8	0.17	0.3	Average	201	176	25	2.7	14	0.30	2.3
♀ 101 lit. 10	241	213	28					♀ 100 lit. 10	218	190	28	1	2	1	5
,, 103 ,, 11	226	194	32					,, 102 ,, 11	210	172	38	2	1	0	0
,, 106 ,, 12	202	174	28					,, 105 ,, 12	237	190	47	2	0	1	0
Sum	669	581	88	7	12	0	5	Sum	665	552	113	5	3	2	5
Average	223	194	29	2.3	4	0	1.7	Average	222	184	38	1.7	1	1.12	1.7
Sum+Sum				20	36	3	6					12	44	4	12
Average	212	185	27	3.3	6	0.17	1		212	180	32	2	6.1	0.71	2

Due to other experiments, the further work in this investigation was carried on during December 18–31, 1938, from 1–5.30 P.M. with 42 female rats at the age of 5–6 months. The time elapsing from separation from males and from having brought forth and fed children, varies a little for the different litters, but as can be seen in Tables 2 and 3, the parallel tests can be compared in these respects as well as in regard to average weight.

Sudden cold weather came on December 18, and it took a couple of days before extra heating could be supplied, so the two first parallel tests had unfortunately to be carried out in a too low temperature. During the rest of the tests, the temperature both in stable and laboratory was kept at 19°C.

It was found, moreover, that the number of drink-trials and approaches-withdrawals was con-

siderably reduced compared with the preliminary experiments. This might be due to the fact that in this experiment *both* connecting-bands were electrified by the same pole, and thus the shock would be more intense than when only every other wire was electrified, as was the case in the preliminary experiment. It was desirable to have both bands electrified by the same pole, then if the rat which tried to drink stood on those wires only which were *not* electrified, the other rats on the grill would also get shocks. This did not happen, in fact in the preliminary experiments but it was a theoretical possibility which we wanted to exclude. The difference mentioned above might also be due to the age difference between the rats concerned. Young growing rats would have a greater need for food and water.[1] It is true that the weight-loss is less for the younger than for the

---

[1] Cfr. Greenman, Milton J. and Duhring, Louise, F. *Breeding and care of albino rats for research purposes*, p. 59 and 62.

older animals. For females alone in Table 1 the average weight-loss is 19 gm, whereas for females alone in Table 3 it is 29 gm. The loss of weight *in per cent*, however, is about the same, 11.95 per cent and 13.04 per cent respectively. Of course, we must be cautious in concluding from loss of weight to need, as the need of water *may* be greater in young rats, even if the percentual loss of weight is not greater.

Due to these various circumstances, we decided to reduce the current from 0.3 to 0.185 milliamperes for the remaining 30 animals to be tested. The results for the first 12 animals are therefore given separately in Table 2 (see page 56).

TABLE 3. *30 Females, 5–6 Months of Age.*
Shock: 500 V. 0.185 Milliampere. Both Main Wires Electrified

	In Groups of Three					in allow.			Alone					in allow.	
	w. I	w. II	d.	sh.	ap.	dr.	ap.		w. I	w. II	d.	sh.	ap.	dr.	ap.
♀ 94 lit. 8	215	189	26					♀ 93 lit. 8	185	163	22	2	5	1	2
„ 96 „ 9	235	209	26					„ 95 „ 9	231	204	27	5	10	1	3
„ 99 „ 9	190	158	32					„ 98 „ 9	230	197	33	4	7	1	0
Sum	640	556	84	14	26	3	0		646	564	82	11	22	3	5
Average	213	185	28	4.7	9	0.09	0		215	188	27	3.7	8	0.65	1.7
♀ 68 lit. 1	238	207	31					♀ 67 lit. 1	236	205	31	3	17	1	2
„ 71 „ 2	200	169	31					„ 73 „ 2	254	218	36	2	1	0	6
„ 72 „ 2	227	196	31					„ 74 „ 2	182	160	22	2	8	1	2
Sum	665	572	93	17	45	3	2		672	583	89	7	26	2	10
Average	222	191	31	5.7	15	0.32	0.7		224	194	30	2.3	9	1.27	3.3
♀ 70 lit. 1	262	228	34					♀ 69 lit. 1	241	212	29	2	12	1	3
„ 76 „ 3	234	203	31					„ 78 „ 3	242	203	39	2	8	1	3
„ 77 „ 3	241	212	29					„ 79 „ 3	253	216	37	2	8	0	3
Sum	737	643	94	14	63	3	0		736	631	105	6	28	2	9
Average	245	214	31	4.7	21	0.09	0		245	210	35	2	10	0.74	3
♀ 107 lit. 13	209	184	25					♀ 108 lit. 13	218	185	33	4	15	1	0
„ 110 „ 14	245	217	28					„ 111 „ 14	268	236	32	3	5	0	0
„ 112 „ 15	270	225	45					„ 113 „ 15	244	211	33	4	18	0	1
Sum	724	626	98	21	66	3	2		730	632	98	11	38	1	1
Average	242	209	33	7	22	0.17	0.7		244	211	33	3.7	13	0.08	0.3
♀ 117 lit. 16	238	212	26					♀ 114 lit. 16	250	211	39	2	14	1	1
„ 119 „ 16	212	180	21					„ 115 „ 16	217	191	26	2	4	1	0
„ 120 „ 16	217	193	24					„ 118 „ 16	202	178	24	2	9	0	4
Sum	656	585	71	9	69	3	5		669	580	89	6	27	2	5
Average	219	195	24	3	23	1.53	1.7		223	193	30	2	9	1.78	1.7
Sum+Sum				75	269	15	9					41	141	10	
Average	228	199	29	5	18	0.44	0.6		230	199	31	2.7	9.6	0.90	30 / 2

It will be seen that for both parallel tests the number of shocks is greater for group-animals than for alone ones. The number of approaches-withdrawals, however, is slightly greater for animals alone (when both parallel tests are taken together). This is due to one animal, female No. 91, litter 8, which has as many as 30 approaches-withdrawals. The group-animals in the first parallel test did not drink in the allowance cage, thus being the only ones not doing so of 27 animals tested in groups.

As previously mentioned, the current was reduced to 0.185 milliamperes for the following tests. It might perhaps have been still more reduced, but if getting to any degree below 0.1 milliamps, some animals will not be frightened away, but keep on drinking just a little at a time. The results of the 30 females are given in Table 3.

The 15 animals in groups have tried to drink and gotten shocks 75 times, while the animals alone have tried only 41 times. The number of approaches-withdrawals is also considerably higher for the grouped animals. The latter have also without exception drunk in the allowance cage, while only 10 of the animals alone have done so. The time elapsing before these 10 rats drank is twice as big as for the group-animals. It is also characteristic that the animals in groups only have 9 approaches-withdrawals before they went ahead drinking, while the animals alone have 30. All other factors are kept constant for the two parallel tests. That the animals alone have lost

2 gm more weight than the animals in groups must be ascribed to the fact that the grouped animals have been allowed to drink until all three in the group had finished (social influence being the object of our study). A rat which has thirsted for some days, will drink several ccm in the course of a few seconds.

Table 4 gives the time elapsing before animals in groups and alone have started to drink.

TABLE 4

	0.15	0.30	0.45	1.00	1.30	2.00	2.30	3.00
In groups	16	18	21	22	22	22	23	24
Alone	4	4	6	9	12	14	15	15

The figures on the top of each column is the time in minutes and seconds, the other figures indicate number of animals. Sixteen of the group-animals drank in the course of 15 sec., as compared with 4 when alone. Twenty-two of the group-animals have drunk within 1 min. as compared with 9 when alone. Only 15 of the animals alone have drunk at all, as compared with 24 of the grouped animals. It should be noted that the 3 group-animals which have not drunk belong to the same group. It seems that if 1 animal reacts negatively to an object, it will spread to the rest. Rats are able to influence each other to a considerable extent. Grey rats of wild parents which are raised by a white tame mother, will become completely tame, and it has been shown that we are dealing with an influence of the tame mother.[1]

DISCUSSION

It might seem superfluous to devote more space to the results. Of course, even if the various tests are heterogeneous in regard to the age of the animals and the shocks received, the results in Tables 1, 2 and 3 can be summarized as the parallel tests are homogeneous. Then we find that the animals in groups have tried to drink and

received shocks 151 times (average 5.6), while the animals alone have only 80 times (average 2.9). The corresponding totals for approaches-withdrawals are 444 and 318. However, the animals giving the results reported in Table 3 must be taken as representative for our study. It seems that for these animals better experimental conditions (i.e. conditions favorable for the observation of social influence) as to the shock, electric arrangement etc., have been obtained than was the case in previous experiments.

As previously stated, the time of observation was 50 min. It might be asked why a limited observation period is used, and what justified the length used. When comparing results from different experiments with the Obstruction Method, this aspect has apparently not been kept in mind. When comparing sexual deprivation and hunger of different duration, etc., as well as when comparing the strengths of the various needs, a definite rank order has been assigned to the drives.[2] Leuba[3] has shown, however, that a completely different rank order might have resulted if a longer period of observation had been used.

It is necessary, therefore, to see how the number of shocks and approaches-withdrawals are distributed during the 50 min. of observation in our experiment (see Table 5).

TABLE 5. *Distribution of Shocks for 15 Females in Groups and 15 Alone in Periods of 5 min.*

	5	10	15	20	25	30	35	40	45	50	Sum
In Groups	55	1	1	6	4	3	5	0	0	0	75
Alone	31	3	3	0	0	2	2	0	0	0	41

[1] E. Wulff Rasmussen. Wildness in rats. Heredity or environment? *Acta Psychologica*, in press.

[2] See Warden, *Op. cit.*

[3] Leuba, C. J. Some comments on the first reports of the Columbia study of animal drives. *Jour. Comp. Psychol.* 1931, 275–79.

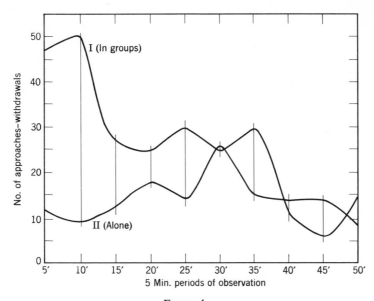

FIGURE 1.

Thus we see that most of the shocks (73 per cent and 76 per cent respectively) are received during the first 5 min. of observation for both categories, whereafter the trials are decreasing, and during the last 15 min. no attempts are made whatsoever by any animal of either category. We have reason to believe, therefore, that the animals have become so negatively conditioned to the dish that they practically without exception would abstain from drinking even if a longer period of observation was used.

The distribution of approaches-withdrawals can be seen from Table 6 and Figure 1.

TABLE 6. *Distribution of Approaches–Withdrawals for 15 Females in Groups and 15 Alone in Periods of 5 Min.*

	5	10	15	20	25	30	35	40	45	50	Sum
In groups	47	50	27	25	30	25	30	12	6	17	269
Alone	11	9	12	18	14	26	15	13	14	9	141

Here, too, most of the observations fall in the first periods. The difference between the group-animals and those alone can easily be seen from the table and the figure. During the first 5 min. period, the group-animals have four times as many movements as the animals alone, during the second period five times as many, and during the third period twice as many. When half of the observation time has passed (25 min.), the animals in groups have thrice as many movements as those alone (179 and 64, respectively). During the second half the number of movements is just about the same for the two categories (80 and 77). This must be explained by the fact that the group-animals will have received many more shocks, and thus have become more negatively con-ditioned to the dish than the animals tested alone. (Cfr. Tables 1, 2 and 3 for the fact that animals in group make many initially more attempts at drinking than animals alone.)

The shock which the rat received when having been drinking for 5 sec. and the current was supplied, was also recorded as a shock. Rats having thirsted will naturally drink whether alone or in groups, but the essential point is whether the rat, after having received a shock and been "frightened" away, *again will approach and attempt drinking.* If we exclude the original shock the difference between the two categories will be seen more clearly. When in groups, they will make renewed attempts at drinking and receiving shocks 124 times (average 4.6), while when alone the corresponding figures will be 52 and 1.9. In other words, the difference turns out more than twice as big. This treatment of the results is most correct for our problem. The original shock has been included in order to control that all rats have been drinking and received a shock.

Theoretically, there should be no possibility with our arrangement for reduced intensity of shock if other rats are present on the grill. To make sure, this has been controlled by connecting up resistances, corresponding approximately to that of the rats, with different spots of the grill, and then take reading on the ampere-meter.

Two of the group-rats (females No. 112 and 120) and three of the rats tested alone (females No. 75, 79, 111) displayed sexual behavior towards a male, thus being in pro-oestrus or in oestrus. We have taken 122 vaginal smears in connection with some other experiments[1] and practically all of them give typical pro-oestrus pictures or intermediates of these two phases, when the female has displayed sexual behavior towards a male.

It would have been preferable if no rats in the oestrus had been tested, as we then must reckon with e.g. increased activity, but for practical reasons it would have been difficult to avoid. In order that all animals should have thirsted for an equal length of time, we must have excluded those parallel tests which were carried out with rats in oestrus, or subjected to thirst a greater number of animals than was necessary for the experiments. There seems to be no reason, however, to make the experiments more difficult than necessary, as the animals in oestrus behaved during experimentation like all the other rats. In other experiments the situation may be different, thus in an experiment on hunger,[1] rats in oestrus were not tested.

MAN AND ANIMALS

One can hardly be cautious enough when comparing man with animals. But when investigations as well as more or less systematic observation and general experience point towards certain facts in man, and the same facts are found experimentally in animals, we can to a greater or less extent, depending upon circumstances, by means of such results in animal psychology support our findings in man.

To understand the drinking of alcohol, the drinking customs are factors of social facilitation which must be taken into account. Few people will drink alone. Abuse of alcohol will give anyone a more or less pronounced feeling of doing something wrong. Drinking together with other people will make it easier for each individual participating. But the factor of social facilitation does not, of course, explain alcoholism. Without the presence of a more or less intense *individual need*, nobody is apt to drink, we suppose, just because other people do—just as rats will not try to drink, alone or in groups, without an individual need of water.

Social facilitation plays an important role for certain criminal offenses. When in groups, criminals are able to commit crimes which they, psychologically speaking, could not do alone.

Mass-psychological phenomena do not seem to be taken sufficiently account of in praxis when evaluating the *reliability of witnesses*. The mutual influence exerted by school children in cases of sexual offenses can hardly be overestimated.

The factor of social facilitation is not to be found only when dealing with less desirable human behavior. Social enterprises, craving tasks of various kinds, are easier to carry out when stimulated by others doing the same thing or working for the same goal.

In general, we may say that many actions are carried out more actively both by human beings and animals when in groups than when alone.

SUMMARY

1. In the present investigation we have showed that rats under certain experimental conditions will react differently when in groups of 3 than when alone. The observed differences in behavior are as follows:

(*a*) Rats, having thirsted for 48 hours and then given an electric shock after 5 sec. of drinking, will try to drink again and consequently get shocks twice as many times when in groups of 3 than when tested alone. In only one of the nine parallel tests have the animals in groups less than *twice* as many attempts at drinking and received shocks than the animals alone.

(*b*) Rats in groups of 3 will approach and withdraw (but without drinking) on the average 16.4 times as compared with 11.8 times for rats alone. In the main experiment (see Table 3) with 30 rats the difference is about twice as large.

(*c*) When placed in a cage immediately after

---

[1] Not yet published. *Acta Psychologica IV.*

the test, 16 of 27 grouped animals will drink in the course of 15 sec., while only 4 of 27 animals tested alone will do so. In the course of 1 min., 22 of the group-animals will drink, but only 9 of those alone. At the end of the observation-period of 3 min., 24 of the group-animals had drunk, but only 15 of the animals tested alone. It is characteristic that the only 3 grouped animals which did not drink, belonged to the same group.

2. Most of the shocks (70 per cent) are received during the first 4 min. of the testing period, and practically all fall within the first half, the complete period of observation being 50 min. The number of approaches-withdrawals is four times as big for group-rats than for rats tested alone during the first 5 min. period, for the second period five times and for the third period thrice as big. At the end of the observation-period, the two categories of animals will not differ in number of approaches-withdrawals. It means that animals in groups will try to drink and consequently get shock and approach-withdraw much more often than animals alone, before they get so negatively conditioned to the risk that they will keep away from this.

3. Other factors, such as the litter of the animal, weight, procedure, etc., are kept constant for the various parallel tests.

4. Female albino rats having thirsted 48 hours (+ time of observation), will lose about one-eighth (13.04 per cent) of their weight. As the rats in groups may have been drinking more or less water in the "allowance" cage after the test, it is more correct only to calculate the loss of weight for the animals tested alone in the main experiment. Their average weight before thirsting was 228 gm, loss of weight by thirsting 29 ± 1.10 gm, and variation 21–45 gm; SD 9.66 gm. The correlation between weight and loss of weight is $+0.42$. It must be remembered that this result is based only on 15 animals.

5. Our experiments seem to a certain extent to support the Gestalt thesis that one reacts to the situation as a whole. The rats will be more negatively conditioned to the dish of water when this is in the experimental box where the shocks have been given than when placed somewhere else. But if the rat has been *strongly* negatively conditioned to the dish in the experimental box, it will not try to drink from it in another cage.

6. Social facilitation plays an important role in every human society. That the same factor has been shown to be operative in animals, lend support to this thesis.

CHAPTER TWO

IMITATION

In the previous chapter one of the observations made by Chen who studied the problem of social facilitation in ants was that nest-building under coaction resulted in a greater uniformity of behavior. Chen observed that the rapid co-worker had an accelerating effect, and the slow co-worker a retarding effect upon the work of the individual ant. These accelerating and retarding effects were obtained in addition to and independently of the overall enhancement of performance which was due to the presence of a coactor.

Obviously the presence of a coactor, or even a spectator, may have considerably more profound and complex effects than simply energizing the subject's dominant responses. Chen's observations of increased interindividual uniformity under grouped conditions indicates that the coactors are rich sources of cues for one another. The fact that rapid workers accelerate nest-building of their partners while slow workers retard it, shows that the presence of another organism not only energizes responses but may trigger them off, as well. Thus, on the one hand the presence of another is a source of nonspecific effects brought about by an increase in general arousal level of the organism. All those responses that would be evoked in the absence of a coactor will also be evoked in his presence, but they will be more vigorous. On the other hand, the coactor may be a source of rather specific effects by providing cues which bring about responses that without him would be un-likely. Thus, the coactor is, at the same time, a source of energizing effects and a source of directive effects.

The study of imitation is concerned primarily with the directive effects upon behavior that can be attributed to the presence and to the behavior of another organism. It is the study of cues which are generated by the behavior of one organism and which benefit the learning and performance of another.

Because imitation occupies a rather important position in the explanation of social life, research and theorizing concerned with it date to the earliest days of social psychology. Gabriel Tarde (1903) attributed to imitation an especially significant consequence, namely that of social uniformity. Imitation, according to Tarde, was indeed the very basis of culture and social uniformity. But, on the other hand, he also maintained that social progress—that is, social change—could be attributed to faulty imitation.

There is only a handful of data on the role of imitation in the spread of what may be "culture-like" habits among animals. It was reported for the first time in 1921 that some birds, probably tits, began opening milk bottles left at the doorsteps of the inhabitants of a small village in southern England. This ingenious way of procuring milk has "become a widespread habit in many parts of England and some parts of Wales, Scotland, and Ireland, and . . . has . . . been practised by at least eleven species of birds." (Fisher & Hinde, 1949). The other form of "culture-like" uniformity was observed by Japanese zoologists. In 1953, while systematically observing a group of monkeys living on the Kosima Island in an almost completely natural environment, a young female was seen washing a sweet potato before eating it—an act previously never recorded. Since then "many monkeys, especially the younger ones show the same behavior, which is expected to become a new cultural habit of this group in the future." (Miyadi, 1958.)

For a long time imitation was regarded as instinctive. In 1896 C. L. Morgan wrote that "The tendency to imitate is based upon an innate constitutional bias to . . . gain satisfaction by reproducing what others are producing" (p. 173). Miller and Dollard (1941), in their classical studies of imitation, gave considerable attention to the question of the learned and instinctive components of imitation. However, they took a rather strong position in favor of the former. Research that followed the pioneering experiments of Miller and Dollard was much less concerned with the genetic basis of imitation. It gave more emphasis to the analysis of imitation as a behavioral process, to the conditions that are critical for its occurrence, to the variables that affect its success, and to the stability and scope of its consequences. More and more attention was given to the behavior of the demonstrator in an attempt to discover what specific elements and aspects of this behavior enhance the imitator's learning. These concerns are reflected in terminological changes in the field that generated such concepts as "observational learning" or "vicarious experience."

The series of articles in this section shows a cross section of research interests in the study of imitation. The paper by Warden, Fjeld, and Koch is concerned with the methodological problems of studying imitation in the laboratory. It describes a method that has been widely used since its inception, and it specifies many experimental conditions which must be met to eliminate artifacts. The important paper by Crawford and Spence focuses upon the question of specifically what it is that the imitator benefits by observing the demonstrator. Their basic concern was with the hypothesis offered earlier by Spence that the main contribution of the demonstrator to the learning of the imitator consists of the enhancement of stimuli that are critical to the imitator's learning. In an earlier paper, in part concerned with imitation among infra-human primates, Spence (1937) inquired if imitation is "a rational process in which an animal perceiving the act of another animal and its results, realizes or understands these consequences and thereupon performs a similar act in order to obtain the same results" or "merely an abridgement of the learning processes . . . as a result of a change in the stimulus . . . conditions . . . ?" (p. 820–1). Would learning be enhanced if the demonstrator could do little to effect such stimulus predifferentiation? And, we see in the excerpt from a study by Turner that stimulus enhancement is indeed a powerful component of imitative behavior, for the "demonstrator" in his study is not a member of the same species of animals, nor, for that matter, of any species.

If the contribution of the demonstrator is primarily to capture the learner's attention and direct it to the relevant stimulus aspects, then the imitator should benefit more by observing a skilled performer than by observing a demonstrator who is in the process of learning. Yet, Herbert and Harsh (1944) in their work with cats have shown that the imitator can profit a great deal by observing errors committed by the demonstrator. The study carried out by Darby and Riopelle also shows this effect under experimental conditions which scrupulously observe all the requirements specified by Warden and his co-workers.

Church's experiment is a good illustration of the classical approach taken by Miller and Dollard, and it extends the work of these authors by showing that the imitator learns more than simply to duplicate the demonstrator's acts. There is a clear indication in Church's study that the stimulus situation which is significant to the imitator, involves not only the behavior of the demonstrator but the entire stimulus complex in which the demonstrator's behavior occurs.

REFERENCES

FISHER, J. and HINDE, C. A. The opening of milk bottles by birds. *British Birds*, 1949, **42**, 347–357.

HERBERT, M. J. and HARSH, C. M. Observational learning by cats. *J. comp. Psychol.*, 1944, **37**, 81–95.

MILLER, N. E. and DOLLARD, J. *Social learning and imitation.* New Haven: Yale Univer. Press, 1941.

MIYADI, D. On some new habits and their propagation in Japanese monkey groups. *Proc. XV Intern. Conf. Zool.*, London, 1958, pp. 857–861.

MORGAN, C. L. *Habit and instinct.* London: E. Arnold, 1896.

SPENCE, K. E. Experimental studies of learning and the higher mental processes in infra-human primates. *Psychol. Bull.*, 1937, **34**, 806–850.

TARDE, G. *The laws of imitation.* New York: Holt, 1903.

IMITATIVE BEHAVIOR IN CEBUS AND RHESUS MONKEYS*

C. J. Warden, H. A. Fjeld and A. M. Koch[1]

A. INTRODUCTION

As shown by the recent summary of Warden and Jackson (8), the early work on imitation in birds and mammals yielded, in the main, negative results. Thorndike found no evidence for imitation in chicks, and the results of Porter on various species of birds are inconclusive. Small decides against imitation in the white rat, whereas Berry argues for a low order of imitation in this type. The negative results of Thorndike on the cat and dog are well known. Berry and Hobhouse insist that cats possess a low order of imitativeness. The tests of Cole and Davis on the raccoon were disappointing. In general, the opinion has steadily gained ground that genuine imitation does not occur among the subprimate forms.

The results to date on monkeys and apes are more or less inconsistent and confusing. Thorndike found little or no evidence for intelligent imitation in monkeys, and the instances cited by Hobhouse are not very convincing. Shepherd concludes that imitation, when present at all, is of little importance in the learning of monkeys. The results of Watson were negative, regardless of whether the imitatee were another monkey or himself. Kinnaman (3) reports two cases of imitation in the rhesus monkey, but these followed a considerable amount of practice on the task. Haggerty (2) offers the best evidence from these earlier studies for imitation in monkeys. Instances of clear-cut imitation were limited to a few of the animals tested, and some of the tasks involved were exceedingly simple for primates. He reports numerous cases of "partial" imitation in which observation apparently aided the imitator somewhat in solving the problem by trial-and-error activities. As a rule, however, the monkeys were allowed practice at the task before being tested for imitation.

Several workers have noted instances of social facilitation and simple imitative behavior in the great apes. Nevertheless, both Yerkes (12) and Köhler (6) seem to feel there is little reason to suspect a high order of imitativeness even in the chimpanzee.

Most of the studies cited above made use of the *Observation-Cage* method which was devised by Thorndike. After the imitatee had been trained, the imitator is allowed to observe him at work from a cage attached to the working cage. When sufficient time has been allowed for observation, the imitatee is removed from the working cage and the imitator transferred thereto. As pointed out by Warden (8, 9) the following criticisms apply to this method: (*a*) A shifting about of the animals may involve a considerable delay between observation and the imitative act; (*b*) the shifting may induce an emotional disturbance in the imitator; (*c*) the shift may disrupt the mental set to imitate. It seems altogether likely, indeed, that the negative results reported on primates by the earlier workers were due in large part to such arbitrary factors which are inherent in the method itself.

In 1930, the *Duplicate-Cage* method was devised by Warden (9) and was later used by Warden and Jackson (8) in testing the imitative behavior of 15 rhesus monkeys. This method involves the presentation of a duplicate puzzle device to the imitator in the observation cage. This makes it unnecessary to shift either of the animals about during a given work period. The imitator may react at once, upon observing the imitatee, since the puzzle device is ready at hand. Moreover, the imitative tendency when aroused is not disrupted by a shift of any sort. As previously observed, this method provides the natural setting for the occurrence of imitative behavior in animals and children. When tested by this method,

* From the *Journal of Genetic Psychology*, 1940, **56**, 311–322. Copyright 1940 by The Journal Press, Provincetown, Mass.

[1] The general method employed was devised by the senior author who is also responsible for the present report. The junior authors carried out the tests and made the statistical computations of the tabular materials.

the rhesus monkey (8) exhibited a marked capacity to learn various problem devices by means of imitation. The present study may be regarded as a continuation of this work to cebus as well as other rhesus monkeys.

B. METHOD AND PROCEDURE

The six monkeys used in this investigation may be classified as follows: two common cebus (*Cebus capucina*), one weeper capuchin (*Cebus apella*), and three rhesus (*Macaca mulatta*). The weeper capuchin (Trader) was supplied by Dr C. R. Carpenter. The other cebus and the rhesus monkeys were secured from local dealers and were 3–3½ years old when tested. The three cebus monkeys and Rhesus No. 2 were males, and Rhesus No. 3 and No. 4 were females. The animals were well adjusted to laboratory work before the imitation tests were begun. All of the monkeys had previously been used in the study on patterned strings (10), and in the study on instrumentation (11) as well. The care of the animals followed the usual Columbia Laboratory routine, as described in detail by Fjeld (1) and Koch (5). Each monkey was caged alone while in the laboratory to prevent the formation of distracting social attachments. Raisins were used as the reward, since they were known to possess a high incentive value for both types of monkeys.

A diagram of the apparatus employed is shown in Figure 1. Each of the two compartments

FIGURE 1. Warden duplicate-cage imitation apparatus. The cages for imitator and imitatee, separated by a wire mesh partition, are similar in every respect. The lights, *A, A* flood the puzzle devices during the test to enhance observational stimulation. The setting is for problem 4 except that the device on the right side is open so the position of the incentive can be seen.

measured 36 in. long, 30 in. wide, and 36 in. high. The dividing screen was made of ½-in. wire-mesh which permitted clear observation of the movements of the imitatee. Preliminary tests have shown that the monkey may be distracted by the presence of the experimenter in the room, hence a one-way light screen was employed. This involved covering the top of the cage with heavy black cloth and the back and sides with beaver board. A screen of rayon cloth through which the experimenter could observe the animals was lowered in front of the cage before making each test. The inside of each compartment was illuminated by an automobile spotlight (32 candle power) focused directly upon the panel bearing the puzzle device. In addition to supplying general illumination, these lights tended to enhance the stimulation value of the puzzle devices. The large central light was not used during the testing period and all lights in the room were turned off before beginning a test. A large electric fan was operated at low speed to serve as a sound screen.

The puzzle devices were mounted on removable panels as indicated in Figure 1. The panels were placed close to the division screen so as to favor ready observation. In this position the imitator was enabled to secure a profile view of the imitatee during his attack upon the problem. The following tasks, arranged in order of difficulty, were employed. (*a*) Pulling a chain which hung down within easy reach in front of the panel, thus raising a door and exposing a raisin in a hole in the panel; (*b*) opening a door in the panel, behind which a raisin had been placed, by manipulating a knob; (*c*) operating a simple latch and then opening the door; and (*d*) operating two latches (upper, lower) arranged as in Figure 1, and then opening the door. As will be seen, the last three problems represented a progressive series of acts. The four problems were first learned by the imitatee and then presented to the several imitators in the order given above. These problems were precisely the same as those in the earlier study (8). It was there shown by testing a control group that each of the problems is above the instinctive level, hence genuine learning is required to solve them. No further control group was necessary, therefore, in the present study. The time required to train the imitatee by trial and error, aided by coaching, was many times as great as that allowed in the series of imitation tests.

The general procedure was the same as that used in the earlier study. The imitator was allowed to adjust to the testing conditions by being placed in the cage for 10 min. daily on from

1 to 4 days as required. During this period practice was prevented by the use of blank panels in the cage. The imitatee was in the opposite side and the lights, light-screen, and the guide-cords (to control the animal) were manipulated as in the experiment proper. A female monkey (Rhesus No. 1), after being trained as the imitatee, had to be discarded at this stage because Cebus No. 1 was very much afraid of her. A male animal (Rhesus No. 5) was trained and used thereafter as the imitatee. So far as could be observed this animal occasioned no emotional upset in any of the imitators tested.

In the experiment proper an animal was given six tests on each problem before being advanced

to the next. As a rule the six tests were made on the same day, although the series was broken by a 24-hour interval in some cases. An interval of 4 days intervened between one problem and another, in order to reduce any possible transfer and interference effects. As will be seen, each animal was given 24 tests of imitation altogether, thus making the group total 144.

The test procedure was precisely the same as that used in the earlier study and may be indicated very briefly. The two animals, with cords attached, were placed in their respective compartments, and anchored in the proper position by the experimenter. The two puzzle devices were set and baited by an assistant[1] stationed behind the

TABLE 1. *Showing the Record of Each Monkey on the Four Problems* (*Six Tests on Each*)

Animal and Problem	Test 1 Sec. R		Test 2 Sec. R		Test 3 Sec. R		Test 4 Sec. R		Test 5 Sec. R		Test 6 Sec. R	
Problem I												
Cebus 1	25	I	20	I	18	I	12	I	14	I	4	I
Cebus 2	60	P	60	P	32	I	60	P+	60	P+	46	I
Trader	60	P	60	P+	22	I	60	P+	8	I	60	P+
Rhesus 2	60	P	25	I	5	I	5	I	3	I	4	I
Rhesus 3	7	I	10	I	9	I	8	I	4	I	3	I
Rhesus 4	60	P+	60	P+	25	I	17	I	6	I	5	I
Problem II												
Cebus 1	60	P	60	P+	57	I	2	I	2	I	2	I
Cebus 2	39	I	14	I	10	I	4	I	12	I	4	I
Trader	2	I	2	I	2	I	2	I	1	I	5	I
Rhesus 2	31	I	20	I	4	I	2	I	2	I	2	I
Rhesus 3	11	I	3	I	9	I	14	I	3	I	7	I
Rhesus 4	21	I	60	P	19	I	4	I	3	I	1	I
Problem III												
Cebus 1	11	I	8	I	3	I	16	I	25	I	49	I
Cebus 2	2	I	14	I	1	I	13	I	3	I	1	I
Trader	60	I	25	I	60	P+	40	I	5	I	40	I
Rhesus 2	60	P	60	P	60	P	60	P	60	P	60	P
Rhesus 3	13	I	27	I	60	P	60	P	15	I	45	I
Rhesus 4	60	P+	5	I	25	I	10	I	5	I	25	I
Problem IV												
Cebus 1	4	I	48	I	60	P	59	I	60	P	60	P
Cebus 2	23	I	13	I	12	I	8	I	7	I	8	I
Trader	9	I	9	I	9	I	19	I	13	I	10	I
Rhesus 2	60	P	60	P	60	P−	60	P−	60	P−	60	P−
Rhesus 3	31	I	60	P	18	I	33	I	48	I	25	I
Rhesus 4	19	I	49	I	19	I	18	I	30	I	7	I

For the meaning of the letter rating under *R*, see analysis of types of imitative behavior in text above.

[1] Thanks are due to Dr. G. M. Gilbert for valuable aid in this connection.

TABLE 2. *Group Averages and Percentages on All Problems*

Degree of Imitation	Total Animals	Prob. 1 Cases No.	%	Prob. II Cases No.	%	Prob. III Cases No.	%	Prob. IV Cases No.	%	Total Cases No.	%
I	6	25	69.4	33	91.7	26	72.2	26	72.2	110	76.4
P+	6	7	19.4	1	2.8	2	5.5	0	0.0	10	6.9
P	6	4	11.2	2	5.5	8	22.3	6	16.6	20	13.9
P−	6	0		0		0		4	11.2	4	2.8
F	6	0		0		0		0		0	

apparatus. The room lights were then turned off. The panel of the imitatee was then illuminated by means of the spotlight and the imitatee allowed to perform. The animal was drawn back to the corner of the cage, the light turned off, and the puzzle device reset. This procedure was repeated five times in rapid succession, the five performances of the imitatee requiring only about 30 sec. Immediately after the fifth performance, the light was turned off in the imitatee's compartment and flashed on in the other compartment. The imitator was released and allowed 60 sec. in which to solve the problem just observed. The imitator was usually leaning forward against the taut cord and often went directly to the panel. If the problem had not been solved within the 60-sec. period, the imitator was drawn back to the corner of his compartment. The exact time required for each solution was taken with a stop-watch.

C. RESULTS

The results have been analyzed into the following degrees of imitativeness, as was done in the earlier study:

I, immediate imitation—solution of task within the 60-sec. test period.

P+, partial imitation—proper act occurred but not exact or forceful enough to work the mechanism.

P, partial imitation—contact made with proper part of mechanism but appropriate movement incomplete.

P−, partial imitation—puzzle device approached and parts inspected but without body contact with it.

F, failure to imitate—no apparent attention to puzzle device.

The individual records for the six tests on each of the four problems are presented in Table 1. The records for a given performance include the time score and a rating as to the degree of imitativeness exhibited. The ratings are assigned in accordance with the letter system indicated above. The time scores vary considerably when imitation (immediate or partial) occurs, depending on the moment when the appropriate act took place. Obviously, the shorter the time the less the opportunity for trial and error activity to enter into the solution.

As indicated in Table 1, no outright failures occurred in the present study. This means that some degree of social facilitation was shown on each of the 144 tests. There were 110 cases of immediate imitation and 34 cases of partial imitation (*P+, P, P−*). The amount of imitative behavior exhibited showed little variation from task to task. The number of instances of immediate imitation on the series of tasks was as follows: 25, 33, 26, 26. The corresponding scores for partial imitation were 11, 3, 10, 10. A summary of the several types of imitative behavior for each problem is given in Table 2. As will be seen, 10 of the 34 cases of partial imitation fall into the *P+* category. This means that the proper pattern of response was copied by the imitator, but not with enough exactitude or force to operate the essential mechanism. The other two types of social facilitation (*P−, P*) involved little more than an enhancement of the stimulus in the setting. It is apparent, from this analysis, that the pattern of response was actually copied in 120 out of 144 tests.

The data bearing on the relative imitative ability of the different animals, in terms of the immediate imitation scores, are given in Table 3. The differences are not marked, except in the case of Rhesus No. 2, who succeeded on somewhat less than half the tests. The other five animals ranged from 19 to 21 successes out of the 24 tests given to each animal. The question of generic differences can hardly arise in view of the small number of each type tested. Nevertheless, it may be noted, except for Rhesus No. 2, the cebus and rhesus monkeys made about the same showing.

TABLE 3. *Showing the Record* (Immediate Imitation) *of Each Animal on the Four Problems in Terms of Average Score*

Animal	Prob. I		Prob. II		Prob. III		Prob. IV		Totals	
	No. Cases	Av. Sec.	No. Cases	Av. Sec.	No. Cases	Av. Sec.	No. Cases	Av. Sec.	No. Cases	% Suc.
Cebus 1	6	15.5	4	15.8	6	18.7	3	37.0	19	79.2
Cebus 2	2	39.0	6	13.8	6	5.7	6	11.8	20	83.3
Trader	2	5.0	6	2.3	5	34.0	6	11.5	19	79.2
Rhesus 2	5	8.4	6	10.2	0	0.0	0	0.0	11	45.8
Rhesus 3	6	6.8	6	7.8	4	25.0	5	31.0	21	87.5
Rhesus 4	4	13.3	5	9.6	5	14.0	6	23.7	20	83.3

As already noted, all instances of immediate imitation involved successful performance within the 60-sec. test period. In many instances, however, the imitative act occurred within a few seconds after observation of the imitatee. The temporal distribution of the 110 cases of immediate imitation is shown in Table 4. In 57 cases (51.8 per cent), imitation occurred within 10 sec., and in 38 cases (34.5 per cent) within 5 sec. Such speedy performance would eliminate the possibility of trial and error activities of any consequence. Moreover, a check of the detailed records indicates that in at least 40 of the 57 cases mentioned above, no such activities occurred. In these instances, the pattern of movement of the imitatee was followed precisely and without excess or fumbling movements. As a matter of fact, 94 of the 110 cases of immediate imitation fell within the first half of the 60-sec. test period. In numerous instances the time score was lengthened because of the fact that the animal hesitated somewhat before approaching the panel. A high time score, therefore, does not necessarily mean a large number of excess or trial and error movements.

TABLE 4. *Showing the Temporal Distribution of Immediate Imitation Scores Within the Test Period* (60 Seconds)

Time Interval (seconds)	Problem I	Problem II	Problem III	Problem IV	All Problems
0– 1	0	2	2	0	4
1– 2	1	10	1	0	12
2– 3	2	3	2	0	7
3– 4	3	4	0	1	8
4– 5	3	1	3	0	7
5– 6	1	0	0	0	1
6– 7	1	1	0	2	4
7– 8	2	0	1	2	5
8– 9	1	1	0	3	5
9–10	1	1	1	1	4
11–15	2	4	5	3	14
16–20	3	2	1	5	11
21–25	3	1	4	2	10
26–30	0	0	1	1	2
31–35	1	1	0	2	4
36–40	0	1	2	0	3
41–45	0	0	1	0	1
46–50	1	0	1	3	5
51–55	0	0	0	0	0
56–60	0	1	1	1	3
Median time	7.3	3.8	13.0	19.0	9.5

TABLE 5. *Showing the Practice Effect (Through Six Tests) on Immediate Imitation*

Problem	Test 1 *I*	Sec.	Test 2 *I*	Sec.	Test 3 *I*	Sec.	Test 4 *I*	Sec.	Test 5 *I*	Sec.	Test 6 *I*	Sec.	Total *I*	Sec.
I	2	16.0	3	18.3	6	15.2	4	10.5	5	7.0	5	12.4	25	12.7
II	5	20.8	4	9.8	6	16.8	6	4.7	6	3.8	6	3.5	33	9.6
III	4	21.5	5	15.8	3	9.7	4	19.8	5	10.6	5	32.0	26	18.7
IV	5	17.2	4	29.8	4	14.5	5	27.4	4	24.5	4	12.5	26	21.1
Total	16		16		19		19		20		20		110	

I refers to the number of cases of immediate imitation for all animals; *Sec.* to the average time in seconds for imitation.

As has been noted, each animal was given six tests on a given problem before being passed on to the next. This schedule raises the question of practice effects within each series of tests, and throughout the experiment as a whole. An analysis of the data on this point will be found in Table 5. The number of cases of immediate imitation, together with the average time of performance, is given for each of the six tests. These data are indicated separately for the several problems. It is obvious that practice effects might be manifested either in a greater number of successes or in speedier reactions as the tests continued. There was no practice effect from Test 1 to Test 2, when the data for the four problems are pooled, as indicated in the last array of Table 5. There is a rise, however, from Tests 3 to 6, in the number of cases of immediate imitation as well as in average speed of response. Such practice effects may have reflected either better observation or a smoother manipulation of the puzzle devices. It is impossible to interpret the results, relative to the effects of practice through the several problems, since the tasks were made more complex as the work advanced.

On the whole, the present results demonstrate about the same level of imitative capacity in monkeys as those of the earlier study by Warden and Jackson (8). This is hardly surprising since the *Duplicate-Cage* method was used in both investigations, and the procedure, schedule of testing, and method of scoring were the same. Moreover, in both cases the monkeys had been previously trained on some other laboratory apparatus, and hence were probably about equally well adjusted to handilng and test conditions. The present results should be compared only with those of the "satisfactory" group of the former study. This group was formed by eliminating six females that were so distracted by the male imitatee that they observed very little and failed on most of the tests. The "satisfactory" group there reported showed 72.4 per cent immediate imitation score and 22.9 per cent partial imitation score. The corresponding values in the present study were 76.4 per cent and 23.6 per cent. In the earlier study the group showed only 4.8 per cent failure scores, whereas no failures occurred in the present study. The slight differences in the results of the two experiments clearly fall within the range of individual differences.

It seems obvious that the old notion that monkeys do not imitate must be abandoned, in view of the results of these studies, involving two generic types (rhesus, cebus) and three species. It should be emphasized, in this connection, that more rigid controls were applied in these studies than had been used by any of the earlier workers. In fact, the generally accepted criteria of imitation (8, 9) were carefully fulfilled for the first time. The tasks employed were sufficiently novel and complex to involve a genuine learning process; the responses classed as immediate imitation appeared promptly after observation of the imitatee; previous practice was excluded by the experimental conditions; the responses of the imitator was substantially that of the imitatee; and a large enough number of instances occurred on the various problems to eliminate the chance factor. Up to the present, no one has suggested a more rigid set of criteria. In view of these findings, it seems altogether likely that the negative results of previous investigators were due in large measure to the arbitrary method (*Observation-Cage*) employed. This suggests the need of retesting such sub-primate mammals as the dog and cat, by the *Duplicate-Cage* method, before drawing conclusions as to the imitative capacity of these types.

D. SUMMARY

1. Six monkeys (three rhesus, two common cebus, one weeping capuchin) were tested for

imitation by means of a newly devised procedure (Warden *Duplicate-Cage* method) on four manipulative problems, six tests being given on each task.

2. Immediate imitation (success within 60 sec.) occurred in 76.4 per cent of the 144 tests. This compares with a score of 72.4 per cent obtained on nine rhesus monkeys in a previous study in which the same method was used.

3. Instantaneous imitation (within 10 sec.) occurred in 57 tests, or approximately one-half of the successes (immediate imitation). In 40 of these 57 cases, the response was made directly without fumbling or trial and error movements.

4. These results demonstrate a high level of imitative capacity in the cebus as well as the rhesus monkey.

REFERENCES

1. FJELD, H. A. The limits of learning ability in rhesus monkeys. *Genet. Psychol. Monog.*, 1934, **15**, 369–537.
2. HAGGERTY, M. E. Imitation in monkeys. *J. Comp. Neurol.*, 1909, **19**, 337–445.
3. KINNAMAN, A. J. Mental life of two macacus rhesus monkeys in captivity. *Amer. J. Psychol.*, 1902, **13**, 98–148, 171–218.
4. KLÜVER, H. *Behavior Mechanisms in Monkeys.* Chicago: Univ. Chicago Press, 1933.
5. KOCH, A. M. The limits of learning ability in cebus monkeys. *Genet. Psychol. Monog.*, 1935, **17**, 163–234.
6. KÖHLER, W. *The Mentality of Apes.* New York: Harcourt, Brace, 1925.
7. WARDEN, C. J. and JACKSON, T. A. A preliminary study of the hunger drive in the rhesus monkey. *J. Genet. Psychol.*, 1935, **46**, 126–138.
8. ———. Imitative behavior in the rhesus monkey. *J. Genet. Psychol.*, 1935, **46**, 103–125.
9. WARDEN, C. J., JENKINS, T. N. and WARNER, L. H. *Comparative Psychology.* Vol. 1, Principles and Methods. New York: Ronald Press, 1936.
10. WARDEN, C. J., KOCH, A. M. and FJELD, H. A. Solution of patterned string problems by monkeys. *J. Genet. Psychol.*, 1940, **56**, 283–295.
11. ———. Instrumentation in cebus and rhesus monkeys. *J. Genet. Psychol.*, 1940, **56**, 297–310.
12. YERKES, R. M. and YERKES, A. W. *The Great Apes.* New Haven: Yale Univ. Press., 1929.

OBSERVATIONAL LEARNING OF DISCRIMINATION PROBLEMS BY CHIMPANZEES*

Meredith P. Crawford and Kenneth W. Spence[1]

INTRODUCTION

A NEW method for studying imitation in primates was put to test in this investigation. The method consisted in giving a chimpanzee, previously trained in discrimination learning, a series of opportunities to observe another animal's differential responses to a pair of stimuli, and then to make irregularly rewarded choices for himself. Any test of imitation requires the imitator to make one or a few repetitions of the demonstra-

tor's response. The present procedure in addition compels the imitator to fixate his response even in the face of irregular or ambiguous rewarding of his choices. Three experimental techniques were used which were designed to bring into relief these two requirements.

Most of the studies already reported on imitation in animals have been made with the problem-box method. These have been critically reviewed by Watson (9), Warden and Jackson (8), and most recently for the primates by Spence (7). In

* From the *Journal of Comparative Psychology*, 1939, **27**, 133–147. Copyright 1939 by Williams & Wilkens Co., Baltimore.
[1] Both authors collaborated in planning the experiment and construction of apparatus: M. P. C. did most of the testing and is responsible for the report in its present form.

elaborating the point of view of Watson, Spence has contended that in the problem-box test the activity of the demonstrator simply enhances certain aspects of the stimulus situation, which the imitator later attacks in his own way. Since discovery of the locus of attack is usually the most difficult part of a problem-box test, the imitator often meets with quick success without need of specific copy, after beginning to manipulate the part last manipulated by the demonstrator.

In contrast to this, the present test procedure starts with an imitator who is already familiar with the relevant aspects of the problem situation, and with the means of manipulating them. He is required to build up, through observation, only a single, well-defined response connection to one of

the stimuli. Thus the first advantage of the present method lies in the delimitation of the required learning to a single differential response, and the elimination of the enhancement factors of a particular locus of attack. A second advantage follows in the possibility of extending the relatively objective methods of analysis of individual discrimination learning to imitative learning (6). Finally, the discrimination method is superior to the problem-box in that a series of problems may be set which are known to be within the individual learning ability of the imitator, and at the same time the essential aspect of each is novel to him, as may be easily tested by a group of free choices preliminary to each test.

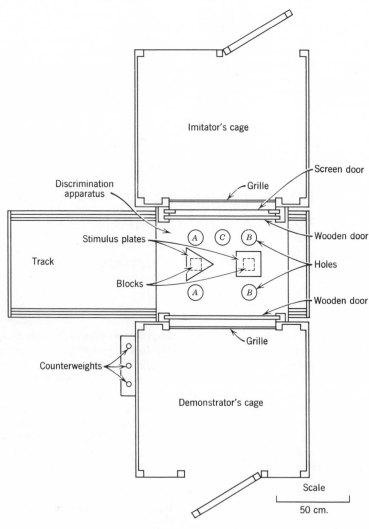

FIGURE 1.

TABLE 1. *The Discrimination Stimuli*

Form	Dimension cm.	Area sq. cm.	Color	Abbreviation
Square	13 × 13	169	White	w.s.
Circle	14.2 (diam.)	158	White	w.c.
Triangle	14 × 18.3	141	White	w.t.
Hexagon	16 (diam.)	168	White	w.h.
Cross	14 (arm)	120	White	w.cr.
Square	12 × 12	144	Red	r.s.
Square	12 × 12	144	Green	g.s.
Block	7.4 × 7.4	55	Bright metallic	b.

In working out this method a rather serious disadvantage appeared in the length of time required for preparation for each test through individual training of both imitator and demonstrator, as well as time consumed in the tests themselves. Possibly this type of test was a rather artificial one in which only exceptionally well motivated subjects might succeed. For this very reason, however, those cases in which learning did take place, although few in number, are the more convincing as exhibitions of imitative learning in chimpanzees.

APPARATUS

A diagrammatic floor plan of the experimental set-up is given in Figure 1. Two restraint cages, the *demonstrator's cage* and the *imitator's cage*, were mounted on platforms and faced each other across a *track*. From the outer end of this track the *discrimination apparatus* could be rolled to a position between the cages at which it is shown in the figure. The sides, floors, and tops of the cages were made of wood, while the rear walls and entrance doors were of hardware cloth on wooden frames. In the front of the demonstrator's cage was a *grille* of iron bars and a vertically sliding *wooden door*, and in corresponding positions on the imitator's cage were a *grille*, a *screen door* (16 gauge wire, 3 mesh to inch), and a *wooden door*. All doors were operated by ropes which led over pulleys in the ceiling of the experiment room to the *counterweights*.

A chimpanzee in either cage might reach through the grille to press down one of the *stimulus plates* on the discrimination apparatus thereby releasing a food cup which would slide under one of the *holes* (A, A, B, B, and C). The mechanisms for releasing food cups were placed beneath the surface of the apparatus and were operated from the surface by means of the *blocks*, to which the stimulus plates might be fastened. The mechanisms could be arranged in different ways so that cups would appear under the holes in various combinations upon depression of a single stimulus.

The seven stimulus plates used in the experiment are listed in Table 1 with their dimensions, colors, and the abbreviations by which they will be designated in the tables below. The red and green squares were of approximately equal brightness. The block is included in the table because in some of the tests only one stimulus plate was used, so that the problem was to discriminate between a stimulus plate and a block. The surface of the discrimination apparatus was painted dull black.

PROCEDURE

On each *demonstration trial* an imitator, through the screen door of his cage, could observe the demonstrator as he made his choice between a pair of stimuli and secured the food reward. A series of demonstration trials was regularly interrupted with one or more *imitation trials*, on which the imitator was allowed to choose between the stimuli and was rewarded without regard to the correctness of his choice, while the demonstrator was hidden from view. Trials of these two types were continued until the imitator mastered the problem, or until it was indicated that he would not profit by further demonstration. All the work done with a pair of subjects on a single discrimination problem constituted a *test* of imitation. Before each test the prospective imitator was given ten irregularly rewarded choices of the stimuli, and was not used if more than half of his choices were correct. The demonstrator, on the other hand, had mastered the problem before each test began, save for a few

TABLE 2. *Results With Technique A**

Number	1 Imit.	2 Dem.	3 Dates	4 Problem	5 Unit of D.T.	I.T.	6 Initial Pref.	7 Number Dem. Trials	8 Results	9 Prev. Learn. by Imitator
1	Helene	Bob	5/18–21/36	w.s.–b.	5	2	5/10	60	5/5	No
2	Helene	Bob	5/21–31/36	w.s.–w.c.	5	2	5/10	323	8/10	Yes
3	Helene	Bob Tom	6/19–20/36	w.cr.–w.t.	5	2	4/10	60	10/10	Yes
4	Bob	Tom	6/12–19/36	w.cr.–w.t.	5	2	6/10	80	10/10	Yes
5	Bob	Tom	6/29–7/2/36	g.s.–w.s.	4	2	5/10	20	10/10	Yes
6	Dick	Tom Bob	6/10–11/36	w.c.–b.	5	2	0/10	320	10/10	No
7	Helene	Tom	4/27–5/11/37	w.t.–w.s.	20	2	5/10	330	5/10	Yes
8	Bob	Tom	4/27–5/9/37	w.t.–w.s.	20	2	5/10	220	10/10	Yes

* On all tests with this technique the demonstrator had mastered the problem before the test began. Double baiting of the food cups on imitation trials was used throughout.

exceptions noted below. With a few exceptions, all the imitators had individually learned other discrimination problems in this apparatus before being used in a test.

Three basic experimental techniques were employed which differed only in the procedure used on demonstration trials. The procedure for the imitation trials remained the same with all three techniques. For each technique the discrimination apparatus had to be specially adjusted, so that cups would appear under the holes in different combinations for each technique.

In *technique A*, when the demonstrator depressed a stimulus, cups appeared under each of the pair of lateral holes nearest the stimulus (*A* and *A* or *B* and *B* in Figure 1). Each cup was baited, so that after the demonstrator had taken his food from the cup nearest him, the imitator was allowed to take food from the cup on his side.

In *technique B*, when the demonstrator depressed a stimulus, a baited cup appeared under one of the lateral holes on the demonstrator's side nearest the depressed stimulus, and another under the central hole (*C* in Figure 1) on the imitator's side, regardless of which stimulus was depressed.

In *technique C*, when the demonstrator depressed a stimulus, a cup appeared on his side under the hole nearest the stimulus, but no cup moved on the imitator's side, so that the imitator received no food.

For each trial the cups were baited at the outer end of the track while all vertical doors were closed. When the apparatus had been rolled into place for a demonstration trial the wooden door on the imitator's cage was raised. When the experimenter had made sure the imitator was looking at the stimuli he raised the door on the

TABLE 3. *Results With Technique B**

Number	1 Imit.	2 Dem.	3 Dates	4 Problem	5 Unit of D.T.	I.T.	6 Initial Pref.	7 Number Dem. Trials	8 Results
1	Helene	Tom	6/23–28/36	g.s.–w.s.	5–4	2	0/10	146	2/10
2	Bob	Tom	6/23–29/36	g.s.–w.s.	5–4	2	0/10	204	5/10
3	Helene	Bob	1/4–5/37	Position of b's	10	5	5/10	15	10/10
4	Helene	Bob	1/11–22/37	w.s.–w.t.	4	2	5/10	180	5/10
5	Helene	Bob	1/26–2/10/37	w.s. and w.t.–b.	4	1, 5	5/10	70	5/10
5a	Helene	Bob	2/15–28/37	w.s. and w.t.–b.	80 40 20	10, 5	5/10	140	10/10
6	Helene	Tom	3/16–4/5/37	w.t.–w.s.	100 20	5, 2	3/10	300	2/5
7	Bob	Tom	3/18–4/6/37	w.t.–w.s.	100 20	5, 2	0/10	310	0/5

* On all tests with this technique the demonstrator had mastered the problem before the test began. Double baiting of the food cups on imitation trials was used throughout. All imitators had learned at least one discrimination problem with this apparatus previous to testing.

demonstrator's cage, and the demonstrator made his choice, correcting it if necessary, and took his food. Then the imitator's screen door was raised, and, if technique A or B was in use, he was allowed to take food. For an imitation trial the lateral cups on the imitator's side were both baited, or were baited irregularly, without regard to the position of the correct stimulus, then the wooden door was raised, and a few seconds later the screen door, while the demonstrator's door remained closed.

children's group included the three males Tom, Bob, and Dick, and the female Helene. Tom and Bob were fairly well adapted to experimental routine, while Dick and Helene showed various types of infantile maladjustment when the present work began. The adolescent group comprised the females Bimba, Bula, Alpha and Kambi, and the male Frank. All of these subjects had served for more than four years in various behavioral experiments and were well adapted to testing routine before this experiment began.

SUBJECTS

Two age groups of chimpanzees served as subjects: four children between 3 and 4 years old, and five adolescents between 7 and 8 years. The

RESULTS

Twenty-six tests, 8 by technique A, 7 by technique B, and 11 by technique C were made during this experiment. The numerical results are pre-

TABLE 4. *Results With Technique C**

	1	2	3	4	5		6	7	8	9	10	11
					Unit of							Cor-
No.	Imit.	Dem.	Dates	Problems			Initial Pref.	Dem. Trials	Results	Dem.'s Status	Reward on I. Trials	rections Allowed I.
					D.T.	I.T.						
1	Tom	Helene	5/18–28/37	g.s.-r.s.	5	2	3/20	99	5/10	D.L.	I.B.	No
2	Bob	Helene	5/20–26/37	g.s.-r.s.	5	2	5/10	178	10/10	D.L.	I.B.	No
3	Bob	Tom	5/29–6/8/37	w.c.-w.h.	5, 4	2	4/10	209	5/10	D.M.	I.B.	No
4	Bob	Tom	6/10–13/37	r.s.-w.s.	4	2	4/10	68	10/10	D.L.	I.B.	No
5	Bob	Tom	6/14–29/37	w.s.-w.h.	4	2	3/10	300	5/10	D.L.	I.B. & D.B.	No
6	Alpha	Bula	10/ 5–19/37	w.t.-w.c.	4, 3, 2	2	3/10	164	5/10	D.M.	I.B. & D.B.	No
7	Kambi	Bula	11/15–29/37	w.t.-w.c.	3, 2	2	3/10	68	5/10	D.M.	I.B. & D.B.	No
7a	Kambi	Frank	1/17–21/38	w.t.-w.c.	2	2	6/10	50	5/10	D.M.	I.B. & D.B.	No
8	Tom	Frank	2/15–4/1/38	w.t.-w.c.	2	2	2/10	150	5/10	D.M.	I.B.	Yes
9	Bimba	Bula	3/ 2–4/8/38	w.t.-w.c.	2	2	4/10	200	5/10	D.M.	I.B. & D.B.	No
10	Bula	Bimba	4/15–5/11/38	g.s.-w.s.	2	2	5/10	130	5/10	D.M.	I.B. & D.B.	Yes
11	Helene	Tom	4/14–5/17/38	w.s.-w.h.	2	2	0/10	200	5/10	D.M.	I.B. & D.B.	Yes

* All imitators had learned at least one discrimination problem with this apparatus previous to testing.

sented for each technique in Tables 2, 3, and 4 respectively. In the first three columns of each table are given the names of the imitator and of the demonstrator, and the inclusive dates of each test. In column 4 the discrimination stimuli are designated by the abbreviations listed above (Table 1), the positive or correct stimulus being entered on the left-hand side of the column. In column 5 is shown the number of demonstration trials which were regularly presented before the group of imitation trials. Column 6 gives the number of times the positive stimulus was chosen by the imitator on the ten preference choices given prior to each test. In column 7 the number of demonstration trials given on each test is found. (The number of imitation trials given on a test may be calculated from the ratio of demonstration to imitation trials given in column 5.) In

column 8 the result of each test is expressed by the number of correct choices made by the imitator during the last ten imitation trials.

A ninth column will be found in Table 2, in which a "yes" is entered if the imitator had learned another discrimination problem before that test, and a "no" if he had not. This column is not included in Tables 3 or 4, since all imitators for techniques B and C had learned another problem previous to testing, as indicated in the footnotes to Tables 3 and 4. In Table 4, columns 9, 10, and 11 indicate conditions which varied for different tests with technique C, but remained constant for testing with other techniques, as indicated by the footnotes to Tables 2 and 3. Column 9 shows whether the demonstrator had already mastered (D.M.) or was learning (D. L.) the problem during the test. Corrections of wrong responses by the demonstrator were allowed as the imitator looked on. In column 10 the method for rewarding the imitator on imitation trials is indicated. Double baiting (D.B.) indicates that both of the imitator's cups were baited so that he received food on every choice regardless of correctness, while irregular baiting (I.B.) refers to the baiting of cups in random order without regard to the position of the correct stimulus. In column 11 a "yes" was entered when, on the irregular baiting trials, an imitator was allowed to "correct" or go to the other cup when food was not found in the cup already moved under the hole, while "no" indicates that the subject was not so permitted.

Qualitative comments must be made on some of the tests. These are presented as notes to the tests listed below. D.T. and I.T. will designate demonstration and imitation trials respectively.

Notes on Table 2

Test 2. Immediately after this test Helene was given the same problem (white square vs. white circle) for individual learning. She made only a chance score, and required 600 trials for mastery. Immediately thereafter she served as imitator in test 3.

Test 3. It should be noted that other investigators (4, 6) have found that the cross seems to have a peculiar initial excitatory value for certain subjects. Perhaps, therefore, this problem was an exceptionally easy one. Tom replaced Bob as demonstrator when the latter was shifted to another problem.

Test 5. This test immediately followed Bob's failure to learn the same problem under technique B (test 2, Table 3).

Test 6. When Bob replaced Tom as demonstrator Dick's I.T. scores began to improve immediately. Dick was more friendly with Bob than with Tom.

Tests 7 and 8. These tests followed the imitators' failures to learn the same problems under technique B (tests 6 and 7, Table 3).

General Note on Table 2. During each test on which the imitator learned (tests 3, 4, 5, 6 and 8) it was observed that he began, on D.T.'s, to orient toward the correct stimulus as he watched through the screen door. As learning became indicated in the imitator's scores it was found that the orientation behavior began to precede the demonstrator's choice movements. Such orientation was also shown in I.T.'s when the imitator was restrained from choice for a few seconds by the screen door.

Notes on Table 3

Tests 1 and 2. Experimentation was unavoidably interrupted before a greater number of demonstration trials could be given.

Test 3. Orientation movements by Helene were observed. The white square and white triangle were paired alternately with the block, in order not to increase the excitatory value of either with respect to the other.

Tests 5 and 5a are divided to indicate the differences in demonstration-imitation trial ratios.

Tests 7 and 8. As previously noted tests 6 and 7 preceded tests 7 and 8 in Table 2.

Notes on Table 4

Test 1. This test was discontinued because Tom refused to work when not allowed to correct I.T.'s.

Test 2. Results must be questioned because it was discovered that the irregular rewarding of imitation choices favored the correct stimulus. The red square was rewarded 10 times and not rewarded 17 times, while the green square was rewarded and not rewarded equally often.

Test 3. During the course of the test the operation of a secondary cue supplied by movements of the experimenter on I.T.'s was suspected, and when it was eliminated Bob's performance dropped from 75 per cent correct to chance.

Test 4. Reward and non-reward on imitation trials was equalized and care was taken to prevent operation of any secondary cues. The learning appeared to be a *bona fide* case of imitation. Gross bodily orientation movements by Bob were not observed on D.T.'s.

Test 5. Conditions were identical with those of test 4 except double baiting was used on I.T.'s during some sessions, to prevent frustration.

Test 6. Work was discontinued after 164 D.T.'s because Alpha refused to watch demonstrator and whimpered throughout the sessions.

Tests 7 and 7a. These two tests should be considered as a single test, unavoidably interrupted, and resumed with a new demonstrator. Kambi was in poor health during test 7a, showed little interest in working and refused to watch the demonstrator, so that experimentation was stopped after only 50 D.T.'s.

Test 8. Tom fell into a position habit which was not broken by repeated D.T.'s on his non-preferred side, by rewarding only the non-preferred side on I.T.'s, or by individual practice with another pair of stimuli from which he chose correctly.

Test 9. Bimba watched Bula very carefully on all D.T.'s, did not develop any position habits, and seemed at all times eager to work. She once made 8 out of 10 correct choices during a single session, but thereafter her choices dropped to chance.

Test 10. Bula watched the demonstrator keenly during the first five sessions and seemed to be very careful on I.T.'s. She made 15–20 correct choices during the I.T.'s of two successive sessions, after which her interest in the problem seemed to lag and she fell into a position habit which was not broken by various procedures.

Test 11. Helene had just finished learning individually three other discrimination problems when this test began. She watched the demonstrator well and did not fall into any position habit.

The imitator learned the problem in 5 of the 8 tests with technique A, the 3 failures being made by the subject Helene, who was known to be a poor individual learner. Only on the 2 easiest of the 7 tests with technique B was learning shown by the imitator. Only 1 clear case of learning occurred with technique C. The positive results with test 2 were questioned because of unbalanced, irregular rewarding. Tests 6, 7 and 7a were unsatisfactory because the imitators did not watch the demonstration trials. While the remaining 7 tests yielded negative results, the imitators Bimba and Bula both showed rapid improvement in their scores during the first few sessions of tests 9 and 10. Each watched her demonstrator's choices with apparent eagerness and used care on her own imitation trials during the first 4 or 5 sessions, making as many as 15 out of 20 correct responses. Thereafter they showed a marked decline in interest in the demonstrator's performance and in care with their own choices, whereupon their scores returned to a chance level.

DISCUSSION

With a few exceptions in technique A, all the imitators had received individual training in discrimination learning in this apparatus previous to imitation tests. They were presumably familiar with the relevant aspects of the stimulus situation, i.e., the stimulus plates, and possessed the response for their manipulation. The problem, with any technique, was simply to condition this manipulation response to the correct one of a pair of stimuli. Since reward was ambiguous or irregular on imitation trials with all techniques, it is assumed that all differential conditioning, or learning by the imitator, resulted from demonstration trials. To account for the differences in results with the three techniques, the following suggestions are offered.

In technique A the delivery of food to the imitator, on demonstration trials, at the side of the apparatus nearest the correct stimulus, served as a primary stimulus for the imitator's approach toward that side. This approach, or even an orientation, in the direction of the correct stimulus, may be considered as a critical act in the group of acts involved in a choice response in this experiment. This orientation was observed in all cases, in which learning took place, to precede the delivery of food by successively longer intervals, so that the imitator's response at one time coincided with, and later preceded, the movements of the demonstrator. It is possible, although not necessary, to assume that the imitator's response became conditioned to the demonstrator's response before being conditioned to the stimulus plate. More simply, the conditioning to the stimulus plate may have begun as soon as the primary food stimulus elicited the orientation response. In any event, the activity of the demonstrator, if effective at all, served only as a type of social facilitation, adding, in some way, excitatory strength to the chosen stimulus. It would have been interesting to arrange for a mechanical operation of the stimulus plates on demonstration trials and to compare learning of the imitator with and without the factor of social facilitation.

In techniques B and C there was no primary stimulus for the orientation behavior of the imitator. Such orientation was, however, observed when learning took place with technique B but not with technique C. The activity of the demonstrator in the absence of the primary food stimulus must have served to elicit this particular segment of the choice response in the imitator, again through the mechanism of social facilitation. That chimpanzees are capable of social facilitation has been shown by Yerkes (10). Humphrey (3) stated the basis for this mechanism in stimulus-response terminology when he said that imitation is based on a conditioned response in which the secondary stimulus and the response are identical. Unfortunately he offers no account of how such a conditioned response might be built up during the life of the organism. The function of the delivery of food in the central cup during technique B is difficult to describe. It doubtlessly served to distract the imitator from observing the demonstrator, but it may also have served as a reward when systematic orientation finally appeared in the imitator's behavior.

Once this response was elicited in the imitator's behavior and was consistently given when the demonstrator made his choice, it should have become conditioned to the correct stimulus plate. There is, however, the possibility that another factor operated to aid in this conditioning. The eating of food by the demonstrator may have served as a *substitute reward* for the imitator. This seems especially likely with technique C in which irregular reward was used on imitation trials, with consequent frustration of correct as well as wrong responses of the imitator.

This capacity for substitute reward might have been built up through earlier social conditioning of the subject, perhaps through the mechanism of second order conditioning suggested by Finch and Culler (2), and in a manner already demonstrated for the dog by Kriazhev (5). This investigator allowed one dog to observe the responses of another dog in a conditioning experiment, in which the primary stimulus was food, and to perceive the secondary stimulus. A conditioned salivary response was built up in the observing animal which had a strength intermediate between a first and second order C.R. With shock as the primary stimulus local defensive responses were not built up in the observing animal.

In considering the results with technique C it appears that only in subject Bob were both the C.R. to the demonstrator's choice (social facilitation) and the capacity for substitute reward present. Perhaps Bimba and Bula possessed only the first type of C.R. and not the capacity for substitute reward, since their scores improved rapidly but then fell away again as the subjects lost interest, possibly because of frustration on imitation trials.

For reasons theoretical as well as practical the present experimental methodology was not carried further into an exploration of the types of social relationships in which the imitative response appears most readily, a field which has been entered by the work of Aronowitsch and Chotin (1). For such a study the test should be simpler, a mere test of social facilitation, without the possible complication of the factor of substitute social reward. This latter factor, entirely unexplored in the primates, deserves study in isolation.

SUMMARY

Exploratory use was made of a new method for studying imitation in primates. A chimpanzee, previously trained in discrimination learning, was given a series of opportunities to observe another animal's discriminations between a pair of stimuli, and then to make critical, irregularly rewarded choices for himself. Three experimental techniques were used which differed in procedure on demonstration trials: In technique A the imitator received food in a cup near the correct stimulus just depressed by the demonstrator. In technique B he received food in the same centrally located cup regardless of the position of the correct stimulus. In technique C he received no food at all on demonstration trials. Technique C was the only real test of imitation; the other two were employed for theoretical interest.

Nine chimpanzees served as subjects in 26 test series. Perfect mastery by the imitator was attained in 5 out of 8 tests with technique A, in 2 out of 7 with technique B and 1 out of 11 with technique C. Partial learning took place in two cases with technique C before clear behavioral indications of loss of interest were shown by the imitator.

In the discussion it was suggested that learning might have taken place in technique A without social facilitation by the demonstrator, while in techniques B or C the imitator's response was elicited solely through social facilitation. In addition, conditioning in technique C may have been aided through the possible function of the demonstrator's eating as a substitute reward for the imitator.

REFERENCES

1. ARONOWITSCH, G. and CHOTIN, B. Über die Nachahmung bei den Affen (*Macacus rhesus*). *Zsch. Morphol. u. Ökol. Tiere*, 1929, **16**, 1–25.
2. FINCH, G. and CULLER, E. Higher order conditioning with constant motivation. *Amer. Jour. Psychol.*, 1934, **46**, 596–602.
3. HUMPHREY, G. Imitation and the conditioned reflex. *Ped. Sem.*, 1921, **28**, 1–21.
4. KLÜVER, H. *Behavior Mechanisms in Monkeys.* Chicago: Chicago University Press, 1931.
5. KRIAZHEV, V. I. [The objective investigation of the higher nervous activity in a collective experiment.] *Vysshaya Nervnaya Deyatel'nost*, 1929, **1**, 247–291.
6. SPENCE, K. W. Analysis of the formation of visual discrimination habits in chimpanzee. *Jour. Comp. Psychol.*, 1937, **23**, 77–100.

7. SPENCE, K. W. Experimental studies of learning and the higher mental processes in infra-human primates. *Psychol. Bull.*, 1937, **34**, 806–850.

8. WARDEN, C. J. and JACKSON, T. A. Imitative behavior in the rhesus monkey. *Jour. Genet. Psychol.*, 1935, **46**, 103–125.

9. WATSON, J. B. *Behavior. An Introduction to Comparative Psychology*. New York: Henry Holt & Company, 1914.

10. YERKES, R. M. Suggestibility in chimpanzee. *Jour. Soc. Psychol.*, 1934, **5**, 271–282.

SOCIAL FEEDING IN BIRDS*

Edward R. A. Turner[1]

THE purpose of this work was to investigate the extent to which social feeding exists in certain species of birds, and to attempt an analysis of the various causal factors involved. The term "social feeding" will be used to describe "feeding resulting from the stimuli provided by another non-prey animal."

Earlier work (see Turner, 1961) has demonstrated that camouflage mechanisms have selective value, but that birds are incredibly keen sighted, and once they have penetrated their prey's camouflage, the latter will suffer heavy casualties. If social feeding plays an important part in a species' normal behavior, then its prey species may become very vulnerable after one member of the predator species has made the initial discovery. This will also apply to edible prey species mimicking aposematic models, and some tentative work was done where a pair of crows (*Corvus corone*) was offered bright red pastry balls. The two birds approached the objects with the utmost caution, and it was with some great hesitation that one of the pair picked up the first model. Having consumed it, he instantly ate another, and only then did his mate join in. It seemed probable that the second bird's feeding resulted from the observation of its mate. Swynnerton (1942) made similar observations upon a pair of swallows (*Hirundo r. rustica*) that were fed with various species of moths. He concluded that one of the swallows was placing considerable reliance on its partner, and would not sample a new insect until the other bird had eaten one first.

It seemed, therefore, that it would be of considerable interest to discover the extent and nature of social feeding.

Some of the possible components of social feeding have been described by Thorpe (1956) and they include social facilitation, local enhancement, conditioning and imitation. Very briefly, social facilitation can be described as contagious behavior where the action of one animal may release identical behavior in another, and Thorpe quotes yawning as a familiar example. Local enhancement is defined as an "apparent imitation resulting from directing an animal's attention to a particular part of the environment." He considers it to be a special form of social facilitation. The term imitation is reserved for the "copying of a novel or otherwise improbable act or utterance for which there is clearly no instinctive tendency."

At the outset it may be convenient to state that a bird that is used in an attempt to release social feeding in another will be referred to as the "actor," and the bird that is thereby induced to feed will be called the "reactor."

From his work on greenfinches (*Chloris c. chloris*), Klopfer (1959) concludes that their power of visual imitation is not well developed. The test that he gave, however, was a somewhat difficult and unnatural one in that it tested their ability to discriminate a seed filled with aspirin from a normal, but otherwise identical seed, that was offered against a different background. The reactor was allowed to see the actor consuming normal seed, and avoiding the aspirin-filled seeds.

* From *Behaviour*, 1964, **24** 1–47. Copyright 1964. E. J. Brill, Publishers, Leiden, The Netherlands.

[1] This work was performed at the Sub-Department of Animal Behaviour of the University of Cambridge during one year's leave from St Luke's. The author is deeply indebted to Dr W. H. Thorpe, F.R.S. and to Dr R. A. Hinde for their warm welcome to Madingley and their subsequent invaluable help in every possible way. This paper represents part of a Ph.D. thesis submitted in 1962.

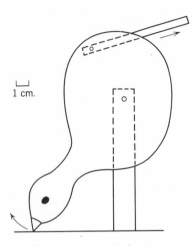

FIGURE 1. The mechanical hen.

If an actor seized either of the seeds it would look identical once the seed had been picked up, and it seems that the test was a difficult one involving microhabitat discrimination. Tits were also tested (Klopfer, 1961) and it was found that single birds or pairs learnt the discrimination at the same rate, whereas greenfinch pairs were often unable to learn the discrimination.

This experiment was designed to find out the effect of a mechanical model hen upon groups of different aged chicks that had been kept in (a) social or (b) isolated conditions.

METHOD. The chicks had been hatched in an incubator on the station, and their ages, to the nearest 4 hours, were known. They were removed from the incubation tray at 4-hourly intervals, and placed in small isolation chambers in the incubator until they were ready to be placed in the larger pens. The social chicks were kept together in groups of five or six individuals and the isolated chicks were kept in solitary confinement. The latter did not see any other birds, and had no experience of observing pecking activities.

The mechanical "hen" consisted of a flat piece of hardboard that was cut out as shown in Figure 1. The beak and eye were indicated in black ink, and the natural color of the hardboard was speckled with blackish spots. The "hen" had a rivet through the middle of the body and could be caused to perform pecking movements by revolving her on her axis by means of the slender piece of wood that was operated by the experimenter. Throughout the trials, the pecking rate was about one per second, but this was deliberately varied from time to time, during the trials,

to give the sort of variation that might be expected in a normal pecking hen. Live hens had, in fact, been trained to feed in front of chicks for investigating social feeding, but the series had to be abandoned because of the extreme variability and unreliability of the hens' performances. The model was much more satisfactory, and in some respects appeared to be more effective than the live hens.

The basic experimental arrangement was that the chicks were separated from the "hen" by wire netting but, owing to the two-dimensional nature of the latter, she could be placed very close to the wire, and the chicks could approach to within 2–3 cm of her. In all cases, the chicks were offered a choice of three orange- and three green-colored wheat grains, that were arranged alternately in a row along the edge of the wire where the "hen" was pecking (see Fig. 2). The trials were arranged so that, in one half of them, the "hen" was pecking at a green grain, and in the other half an orange grain was substituted. In all cases the grains were firmly glued down so that neither the "hen's" or the chick's pecks would loosen them. This ensured that the chick did not receive positive reinforcement from its pecking activity.

The individual trials lasted 20 mins., and a chick was used once only. The following records were made:

1. The total number of pecks delivered by the chick at any object present in the arena.

2. The total number of pecks aimed at (a) the green grains, and (b) the orange grains. (Orange and green grains were selected because a paper by Hess (1956) indicated that these two colors would have an approximately equal pecking attraction.

FIGURE 2. The mechanical hen pecked at a green colored grain and the chick had the choice of green or orange grains. All grains were firmly glued down.

3. The chick's latency before approaching within 10 cm of the "hen."

4. The food latency of the chick, i.e. the time that elapsed before the chick took a peck at either a green or an orange grain.

From these figures it was possible to calculate:

5. The percentage of pecks that were aimed at the grains as opposed to all the other possible objects in the chicks' arena, i.e. pecks on the floor, sides of the arena, wire netting, nails, etc.

6. The percentage of "correct" grain pecks. A peck was considered to be correct if it was aimed at a grain of the same colour as that being pecked at by the "hen." It should be noted that if any color preferences were being indulged by the chicks, then the effect upon the results would be cancelled out by the technique of alternating the color of the "hen's" grain for different trials.

Experiments with live hens had shown that the chicks would not approach very readily unless the heat source was also in the vicinity of the hen. The experiments were performed in a section of a large room, and it was impossible to keep the room at a constant temperature. This problem was partially solved by using a brooder lamp over the experimental area where the "hen" was pecking. The temperature immediately under the lamp was close to 100°F, and it did not fluctuate to any great extent in this limited area. In preliminary work the lamp was placed over the centre of the arena, and the chicks showed a marked tendency to remain in the middle under the lamp. They would only approach the "hen" for short periods before returning to the lamp again. If the chicks were cold, e.g. if no lamp was used at all, they showed much more interest in the "hen," thus a temperature gradient will play an important part in the elicitation and reinforcement of the approach behavior.

In view of this aspect of the preliminary work, the lamp was hung immediately over the "hen," and for the younger age-group chicks, control experiments were performed to find out the behavior towards an immobile "hen" with a lamp placed immediately above her.

Several different age groups were tested (see headings).

I. Experiment to determine the effect of a mechanical "hen" upon naive chicks of ages between 16 and 30 hours that had been kept in isolation.

Results

1. The latency towards a moving "hen" was lower than that shown by control chicks towards a stationary "hen." (Means 1.75 and 4.1 minutes respectively. U-test show differences significant at $p < 0.01$ level.)

2. The total number of pecks delivered at all objects in the arena was greater than that for the control group. (Means 44 and 9 respectively. U-test, $p < 0.01$.)

3. The total number of pecks aimed at the green and orange grains was greater than that of the control group. (Means 15 and 1 respectively. U-test, $p < 0.01$.)

4. The food latency was lower than that for the controls. (Means 7.4 and 18 min. respectively. U-test, $p < 0.01$.)

5. More pecks were aimed at the "correct" colored grain than at the "incorrect" one. Equal number of tests were made with the "hen" pecking at (a) green and (b) orange grains and the results were tested with the Wilcoxon Matched Pairs test. Fifteen chicks gave more correct than incorrect and seven gave more incorrect than correct. Total correct pecks 228; total incorrect pecks 121 ($p < 0.01$).

Conclusions

The effect of the pecking "hen" upon the very young naive chicks was as follows:

(a) Attraction to the site of the "hen."

(b) Pecking near the "hen's" activity.

(c) Preferential pecking at a class of objects similar to those that were being pecked at by the "hen." In two cases this was strikingly obvious, and the chicks actually pecked at three correct grains one after the other. As can be seen from Fig. 2, this entailed missing out the three incorrect grains in the line.

II. The effect of the mechanical "hen" upon 30-hour chicks that had been kept in (a) isolated and (b) social conditions.

Results

1. The social birds delivered more pecks than did the isolated birds. (Means 63 and 48 respectively. U-test, $p < 0.03$.)

2. A control group of socialised chicks who were tested with a stationary "hen" also produced more pecks than a control group of isolated chicks under the same conditions. (Means 58 and 8 respectively. U-test, $p < 0.02$.)

3. The food latency of the experimental social chicks was less than for the experimental isolates. (Means 2.75 and 9 min. respectively. U-test, $p < 0.05$.)

4. The food latency of the control social group was lower than the control isolates. (Both groups

with stationary "hen.") (The means were 10 and 18 min. respectively. U-test, p < 0.05.)

5. The "hen" latencies, between the experimental socials and isolates, were not quite significantly different, but approached it fairly closely. From Guiton's experiments (1959) a difference would be expected. (Means 1.75 and 2.25 min. respectively.)

6. The socialized control chicks, with a stationary "hen" either produced more, or else less pecks than did the experimental socialized group with the pecking "hen." Out of 5 social control chicks, 2 delivered more and 2 less pecks than did any of the 12 experimental social birds (p = 0.01, Fisher Yates test).

Conclusions and Discussion

(a) Socialization produces a general increase in pecking.

(b) It is not clear whether socialization for a short period will also increase the pecking responses of the chicks to the effects of movement. In this experiment the issue was confused by the fact that the chicks were responding to the "hen" and were expending a good deal of time and effort in trying to get close to the "hen" instead of, as in the case of the controls, having the whole of the time available for pecking activities. This is indicated by the "all or none" effect shown in the socialized control group where the "hen" was motionless. The results show that the experimental social group took more pecks than the social controls but the differences were nonsignificant. It is therefore suggested that there is an increased tendency for the socialized chicks to deliver more pecks when stimulated by a moving model, but that in this experiment the effect was masked by the waste of time in efforts to approach the inaccessible moving model.

III. The effect of the mechanical "hen" upon 7-day-old chicks that had been kept in (a) isolated or (b) social conditions.

Results

1. Isolated chicks produced more pecks than did the social group. (Means 64 and 19 respectively. U-test, p < 0.01.)

2. Their food latency was also lower than that of the social group. (Means 5.5 and 11.75 min. respectively. U-test, p < 0.02.)

3. Their "hen" latency was also lower than that of the social group, but not significantly so. (Means 2.5 and 4.5 min. respectively.)

Conclusions

(a) Socialized 7-day-old chicks were apparently less responsive, from a pecking point of view,

towards the mechanical "hen" than are the isolated birds.

(b) Guiton (1959) showed that the approach response to new objects diminished under periods of socialization, and these results seem to agree with the observations above.

(c) The effects of 7 days of socialization as opposed to 7 days of isolation are thus shown to be the reverse of those demonstrated by the 30-hour social and isolate birds. From these experiments it is not possible to ascribe a reason for this, but it could be explained by the growth of fear of strange objects under socialization, or by distress caused by being parted from their companions.

IV. The effects of the mechanical "hen" upon 21-day chicks that had been kept in (a) isolated or (b) social conditions.

It was only possible to perform a limited number of tests in this series, and only five isolate and eight social birds were used. It was thus not possible to produce statistically acceptable results. The results for the 21-day birds were in close agreement with those obtained from the 7-day-old chicks.

Results

1. The isolated chicks gave 42 mean pecks per trial as against a figure of 9.6 mean pecks for social chicks.

2. The mean food latency for isolates was 7.0 min. as against 12.8 min. for the social birds.

3. The "hen" latency for the isolates was 2.5 min. as against 6.75 min. for the social birds.

GENERAL DISCUSSION AND CONCLUSIONS

Development of Social Feeding

The experimental evidence has shown that the social feeding behavior of chicks is well developed and may play an important part in their feeding behavior. Some evidence has been produced for the nature and the origin of the behavior.

Young naive chicks are attracted by moving objects, move towards them, and when in their proximity appear to be contented and start pecking at a wide range of objects. They deliver less pecks if the object is not moving. Pointed objects, and particularly objects in juxtaposition to sharp objects release very strong pecking responses in chicks. The young chicks will preferentially peck at objects that are similar to those being pecked at by a mechanical "hen."

Although chicks that have been in the company of contemporaries for 7 days or more are less attracted by a mechanical "hen" than are similar aged chicks that have been kept in isolation, they are very strongly attracted by their own companions. Thus rapid learning has channelled a generalized approach into a specific one (see also Guiton, 1959). Both the isolated and socialized chicks become much more specific in their pecking with increasing age and cease to peck at their excreta and other objectionable or irrelevant objects. This is presumably due to conditioning of a positive or negative nature. In the case of irrelevant objects producing no reward this behavior could be called habituation, but it is possible that no reward is equivalent to punishment.

REFERENCES

GUITON, P. (1959). Socialisation and Imprinting in Brown Leghorn Chicks.—*Animal Behaviour*, 7, p. 26–34.

HESS, E. (1956). Natural Preferences of Ducklings and Chicks for objects of different colours.—*Psychological Report Univ. of Chicago*.

KLOPFER, P. H. (1959). Social interactions in discrimination learning with special reference to feeding behavior in birds. *Behaviour*, 14, p. 282–299.

——— (1961). Observational Learning in Birds: the establishment of Behavioral Modes.—*Behaviour*,

17, p. 71–80.

SWYNNERTON, C. F. M. (1942). Observations and Experiments in Africa by the late C. F. M. Swynnerton on Wild Birds eating Butterflies and their preferences Shown.—*Proc. Linn. Soc. London*, 1, p. 10–16.

THORPE, W. H. (1956). *Learning and Instinct in Animals*.—Meuthen, London.

TURNER, E. R. A. (1961). Survival Values of Different Methods of Camouflage as shown in a Model Population.—*Proc. Zool. Soc. Lond.*, 136, p. 273–284.

OBSERVATIONAL LEARNING IN THE RHESUS MONKEY*

C. L. Darby and A. J. Riopelle

NUMEROUS attempts have been made to demonstrate that species of animals other than man are capable of learning by observation or imitation. An especially important study for our present purposes is that by Crawford and Spence (1939), who, using two opposing cages with a stimulus tray between them, trained a chimpanzee to observe another depress one of three stimulus objects. After several trials by the demonstrator, the observer was given a test trial to determine if the observer would depress the same object. Both demonstration and test trials continued alternately until the observer either learned the problem or evidenced inability to learn. Results were ambiguous, with only one of eight chimpanzees learning a "pure" imitation problem. The ambiguity of the results suggested that although general procedure was correct, the stimulus-display conditions were not conducive to rapid discrimination learning.

Data collected since have defined the conditions necessary for rapid discrimination learning by primates (Harlow, 1944; Jarvik, 1953; Jenkins, 1943; McClearn and Harlow, 1954), and it was thought that by utilizing the results of these recent researches, unambiguous data might be collected on observational learning. Because of these considerations, we were motivated to try a

* From the *Journal of Comparative and Physiological Psychology*, 1959, **52**, 94–98. Copyright 1959 by the American Psychological Association. Supported in part by a grant (M-589) from the National Institute of Mental Health of the National Institutes of Health, Public Health Service.

This paper is based in part on a doctoral dissertation submitted to the Graduate School of Emory University (Darby, 1956).

demonstration of observational learning under more efficient conditions and to learn what we could about factors contributing to successful performance.

Two studies were conducted. In the first, described in detail elsewhere, Darby (1956) placed two monkeys that could solve object-quality discrimination problems in a single trial (i.e., had formed learning sets) side by side in the Emory version (Riopelle, 1954) of the Wisconsin General Test Apparatus (Harlow, 1949). One animal, the demonstrator (D), was given 1, 2, or 3 trials on an object-quality dsicrimination problem, then another animal, the observer (O), was given 3 trials on the same problem. The Os made approximately 65 per cent correct responses on their first trials of the 400 problems given. Although intraproblem performance improved rapidly, interproblem performance remained unchanged throughout the series of problems. Also, O's performance was no better after 3 than after 1 or 2 demonstration trials. The results suggested that the display aspects of the stimulus presentation were probably still inefficient. Nevertheless, the results were encouraging, and we undertook the present investigation hoping that, with a few additional alterations in the procedures, the efficiency might be high enough to permit the use of inexperienced monkeys, thereby permitting us to trace the development of proficiency in observational learning.

METHOD

SUBJECTS. Four experimentally naive adolescent and young adult rhesus monkeys (Nos. 58, 61, 65, 70) were used as Ss in this experiment. They were subjected to a standardized training and test-adaptation procedure in which they were accustomed to the apparatus, to E, and to the displacement of a single object from the test-tray in order to obtain a morsel from a foodwell.

The animals were grouped in pairs, each pair constituting an experimental unit. The pairings were maintained throughout the experiment. On a particular problem, one member of the pair served as D and the other as O.

APPARATUS. The device, illustrated schematically in Figure 1, consisted of two restraining cages and a sliding test-tray between them. This tray was retracted from the testing location during the baiting and placing of the test objects, and it was returned at the beginning of each trial.

Two foodwells, 1.5 in. in diameter and 10 in. apart, were drilled in the 12- by 18-in. test board. The distance from the foodwells to the bars of the restraining cages was approximately three-fourths of the maximum reach of the animals.

The cages were completely enclosed with sheet metal with the exception of the side facing the test surface; this was barred. The opaque walls served to restrict observation to the quadrant containing the test tray. Immediately in front of the barred side of the cage was a double screen consisting of an inner, opaque screen, constructed of $\frac{1}{4}$-in. plywood, and an outer, transparent screen, made of $\frac{1}{4}$-in. transparent plastic.

The screens were so arranged that the opaque screen could be raised, leaving the transparent screen in place. This arrangement constituted the "observing" position, for the anmal could see but not touch the test objects. By raising the opaque screen further, the transparent screen was carried with it, allowing the animal access to the test board. This constituted the "testing" position.

The problems used in testing were drawn at random from the laboratory's random assortment of over 1,000 mutually different stimulus objects.

A one-way screen, fixed at one side of the apparatus, prevented the monkeys from seeing E.

PROCEDURE. Testing was initiated after both members of a test pair had entered the restraining cages. All screens were lowered, and the test tray was baited. Depending on whether or not the first, or demonstration, trial (D trial) was to be rewarded, food was placed in both or neither (each precisely 50 per cent of the trials) of the foodwells and covered with the stimulus objects. The tray was then moved to the testing location. The screens were then raised to the observing position and held there for approximately 1 sec. The screen in front of D was then raised to the testing position to permit a choice. The other

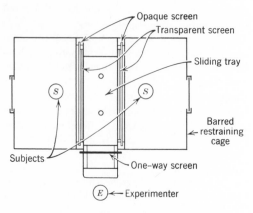

FIGURE 1. Schematic diagram of apparatus.

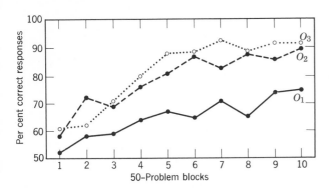

FIGURE 2. Development of proficiency in observational and discrimination learning. Curve O_1 denoted performance by observer after a single demonstration trial. Other curves denote performance on second trials of observer and demonstrator.

monkey, O, was not allowed a response on this trial.

As soon as D displaced an object, all screens were lowered and the test tray was retracted and baited for the first O trial. If the D trial had been rewarded, food was placed under the object which had been displaced and it was removed from under the other object. If the trial was not rewarded, food was placed under the object which had not been displaced. The O was then given its first trial, the task being to select the object displaced by the demonstrator on trial D_1 if D had received reward or to select the nondisplaced object if D had received no reward. Four additional trials completed the problem. The above procedure was repeated in each subsequent problem in the experiment.

Six trials were allowed on each problem, 2 for D and 4 for O. The D was allowed a single trial, and then O was given 1, 2, or 3 trials. Then D was permitted a second opportunity to respond. On each O trial, D was allowed to observe O's response. Following this, O was allowed as many more trials as were necessary to make 4 trials on that particular problem.

The order of presentation, the side on which reward was presented, the sequence in which the observer role was played, and the side on which the animal was placed were all randomly determined.

Between 10 and 15 problems were allowed per day for each pair of animals. Training was continued until each animal had served as O on 500 problems. Necessarily, each animal also served as D on 500 problems. Therefore, every monkey received some training on 1,000 problems.

RESULTS

Figure 2 shows the performance of O on its first trial (O_1) and on its second trial (O_2). Also

shown is the performance on D's second trial (D_2), that for D_1 having been fixed precisely at 50 per cent. The most important curve is that labeled O_1; it is a measure of how much information O gained by the single demonstration trial. The curve starts at 52 per cent and rises gradually to 75 per cent correct. The linear trend of this curve rose significantly (0.01 level), and it shows that as the experiment progressed, the Os learned increasing amounts from the single observation trial on each problem.

Curve O_2 contains the usual learning-set performance; it is superior to O_1 and rises to 90 per cent correct. Curve D_2 shows the same thing for the demonstrator. The slight but consistent superiority of curve O_2 over D_2 suggests that it is more efficient to have the first observation trial precede a test trial rather than succeed it. Both curves are higher than curve O_1. This superiority of discrimination learning-set performance over observational-learning performance shrinks but does not disappear when the learning-set curves are combined and extended over 1,000 problems rather than over the 500 problems as plotted in Figure 2.

What are some of the factors determining observational proficiency? According to Spence (1937), "stimulus enhancement," by which he means that O will respond to those aspects of the problem manipulated by D, is an important factor, especially in problem boxes. If we apply this argument to the discrimination-learning situation, we would predict that O will tend to displace the object displaced by D, regardless of whether or not D obtained reward. If so, the percentage of correct responses made on trial O_1 when D_1 was rewarded should be higher than that for responses when D_1 was wrong (unrewarded), for in the former case O chooses the same object as was chosen by D, and in the latter case it chooses the other object to obtain reward. Figure 3 shows these data. In seven out of ten points on the graph, performance on O_1 after a nonrewarded D_1 surpasses that after a rewarded D_1.

TABLE 1. *Summary of Analysis of Variance*

Source of Variation	df	MS	F
Animals (A)	3	50.28	4.45*
Blocks (B)	9	25.37	2.25*
Reward (R)	1	115.20	10.21**
(A) × (B)	27	5.66	
(A) × (R)	3	11.30	
(B) × (R)	9	17.95	1.59
(A) × (B) × (R)	27	11.28	
Total	79		

* $P \gtrsim 0.05$.
** $P \gtrsim 0.01$.

The data from which Figure 3 was derived contained all the information relevant to the major purpose of the investigation. Those data were therefore subjected to analysis of variance, a summary of which appears in Table 1. In addition to the ever-present individual differences, we found a significant difference among blocks of problems, reflecting the over-all improvement in performance. Also, whether or not the D trial was rewarded was a significant factor in determining O's level of proficiency. The absence of significant interaction effects suggests that the superiority of a nonrewarded D trial over a rewarded D trial holds for most monkeys and throughout an important portion of the acquisition of proficiency in observational-learning performance.

DISCUSSION

The results of this investigation clearly show that rhesus monkeys can acquire information about the solution of an object-quality discrimination problem simply by watching another monkey execute a single trial on the problem. Thus, we have demonstrated that they can acquire some degree of proficiency in observational learning. Doubtless, had the experiment continued, even greater proficiency would have obtained. Moreover, the method used, which involves multiple presentations of simple problems, permits repeated demonstrations of the phenomenon, and is thus useful for further systematic experimental analysis.

What is the nature of O's response on its first test trial? Clearly, O is not "imitating" in the "matched-dependent" sense of Miller and Dollard (1941), for O's response does not consistently match (or oppose) D's response nor does it occur simultaneously with it. Also, O is not imitating via a mediating process of "identification," as has been proposed by Mowrer (1950) in the case of the "talking" birds, for O does not reproduce D's actions and O derives no reward from D, so essential for the development of the identification. Indeed, D frequently competes with O for food.

Spence's notion of stimulus enhancement (1937) does not explain O's behavior either, also for the reason that O can correct for D's mistakes. Indeed, O's performance is better if D makes a mistake than it would be if D made a correct guess. This latter finding is not without precedent (Moss and Harlow, 1947; Riopelle, 1955) and perhaps results from a distracting effect of seeing food, Thus, when D obtains food, O's attention is distracted from the critical stimuli. The desirability of witnessing of errors by O in a problem-box situation has been emphasized by Herbert and Harsh (1944).

The present experimental arrangement satisfies all the requirements of a nonspatial delayed-response test. Although this test is admittedly of

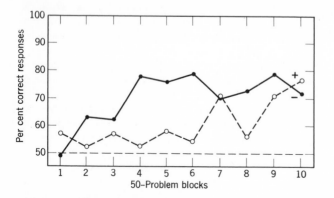

FIGURE 3. Observational-learning performance after a rewarded (+) and after a nonrewarded (−) demonstration trial.

great difficulty, even for chimpanzees, when conducted under the usual procedures (Yerkes and Nissen, 1939), successful performance has been demonstrated previously for rhesus monkeys in an observational-learning situation (Harlow, 1944). Highly successful delayed imitation has been demonstrated by the Hayeses (1952) with their home-raised chimpanzee, Viki.

In the present situation, O simply uses D as a sign stimulus to denote the object O must displace to obtain food. It is interesting to note also that this response satisfies Maier's requirements for a test of reasoning.

SUMMARY

Four monkeys were each given 500 object-quality discrimination problems to solve. Prior to their first trial on each problem they witnessed another monkey execute a single demonstration trial. In the 500 demonstration trials viewed by each observer, exactly half were "correct," i.e., the demonstrator received reward for its selection. The observer was never rewarded on this trial. Its reward came whenever it made a response to the appropriate stimulus object. Appropriateness was defined as a response to the object selected by the demonstrator if the demonstrator had been rewarded or a response to the non-selected object if the demonstrator had received no reward.

All observers showed improvement in their ability to derive information from the observation trial; the observers' first test-trial performance rose from chance level to 75 per cent correct and was still rising at the end of the experiment.

Performance on later trials in the same problems was even higher. It was also found that observational-learning performance was higher if the demonstrator had made a "mistake" than it was if it had made a "correct" response.

The similarity between the procedure of this experiment and that used to test non-spatial delayed response was emphasized.

REFERENCES

CRAWFORD, M. P. and SPENCE, K. W. Observational learning of discrimination problems by chimpanzees. *J. comp. Psychol.*, 1939, **27**, 133–147.

DARBY, C. L. Observational learning in the rhesus monkey. Unpublished doctoral dissertation, Emory Univer., 1956.

HARLOW, H. F. Studies of discrimination learning by monkeys: II. Discrimination learning without primary reinforcement. *J. gen. Psychol.*, 1944, **30**, 13–21.

HARLOW, H. F. The formation of learning sets. *Psychol. Rev.*, 1949, **56**, 51–65.

HAYES, K. J. and HAYES, C. Imitation in a home-raised chimpanzee. *J. comp. physiol. Psychol.*, 1952, **45**, 450–459.

HERBERT, M. J. and HARSH, C. M. Observational learning by cats. *J. comp. Psychol.*, 1944, **37**, 81–95.

JARVIK, M. E. Discrimination of colored food and food signs by primates. *J. comp. physiol. Psychol.*, 1953, **46**, 390–392.

JENKINS, W. O. Spatial factors in chimpanzee learning. *J. comp. Psychol.*, 1943, **35**, 81–84.

McCLEARN, G. E. and HARLOW, H. F. The effect of spatial contiguity on discrimination learning by rhesus monkeys. *J. comp. physiol. Psychol.*, 1954, **47**, 391–394.

MILLER, N. E. and DOLLARD, J. *Social learning and imitation.* New Haven: Yale Univer. Press, 1941.

MOSS, E. M. and HARLOW, H. F. The role of reward in discrimination learning in monkeys. *J. comp. physiol. Psychol.*, 1947, **40**, 333–342.

MOWRER, O. H. On the psychology of "talking birds" —A contribution to language and personality theory. In *Learning theory and personality dynamics.* New York: McGraw-Hill, 1950.

RIOPELLE, A. J. Facilities of the Emory University primate behavior laboratory. *J. Psychol.*, 1954, **38**, 331–338.

RIOPELLE, A. J. Rewards, preferences, and learning sets. *Psychol. Rep.*, 1955, **1**, 167–173.

SPENCE, K. W. Experimental studies of learning and the higher mental processes in infra-human primates. *Psychol. Bull.*, 1937, **34**, 806–850.

YERKES, R. M. and NISSEN, H. W. Pre-linguistic sign behavior in chimpanzee. *Science*, 1939, **89**, 585–587.

TRANSMISSION OF LEARNED BEHAVIOR BETWEEN RATS*

Russell M. Church

THORNDIKE (14) tried to determine whether or not animals are able to learn by observation, i.e., whether or not animals can learn to solve problems by watching a trained animal perform the solution. His results were negative, and his conclusions were that animals learn by trial and error rather than by observation. Thorndike's experiments on observational learning led to a great number of similar experiments. Bird has reviewed most of these studies up to 1940 (2), and Spence has reviewed most of the primate studies up to 1937 (11). A number of the investigators agreed with Thorndike's conclusion that there is no learning by observation in animals, but others claimed to have obtained evidence for this phenomenon. In the latter experiments, the demonstration of the correct response may have increased the activity level of the S, or it may have directed his attention to the crucial part of the apparatus (11). Without control of these extraneous variables it is not possible to demonstrate transmission of learned behavior by the observational learning experiment.

Miller and Dollard (7) adopted the position that observation is not sufficient for the transmission of learned behavior between animals but that imitative behavior, like any other behavior, is acquired on the basis of the principles of trial-and-error learning. They demonstrated that hungry rats can learn to follow a leader in a T maze to secure food. These results, also reported by several other experimenters (1, 9), demonstrate that animals can learn to imitate. It was not determined in any of these experiments whether the followers learned the cues to which the leaders had been responding or whether they merely learned to follow the leader. If the followers would respond consistently to the cues to which the leaders previously had been responding, we may speak of transmission of learned behavior.

The purpose of this experiment is to demonstrate transmission of learned behavior between rats. The design of the experiment treats this as a special case of incidental learning of cues, in which the primary cue is the leader rat. After S learns to follow a leader (the primary cue), the leader responds consistently to some new cue (the incidental cue) with S still following him. After a number of such trials S is placed in the situation *alone*, i.e., without the primary cue of the leader. The consistency to which he then responds to the incidental cue is the index of degree of transmission of learned behavior from leader to follower. On the basis of this analysis, it is clear that this experiment on transmission of learned behavior falls into the class of experiments on incidental learning (3, 4, 6).

METHOD

APPARATUS. The apparatus was an elevated T maze (cf. Fig. 1). The stem of the T and each of the arms was 36 in. long. The maze was 32 in. high. The path was wire mesh covering the wood of the maze and was 1.5 in. wide. Two Plexiglas starting gates on the stem were located 12 in. and 24 in. from the foot. Two Plexiglas goal gates on each arm were located at 12 in. and 24 in. from the choice point. The gates were 12 in. high and 8¾ in. wide. Including the frame into which they were set, the gates extended 4 in. to each side of the path and 4 in. below it. On each of the two arms square metal containers 1.5 in. on a side were sunk 19 in. and 33 in. from the choice point. The lids of these containers were flush with the path. When the lids were raised a thimble of water was exposed from which Ss could drink.

Two pairs of 7½-watt red lights were used as discriminative stimuli. One pair was located 5 in.

* From the *Journal of Abnormal and Social Psychology*, 1957, **54**, 163–165. Copyright 1957 by the American Psychological Association. This experiment formed a portion of a dissertation submitted to the faculty of Harvard University in partial fulfilment of the requirements for the degree of Doctor of Philos-

ophy in Social Psychology (5). The research was facilitated by the Laboratory of Social Relations, and it was supported by a Ford Foundation Grant-in-Aid to Professor Richard L. Solomon, to whom the author is indebted for aid and advice.

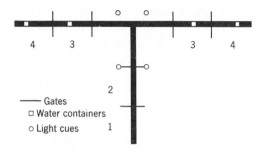

FIGURE 1. Schematic drawing of the elevated T maze. *Key:* 1 = follower's starting section; 2 = leader's starting section; 3 = follower's goal sections; and 4 = leader's goal sections.

from the choice point on each arm, and the other pair was located on the front starting gate, level with the path.

The room which contained the apparatus was 6 ft. by 10 ft. The entire maze was painted gray, and it was surrounded by a cheesecloth curtain. The room was illuminated by a 10-watt bulb hung 36 in. above the rear starting gate. There was another 10-watt bulb in a desk lamp on *E*'s desk which was located behind the foot of the apparatus. During all experimentation a continuous white noise was used to mask extraneous auditory stimulation.

SUBJECTS. Twelve naive male albino rats, about 90 days old, were used in this experiment.

PROCEDURE. *Training of the leaders.* Three *S*s were trained to go to the left arm of the T maze and three *S*s were trained to go to the right arm of the T maze. On each trial *S* was placed in the leader's starting section and the front starting gate was raised. If *S* went to the leader's goal section on the correct arm, he found water in the thimble; if he went to the incorrect arm, he found no water in the thimble. Regardless of the arm he chose, the front goal gate closed behind him (noncorrection method) and he was removed in 30 sec. No *S* was used as a leader who did not consistently reach the correct goal section within 2 sec. after reaching the starting gate.

Training of the Followers. On each trial *S* was placed behind the rear starting gate and a leader was placed behind the front starting gate. Both starting gates were raised simultaneously. The leader would run quickly to the choice point, turn to the left (right) arm, and run to the leader's goal section. At that time the front goal gate on both arms would be closed. If *S* went to the same arm as the leader, the lid of the container with water was opened; if he went to the opposite arm, it remained closed. In either case, as soon as *S*

reached one of the follower's goal sections, the rear goal gates on both arms would be closed (noncorrection method) and the rat would be returned to his home cage in 15 sec. Each *S* was given 150 trials, 10 trials a day spaced at about 3-min. intervals. Leaders were selected according to a systematic order, such that the sequence of correct trials was LRLLRRLRRL, RLRRLLR-LLR. The *S*s were housed three in a cage with consequent artificial illumination. They were given ad lib food in the home cage, and 30 min. after running each day they were given access to water for 90 min.

Training on the Incidental Cue. One hundred additional trials were run in which either the two lights on the left of the apparatus or the two lights on the right of the apparatus were lighted. For a given *S* the side on which the light appeared varied in a systematic order. For half the *S*s leaders were selected who, because of their previous position training, would go to the arm with the light on; for the other half, leaders were selected who would go to the side with the light off. On each trial when *S* reached *either* of the follower's goal sections, the lid of the container with water was opened and *S* was permitted to drink for 15 sec. In all other respects, the procedure for training on the incidental cue was identical to that used in training to follow.

Test for Learning of the Incidental Cue. To test the degree of learning of the incidental cue, the procedure for the training on the incidental cue was continued *except that there was no leader.* Thus, on each trial the lights were on the left (or right) and the follower was placed in the rear starting section. The gates were raised, and regardless of the arm to which *S* went, he was given water for 15 sec. in the follower's goal section. Ten such test trials were run on each of 2 days. The number of times *S* chose the arm marked by the cue which had been positive for the now-absent leader was recorded as the measure of the amount of transmission of learned behavior.

RESULTS

On the initial ten trials there was no significant tendency for *S*s to go to the same arm of the T maze as the leader or to go to the opposite arm. The *S*s gradually learned to follow the leader. On the last 20 trials the six *S*s followed the leader on all but one trial (cf. Fig. 2).

On the initial trials after the introduction of the incidental cue there was a slight decrease in the percentage of following responses (external inhibition?). Despite the fact that Ss could obtain water on either arm of the maze during this period, they followed the leader on 95 per cent of the trials. A few Ss developed side preferences during this treatment, thus slightly decreasing the total percentage of following responses.

After the leaders had been omitted, Ss went to the arm marked by the cue to which the leader had been going on 77 per cent of the 20 test trials ($t = 4.4$; $p < 0.01$). These test trials demonstrate that Ss learned the cues to which the leaders were responding.

DISCUSSION

Miller and Dollard (7) have demonstrated that imitative behavior may be acquired on the basis of trial-and-error learning. The present experiment further confirms this hypothesis, and it also presents evidence that such imitative behavior can become *independent* of the behavior of the leader. In this experiment rats acquired a discrimination habit in a T maze after following leaders who had acquired that habit. Such transmission of learned behavior is to be expected on the basis of a continuity theory of discrimination learning (12). According to the continuity theory, a set to respond to one aspect of the stimulus situation does not interfere with the learning of other effective cues preceding the reward. Thus, in the present case, a set to follow the leader should not interfere with the learning of the incidental cue preceding the reward. Although

there is some evidence to the contrary (6) most evidence supports the hypothesis that incidental cues may be learned (3, 4, 10).

The present laboratory demonstration does not duplicate the naturalistic situation, but it indicates a mechanism through which social transmission of behavior could readily occur. The positive findings encourage speculation about the possibility of a nonpurposive transmission of culture among animals. According to the hypothesis, a rat who has learned to follow a leader will learn to respond to the cues to which the leader has responded. If another rat learned to follow him, he, too should learn to respond independently to these cues. In this case, behavior learned by an original leader would have been transmitted through two "generations." Although the necessary condition of following behavior has been demonstrated in certain animal species (8, 13), experiments in field situations have not yet been performed to determine whether transmission of learned behavior occurs in nature.

SUMMARY

Six rats were trained to follow leader rats in an elevated T maze to secure a reward of water. An incidental cue of two lights was then introduced such that the leaders responded consistently with respect to it. After 100 trials of following leaders who were responding to the incidental cue, Ss were given 20 trials alone. On 77 per cent of these test trials Ss went to the arm marked by the cue to which the leader had been going. This finding

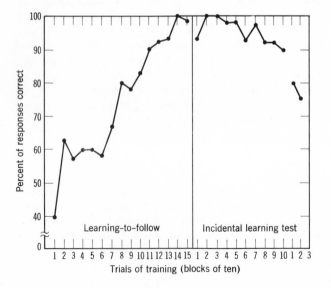

FIGURE 2. Transmission of learned behavior; percentage of responses correct during training to follow the leader, after addition of the incidental cue, and after elimination of the leader.

was interpreted as a demonstration of transmission of learned behavior between animals, and it was explained on the basis of a continuity theory of discrimination learning.

REFERENCES

1. BAYROFF, A. G. and LARD, K. E. Experimental social behavior of animals III. Imitational learning of white rats. *J. comp. physiol. Psychol.*, 1944, **37**, 165–171.
2. BIRD, C. *Social psychology.* New York: Appleton-Century, 1940.
3. BLUM, R. A. and BLUM, J. S. Factual issues in the "continuity" controversy. *Psychol. Rev.*, 1949, **56**, 33–40.
4. BRUNER, J. S., MATTER, J. and PAPANEK, M. L. Breadth of learning as a function of drive level and mechanization, *Psychol. Rev.*, 1955, **42**, 1–10.
5. CHURCH, R. M. Factors affecting learning by imitation in the rat. Unpublished doctor's dissertation, Harvard Univer., 1956.
6. LASHLEY, K. S. An examination of the "continuity theory" as applied to discrimination learning. *J. Gen. Psychol.*, 1942, **26**, 241–265.
7. MILLER, N. E. and DOLLARD, J. *Social learning and imitation.* New Haven: Yale Univer. Press, 1941.
8. SCOTT, J. P. Social behavior, organization, and leadership in a small flock of domestic sheep. *Comp. psychol. Monogr.*, 1945, **18**, No. 4.
9. SOLOMON, R. L. and COLES, M. R. A case of failure of generalization of imitation across drives and across situations. *J. abnorm. soc. Psychol.*, 1954, **49**, 7–13.
10. SPENCE, K. W. An experimental test of the continuity and noncontinuity theories of discrimination learning. *J. exp. Psychol.*, 1945, **35**, 253–266.
11. SPENCE, K. W. Experimental studies of learning and the higher mental processes in infra-human primates. *Psychol. Bull.*, 1937, **34**, 806–850.
12. SPENCE, K. W. The nature of discrimination learning in animals. *Psych. Rev.*, 1936, **43**, 427–449.
13. STEWART, J. and SCOTT, J. P. Lack of correlation between leadership and dominance relationships of goats. *J. comp. physiol. Psychol.*, 1947, **40**, 255–264.
14. THORNDIKE, E. L. Animal intelligence. *Psychol. Rev. Monogr. Suppl.*, 1898, **2**, No. 4.

CHAPTER THREE

SOCIAL STIMULATION AND SOCIAL DEPRIVATION

THE previous chapters have considered evidence showing that the presence of others *during* the course of learning and performance of specific response patterns has pronounced effects. The presence of another acts to increase the general arousal level of the organism, and at the same time, it presents it with a rich source of directive cues. Naturally, we would expect that if a given animal has no social contact with others of his species for a considerable length of time, a variety of his responses will be deeply affected. These effects should be particularly dramatic if social deprivation occurs early in life when most of the learning that has important adaptive functions takes place. To the extent that the presence of other animals that serve as models facilitates the acquisition of adaptive behavior patterns by the young animal, withholding contact with such models should demonstrate to the researcher what it is that the animal acquires from those around him; and what, in short, are the specific outcomes of socialization.

In the chapter on imitation we noted from Turner's study that when chicks are isolated for at least a week they become considerably more influenced by a mechanical "hen" than when they spend this period under social conditions. Isolated chicks show vastly more pecking in response to the mechanical "hen," and their first pecking response following the introduction of the mechanical model occurs much sooner. Turner observed that as the chick grows older it begins to pay more attention to *appropriate* models (that is, to other chicks and hens) than to inappropriate ones, and his data suggest therefore that early social experience helps the animal to acquire an ability to discriminate between appropriate and inappropriate models—an ability of immense importance for survival. For, if the animal acquires much of his adaptive behavior by means of imitation, it would be disastrous if it were equally prone to imitate his predators as his peers. It is true perhaps that the imitation of an appropriate model is more likely to be rewarding than imitation of an inappropriate model. There is, however, no evidence at the present to show that the animal's preference for a member of his own species over, let us say, a mechanical device would persist even though the former model were less effective in assuring reinforcement than the latter.

The studies in this chapter show the pronounced effects which the separation of an animal from others of his species has on its behavior. Mason's study examines these effects across an entire variety of behaviors, and demonstrates how widespread these effects of separation can be for rhesus monkeys. The experiment with rats performed by Uyeno and White illustrates the specifically *social* effects of isolation quite clearly. Were we to anthropomorphize about these rats, we would regard the behavior of the isolated animals as inconsiderate and selfish. The third experiment in this chapter, by Morrison and Hill, also using rats as subjects, again concentrates upon a specific behavioral pattern, the avoidance response under threat. It clearly shows that this response is dramatically affected by a period of isolation.

In the Uyeno–White and Morrison–Hill studies isolation of the animals was accomplished by placing them in separate wire cages but in the same room. These animals, therefore, did have visual, auditory, and olfactory contact with each other. Nevertheless, strong effects upon behavior were observed. In the experiment performed by Thiessen, Zolman and Rodgers mice were housed so that visual interaction was also prevented. Their isolation, however, was not begun until they were 48 days old. Again, strong behavioral effects were observed with the socially isolated animals showing lesser response vigor. Moreover, endocrine functions which are fair indicators of the general activation level, also showed consistent differences in favour of the grouped mice. The results of the Thiessen–Zolman–Rodgers study add weight to the assumption discussed in the chapter on social facilitation that the presence of others contributes to the elevation of the animals' general arousal level.

There are many methodological and theoretical problems in comparing separation effects across species and genera. When we are isolating a *young* monkey from others of his troup we are simultaneously separating it from its mother—for a mammal, an extremely rich source of social effects. In fact, we have not as yet developed techniques that would allow us to separate a mammal at birth. In all studies using mammals as subjects the experimental isolation begins only after some social experience has already taken place. Mason separated his rhesus monkeys at the age of one month; Uyeno and White separated their rats at 17 days, and Morrison and Hill at 25 days. Complete isolation from birth can be achieved with birds, however, and Turner's chicks were so isolated. The question, however, arises if fair comparisons can be made between the effects of isolation on birds, rats, monkeys, and other animals. It is thus important to keep in mind that in separating a young mammal from social contact we are taking it away from its mother. The role of the mother in socialization of nonmammals is of course much smaller. The study of Seay, Hansen and Harlow shows, in contrast, that young rhesus monkeys (six to seven months old) are deeply affected by the separation from their mothers, even if they retain social contacts with their peers.

The final paper in this chapter deals with a form of social behavior, close in form to imitation, which has generally been regarded as instinctive, in the modern sense of the word. But Guiton shows in his experiments that even imprinting is subject to influences depending on the animal's social experience.

THE EFFECTS OF SOCIAL RESTRICTION ON THE BEHAVIOR OF RHESUS MONKEYS: I. FREE SOCIAL BEHAVIOR*

William A. Mason

THE present researches are part of a series of experiments investigating the effects of restricted social experience on the social behavior of rhesus monkeys. The principal comparisons in these experiments are between monkeys living under free-ranging conditions until captured, and laboratory-born monkeys separated from their mothers in early infancy and raised under conditions which limited the nature and extent of early social experience.

To insure uniform and controlled living conditions and provide ready access to infant Ss for testing, early maternal separation is virtually essential to the effective conduct of a major psychological research program utilizing infant monkey Ss. Such a program has been in progress at the University of Wisconsin since 1953, and a large number of infant monkeys have been separated from their mothers and housed in individual living cages at the Primate Laboratory.

When maintained in accordance with the general recommendations of van Wagenen (1950), there is no indication that early maternal separation adversely affects the growth and viability of the infant macaque. However, the reduction in intraspecies social contacts and the routine and relatively impersonal nature of the caretaking methods employed in the laboratory create an impoverished social environment which approaches the more extreme forms of institutional environments for human children.

Investigations assessing the effects of institutionalization on human personality development and social behavior have been summarized by Bowlby (1952), and the findings indicate that deprivation of normal socialization experiences in human children results in a wide range of personal and social deficiencies and aberrations including affective disorders, limited capacity for sustained and effective social relationships, and psychopathic tendencies.

Although there exist no comparable data on the effects of similar restrictions on the development of social behavior in the nonhuman primates, field workers have generally assumed that prior socialization experience is of fundamental importance in the development of orderly and efficient patterns of social interaction characteristic of nonhuman primate societies (Carpenter, 1942; Imanishi, 1957). Because of formidable practical difficulties, however, careful longitudinal studies have not been completed under field conditions, and the process of socialization in free-ranging primates has not been fully described. The available evidence indicates that under natural conditions there is ample time and opportunity for complex social learning to occur. The period of infantile dependency in Old World monkeys is relatively long, extending in most forms for approximately two years. Following this there is a period of several years during which the tie to the mother gradually weakens and the individual has the status of a juvenile which associates with other young monkeys and does not participate fully in adult functions and activities. Adult status is probably not attained by the male before its sixth year; females mature somewhat earlier.

During these early years the young monkey has considerable social mobility, providing a wide range of social contacts and many opportunities to observe and participate in a variety of activities with peers and with adults of both sexes. It is probable that basic social attachments binding the individual to the group are established and strengthened during this time and

---

* From the *Journal of Comparative Psychology*, 1960, **53**, 582–589. Copyright 1960 by the American Psychological Association. Support for this research was provided through funds received from the Graduate School of the University of Wisconsin, Grant G6194 from the National Science Foundation, and Grant M-722 from the National Institutes of Health.

many of the essential patterns of social intercourse are developed and refined. Thus, restriction upon the nature and extent of early socialization experience might be expected to produce inadequacies in subsequent relations.

The present research investigated differences in the form and frequency of spontaneous social interactions of feral and laboratory-raised rhesus monkeys.

EXPERIMENT 1: INTRAGROUP COMPARISONS OF FERAL AND SOCIALLY RESTRICTED MONKEYS

Method

Subjects. The *S*s were two groups of six adolescent monkeys. One group of three males and three females, the Restricted group, was born in the laboratory. They were separated from their mothers before the end of the first month of life and were housed in individual cages which allowed them to see and hear other young monkeys, but which prevented physical contact between them. Opportunities for more extensive intraspecies social contacts were confined to a few brief periods during the first year of life. Interspecies social experience was limited almost entirely to daily contacts with human beings in connection with routine caretaking and testing activities. At the start of the experiment three *S*s were 28 months of age and three were 29 months old.

The second group of three males and three females, the Feral group, was captured in the field. Immediately following their arrival in Madison they spent approximately 3 months on a zoo monkey island with about 20 other monkeys of similar age and background and the next 3 months with the same animals in a large group-living cage in the basement of the zoo's primate building. This section is not on exhibition. The entire group was shifted to the laboratory when the six *S*s later used in this study were about 20 months old. They were housed in pairs for 8 months. At the time of separation, 1 month before the start of the experiment, none was caged with any other member of the group of six. During the month before the experiment began and throughout the period covered by this research, the *S*s lived in individual cages. They were selected from the larger group to match the *S*s of the Restricted group in sex, weight, and dentition. It is estimated the age differences among *S*s were no greater than 1 month.

Apparatus. The test chamber, constructed of gray plywood, measured 6 ft. by 6 ft. by 6 ft. and was illuminated from above by a 150-watt bulb. Two opposing walls contained one-way observation panels, and in each of the other walls was a large hinged door fitted with a smaller sliding door through which the *S*s entered the chamber from an adjacent carrying cage. Behavior was recorded on a multiple-category keyboard which operated the pens of an Esterline-Angus recorder, giving a continuous record of the frequency, duration, and temporal patterning of social interactions. By appropriate wiring of the keyboard a total of 11 response categories were available for each member of a test pair. (See Table 1.)

TABLE 1. *Definitions and Reliability of Response Categories*

1. Approach: Moves to within 6 in. of other monkey (75 per cent).
2. Aggression: Includes intense vocalization (barks, growls), biting, pulling. Threat or bluff (Chance, 1956) was not scored as aggression. No occurrence during reliability check.
3. Groom: Systematically picks through another's fur with hands (95 per cent).
4. Mount: Characteristically, grasps the partner's hips with hands, and feet clasp her legs (96 per cent).
5. Play: Tumbling, mauling, wrestling, and nipping. Less vigorous and intense than aggression, is not accompanied by intense vocalization, and rarely elicits squealing or other evidence of pain in partner (79 per cent).
6. Sexual presentation: Assumption of female mating posture. Hindquarters are elevated and turned toward the partner (60 per cent).
7. Social facilitation of exploration: Activity of one animal with respect to some inanimate feature of the room elicits approach, observation, or display of similar behavior from another (59 per cent).
8. Social investigation: Close visual, manual, and/or oral investigation of the partner. Particular interest toward apertures. No occurrence during reliability check.
9. Thrusting: Piston-like movements usually accompanying mounting (95 per cent).
10. Visual orientation: Passive observation of other (66 per cent).
11. Withdrawal: Abrupt movement away from other. Not scored during play or aggression unless it terminated interaction (65 per cent).

Procedure. Once a day for 14 days each *S* was individually placed in a carrying cage at the entrance to the chamber and was allowed 3 min. in which to enter and explore the room. This was followed by seven daily 3-min. sessions in which entry was forced for all *S*s not entering the room during the first few seconds of the session.

In the social testing which followed these individual adaptation sessions, Feral and Restricted monkeys were paired only with the other members of their group. Each of the 15 pairs obtainable from each group was observed in a series of 16 social test sessions each 3 min. long. To initiate a session, the sliding door on the *O*'s right was raised and when the *S* had entered the room, the second sliding door was raised. When both *S*s were in the chamber, the Esterline-Angus recorder was started. A timer automatically stopped the recorder at the end of the test period. The testing schedule was so arranged that all pairs were tested once every two days with the restriction that no *S* participated in two successive test sessions. For each pair the order of testing and the side from which the *S*s entered the chamber were counterbalanced over sessions.

The behavior of each *S* was separately recorded by depressing the appropriate response-keys during the period of behavioral occurrence. To provide an estimate of the reliability of the method, two *O*s were present and independently recorded the social interactions of 15 pairs on a total of 90 sessions. Table 1 presents definitions of the response categories and their corresponding reliability values for frequency measures. Reliability is expressed in terms of per cent agreement:

$$\frac{\text{Frequency of agreements}}{\text{Frequency of agreements} + \text{disagreements}}$$

Results

To conserve space only the response categories of play, grooming, aggression, and sexual behavior are analyzed in detail. Among the other response measures, differences between groups in the frequency of approaches, withdrawals, and social investigation were not statistically significant. The frequency of visual orientation responses was higher among Feral *S*s, and socially facilitated exploration occurred more frequently in the Restricted group. These differences were statistically significant, *p* being 0.01 for each of these comparisons. Unless otherwise indicated these and all subsequent statistical comparisons were based on *t* tests performed on total individual scores.

Play. Although the form of play behavior was similar in both groups, play occurred more frequently among Restricted *S*s. The mean total incidence of play was 342.0 and 179.8 for Restricted and Feral groups, respectively. Although the difference between groups is substantial, it falls short of significance at the 0.05 level, principally because within each group there were large and consistent sex differences in the incidence of play. Both males and females of the Feral group had lower scores than like-sexed *S*s in the Restricted group, and all females showed less play behavior than did males. Mean play scores for females were 115.7 and 29.0 in Restricted and Feral groups, respectively, whereas corresponding values for males were 568.2 and 330.7. Play behavior did not occur in female pairs in the Feral group and occurred in female pairs of the Restricted group in only six sessions. The difference between sexes in the frequency of play for the combined groups is significant at the 0.001 level.

Grooming. The incidence of grooming was relatively low in both groups occurring in three *S*s in the Restricted group in 10 sessions and in four *S*s in the Feral group in 25 sessions. Grooming was observed in seven Feral pairs and in five Restricted pairs. These differences were not statistically significant.

Grooming episodes tended to be substantially longer in Feral pairs and occasionally occupied nearly the entire test period. The mean duration of grooming behavior, determined by dividing total duration by number of occurrences, was 1.6 sec. and 25.3 sec. for Restricted and Feral groups, respectively, and this difference is significant at the 0.05 level.

Qualitative differences between groups were observed in the response to grooming. While being groomed members of the Feral group were characteristically passive and immobile, whereas the *S*s in the Restricted group showed no specific or consistent complementary responses.

Aggression. The mean total frequency of aggression for Restricted and Feral groups was 11.7 and 2.0, respectively, and this difference is significant at the 0.05 level. Eleven pairs in the Restricted group engaged in aggression in at least one session and seven pairs fought in more than one session. In the Feral group, aggression occurred in six pairs, and in only two of these pairs did fighting occur in more than one session.

Aggressive episodes tended to be longer in the Restricted group, the mean duration being 4.87 sec. for these *S*s, as compared with 1.21 sec. for

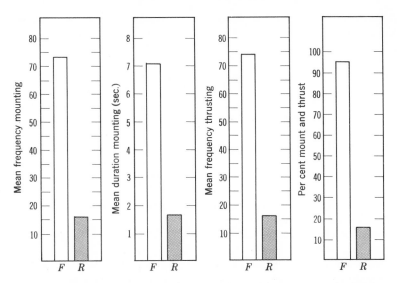

FIGURE 1. Comparisons of sexual behavior of Restricted (*R*) and Feral (*F*) males.

Feral *S*s. This difference, however, was not statistically significant. Although there were no apparent differences between groups in the form of individual aggressive responses, it was characteristic of the *S*s in the Restricted group, particularly during the early phases of the experiment, to respond with aggression when attacked, whereas Feral *S*s generally withdrew from attack or submitted passively and without retaliation.

Sexual Behavior. A striking difference was apparent between Feral and Restricted males in the frequency and integration of sexual behavior. Mean total frequencies are presented in Figure 1, and it is evident that Feral males showed more mounting and thrusting and had sexual episodes of substantially longer duration. All differences were significant at the 0.05 level. The incidence of mounting was negligible among females in both groups.

Gross qualitative differences in the organization of the male sexual act were present. Males in the Restricted group never clasped the partner's legs with their feet during mounting and would frequently assume inappropriate postures and body orientation. Many responses by these males could not be categorized as sexual until thrusting was observed, hence mounting was not scored for 64 per cent of the total thrusting responses. Inasmuch as the presence of thrusting presumably implies some attempt to mount, the mounting scores of the Restricted males were subsequently increased by the number of thrusting responses in which mounting was not recorded. This raised their mean total mounting score from 16.0 to 23.3. The difference between groups was still statistically significant ($p = 0.05$).

Ejaculation was not observed in either group. Menstrual cycles in the females were absent or irregular, and there was no apparent relationship between male sexual behavior and cyclic activity of the females.

With the exception of one highly dominant female in the Feral group, assumption of the female sexual posture (presentation) occurred in all *S*s. Mean total frequency of sexual presentation was 23.5 and 12.2 in Feral and Restricted groups, respectively. Males in both groups made this response less frequently than females, accounting for 28 per cent and 10 per cent of total presentations in Feral and Restricted groups, respectively. Differences between groups and between sexes in the frequency of sexual presentation were not statistically significant.

Sexual presentation was notably more stereotyped for Feral animals of both sexes. In this group presentation was occasionally preceded by gazing intently at the partner while making rapid movements of the lips, and it was virtually always accompanied by postural adjustments, including flexing of the legs as the partner mounted. These behaviors were never observed among Restricted *S*s.

To provide further evidence on the nature of the differences between groups in the male sexual pattern, a second experiment was run in which the males from Feral and Restricted groups were tested with the same socially experienced females, thus eliminating differential social experiences of the sexual partner as a factor contributing to differences in male sexual performance.

EXPERIMENT 2:
BEHAVIOR OF RESTRICTED AND FERAL MALES WITH SOCIALLY EXPERIENCED FEMALES

Method

Subjects. The Ss were the three Feral and three Restricted males previously described, and three adolescent females captured in the field and without prior contact with any of the males. During the 6-month period between experiments 1 and 2, each male participated about 10 hr. in the social tests of gregarious tendencies and food competition. The males lived in individual cages throughout the period covered by the preceding experiments, and all Ss were individually housed during the present research.

Apparatus and Procedure. The apparatus and the testing and recording procedures were the same as those described in Experiment 1. In the present experiment, however, only male-female pairs were tested. Following five individual 3-min. adaptation sessions for all Ss, each male was tested in ten 3-min. sessions with each female. Males participated in one test session a day and each female was used in two sessions. No female was tested in two successive sessions. The response categories used in Experiment 1 were extended and modified to provide further information on the integration of the male sexual pattern. Response categories retained from Experiment 1 without modification included aggression, approach, groom, play, sexual presentation, thrusting, and withdrawal. Measures of visual

orientation and social facilitation were not obtained in the present experiment. Mounting was classified as follows: "Appropriate mounting orientation"—longitudinal axes of the bodies are aligned during mounting, with the Ss facing in the same direction. "Inappropriate mounting orientation"—all attempts at mounting not scored as appropriate mounting. Additional categories included: "Hip clasp"—within 5 sec. before mounting the male places both hands on partner's hips. "Foot clasp"—male grasps female's legs with both feet during mounting. "Anogenital investigation"—visual, manual, and oral investigation of partner's anogenital region.

Results

Mean and individual totals for frequency of nonsexual responses in males are presented in Table 2. None of the differences is significant, although what differences there are tend to be in the same direction as in Experiment 1.

The sexual responsiveness of the Restricted males showed a striking increase relative to their performance levels in Experiment 1. In the present experiment the mean incidence of mounting per session, including both appropriately and inappropriately oriented responses, was 2.44 as compared with 0.20 in Experiment 1. Similarly, the mean frequency of thrusting increased from 0.13 per session to 1.84. These changes are significant at the 0.02 level as determined by t tests for correlated measures. Although the higher frequency of mounting behavior might conceivably be related to modification of scoring categories between Experiments 1 and 2, the same interpretation would not apply to measures of thrusting. A more reasonable hypothesis is that the behavior of the socially experienced female partner and/or experience gained by Restricted males in the social tests intervening between

TABLE 2. *Nonsexual Responses of Feral and Restricted Males in Experiment 2*

Response	Feral S				Restricted Ss			
	331	336	337	Mean	3	4	5	Mean
Aggression	1	1	0	0.7	0	0	0	0.0
Approach	368	293	38	233.0	641	228	358	409.0
Groom	9	14	0	7.7	0	1	0	0.3
Dur. groom (sec.)	4.8	15.4	0	6.7	0	1	0	0.3
Play	114	47	0	53.7	144	79	129	117.3
Withdrawal	26	7	15	16.0	14	2	21	12.3

Experiments 1 and 2 contributed to this increase in sexual responsiveness.

Comparative data on the sexual performance of Restricted and Feral males in the present experiment are presented in Table 3. The differences between groups in the frequency of mounting and thrusting were not statistically significant. Interpretation of this outcome is complicated by the fact that Feral Male No. 337 mounted only once in this test series, as compared with 83 mounts in Experiment 1. The remaining animals in this group showed no evidence of lowered sexual responsiveness relative to their performance in Experiment 1. There is no indication, however, that these animals differed reliably from Restricted males with regard to frequency of thrusting and mounting (appropriate and inappropriate mounting orientations combined), which provides further evidence of enhanced sexual responsiveness of Restricted males in the present experiment.

In spite of more frequent sexual responses the sexual performance of the Restricted males was poorly integrated (see Fig. 2). Only 33 per cent of total mounts in this group were appropriately oriented, 76 per cent of total mounts were accompanied by thrusting and only 3 per cent included clasping the partner's legs. Comparable values for the Feral males exceeded 98 per cent and the differences between groups were significant at the 0.01 level. The duration of mounting was again substantially shorter among Restricted males, mounting was less frequently preceded by grasping the partner's hips and the incidence of anogenital investigation was lower. These differences, however, were not statistically significant. Ejaculation was not observed in either group.

The behavior of the females clearly suggested that they were not responding equivalently to the two groups of males. Females frequently failed to assume appropriate receptive postures in response to sexual advances of Restricted males (see Fig. 2D), and made fewer approaches to these animals. Furthermore, the high incidence of cowering and grimaces in the presence of Restricted males suggested that the females were afraid. This impression receives support from the finding that their withdrawal scores were higher with Restricted partners. (Mean totals: Restricted, 210.7; Feral, 66.0.) This difference was significant at the 0.01 level as determined by Wilcoxon tests performed on individual trial totals.

DISCUSSION

The results of Experiment 1 indicate that restriction of intraspecies social experience of rhesus monkeys during the first two years of life retards the development of integrated social responses and orderly patterns of social interaction. Fighting was more frequent and prolonged among Restricted monkeys. They groomed less frequently, and grooming episodes were shorter. Sexual behavior in Restricted Ss was brief and showed gross deficiencies in organization, which were particularly evident in the behavior of males. The extent to which the performance of the Restricted monkeys was influenced by the brief social contacts provided them during the first year of life cannot be determined, but unpublished data on animals whose social experience was even more severely restricted suggest that this early social experience had some effect. Although the members of the Feral group in Experiment 1 were not strangers, the absence of any major changes in the social behavior of Feral

TABLE 3. *Sexual Responses of Feral and Restricted Males in Experiment 2*

Measure	Feral S				Restricted S				p
	331	336	337	Mean	3	4	5	Mean	
Approp. mount	64	53	1	39.3	25	30	18	24.3	ns
Inapprop. mount	0	0	0	0.0	60	40	47	49.0	.01
% approp. mount	100	100	100	100.0	29	43	28	33.3	.01
Mounting dur. (sec.)	4.2	5.3	10	6.5	2.4	2.5	2.7	2.5	ns
Thrusting	65	53	1	39.7	69	55	42	55.3	ns
% mount + thrusting	100	98	100	99.3	81	79	68	76.0	.01
Foot clasp	63	53	1	39.0	0	7	0	2.3	ns
% mount + foot clasp	98	100	100	99.3	0	10	0	3.3	.01
Hip clasp	49	47	0	32.0	24	30	14	22.7	ns
% mount + hip clasp	66	81	0	49.0	16	34	9	19.7	ns
Anogenital investig.	27	21	0	16.0	4	2	5	3.7	ns

males between Experiments 1 and 2 strongly suggests that possible pre-experimental contacts among members of the Feral group did not appreciably influence the present results.

The data on sexual behavior of Restricted males suggests that the components of the male copulatory pattern are differentially dependent upon social experience. The tendency to approach and bring the genitalia in contact with the partner was present from the first tests, as evidenced by the fact that all Restricted males attempted to mount on the first day of Experiment 1. Grasping the partner with the hands also appeared early in testing, but throughout the present experiments this response was less stereotyped and precise among Restricted males and was often accompanied by nipping, tugging, or other playful behaviors. Foot clasping was particularly deficient in Restricted males. This response was never observed among Restricted males in Experiment 1. In Experiment 2, only one of these males ever succeeded in grasping the partner's legs with both feet simultaneously, and

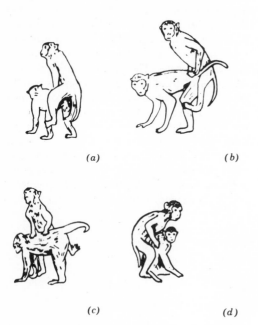

(a)   (b)

(c)   (d)

FIGURE 2. Sexual behavior of Feral and Restricted males with socially experienced female partners. *A.* Rear view of Feral male in typical copulatory position. *B.* Side view of Feral male in typical copulatory position. *C.* Restricted male attempting to . mount from the side. Note elevation of left foot. *D.* Sexual behavior of a Restricted male. Although mounting orientation is appropriate (as defined herein), the hands are placed high on the female's trunk and she remains sitting. The male is thrusting against the female's back. All figures were traced from moving-picture film.

this occurred only 7 times in 70 mounting attempts. A second male in this group occasionally raised its feet alternately as though attempting to place them, but never grasped with both feet simultaneously (see Fig. 2C). Closely related to this deficiency was the absence of efficient and appropriate postural orientation with regard to the sexual partner.

The data on sexual behavior are consistent with previous observations (Bingham, 1928; Foley, 1935; Maslow, 1936; Yerkes and Elder, 1936), and support the generalization that among nonhuman primates social experience is relatively more important to male than to female sexual behavior (Ford and Beach, 1952). Had more sensitive measures of female sexual behavior been used, however, there is little doubt that deficiencies would also have been demonstrated in the performance of Restricted females, although it is unlikely that these were sufficient to prevent effective coitus with an experienced male.

Social organization among nonhuman primates is characteristically orderly and efficient. Presumably, regular social relationships are dependent upon stable interindividual stimulus-response tendencies. Sexual presentation, presentation for grooming, grasping the hips preparatory to mounting, and the threat pattern are a few of the highly stereotyped responses described for rhesus monkeys which ordinarily function as social cues, eliciting appropriate reciprocal responses from other animals. These stimulus-response relationships form the basis for social coordination, communication, and social control in feral groups (Carpenter, 1942; Chance, 1956; Maslow, 1936). Insofar as the present findings bear on this problem, they suggest that among animals whose socialization has been restricted, the cue function of many basic social responses is poorly established if not absent altogether.

SUMMARY

1. Comparisons were made of the spontaneous social interactions of monkeys raised in a socially restricted laboratory environment and Feral monkeys captured in the field.

2. Pairs of Restricted monkeys showed more frequent and prolonged fighting and fewer and less prolonged grooming episodes than Feral pairs. Differences between groups were found in the frequency, duration and integration of sexual behavior, which were particularly evident in the behavior of males.

3. Restricted and Feral males were subsequent-

ly tested with the same socially experienced females, thus eliminating inadequacies in the sexual partner as a factor contributing to the differences in male sexual performance. Gross differences in the organization of the male copulatory pattern were still apparent.

4. In addition to differences between groups in the form and frequency of these basic social responses, the data suggest that responses to social cues are poorly established in monkeys with restricted socialization experience.

REFERENCES

BINGHAM, H. C. Sex development in apes. *Comp. psychol. Monogr.*, 1928, **5**, 1–165.

BOWLBY, J. *Maternal care and mental health.* Geneva: World Health Organization, 1952.

CARPENTER, C. R. Societies of monkeys and apes. *Biol. Sympos.*, 1942, **8**, 177–204.

CHANCE, M. R. A. Social structure of a colony of *Macaca mulatta. Brit. J. anim. Behav.*, 1956, **4**, 1–13.

FOLEY, J. P., JR. Second year development of a rhesus monkey (*Macaca mulatta*) reared in isolation during the first eighteen months. *J. genet. Psychol.*, 1935, **47**, 73–97.

FORD, C. S. and BEACH, F. A. *Patterns of sexual behavior.* New York: Harper, 1952.

IMANISHI, K. Social behavior in Japanese monkeys, *Macaca fuscala. Psychologia*, 1957, **1**, 47–54.

MASLOW, A. H. The role of dominance in the social and sexual behavior of infra-human primates: III. A theory of sexual behavior of infra-human primates. *J. genel. Psychol.*, 1936, **48**, 310–338.

VAN WAGENEN, G. The monkey. In E. J. Farris (Ed.), *The care and breeding of laboratory animals.* New York: Wiley, 1950. Ps. 1–42.

YERKES, R. M. and ELDER, J. H. Oestrus, receptivity, and mating in chimpanzee. *Comp. psychol. Monogr.*, 1936, **13**, 1–39.

SOCIAL ISOLATION AND DOMINANCE BEHAVIOR*

Edward T. Uyeno and Margaret White

SEVERAL investigators have demonstrated that animals raised in social isolation were inferior to those raised in a group. For example, Andervont (1944) and Muhlbock (1950) showed that susceptibility to tumors in certain strains of mice was greater in isolates than in nonisolates. Reynolds (1963) found that shock escape performance of isolated Sprague-Dawley rats was inferior to that of their nonisolated littermates. Moreover, Harlow and Harlow (1962) and Mason (1960) reported that isolated rhesus monkeys showed inability to breed and, when conception was achieved, inappropriate rearing behavior.

There have been several studies of aggression and dominance in isolated animals. Hutchinson, Ulrich, and Azrin (1965) noted that aggressive behavior to foot shock occurred less frequently in

isolated rats than in nonisolated Ss. Since isolated C57BL/10 mice were significantly slower to fight than nonisolated ones, King and Gurney (1954) concluded that the former were significantly less aggressive than the latter. Rosen and Hart (1963) reported that, when isolated and nonisolated male deermice (*bairdii*) competed for water, the former were less dominant than the latter. Thompson and Melzack (1956) found that in competing for bones, isolated dogs were less dominant than those raised as pets.

On the other hand, some studies have indicated that socially isolated animals were more aggressive than nonisolated ones. Kuo (1960) found an appreciably higher percentage of aggressive quail in isolated groups than in nonisolated groups. Ginsburg and Allee (1942)

* From the *Journal of Comparative and Physiological Psychology*, 1967, **63**, 157–159. Copyright 1967 by the American Psychological Association. This

investigation was supported by Public Health Service Research Grant MH-06655 from the National Institute of Mental Health.

reported that the fighting tendencies of isolated albino and black strains of mice were greater than those of their nonisolated counterparts. Such isolation-induced fighting behavior of mice was also observed by Janssen, Jageneau, and Niemegeers (1960), Uyeno and Benson (1965), and Yen, Stanger, and Millman (1958).

It was considered important to obtain data on the results of competition under more stringent and demanding conditions than those in the works cited. The present experiment was designed to determine, under conditions involving survival motivation, whether or not rats socially isolated for 5 weeks beginning at the age of 17 days would be more dominant than their nonisolated littermates.

METHOD

SUBJECTS. Fifteen litters of male Wistar rats (four rats per litter) were weaned at 17 days of age and earmarked. Two rats from each litter were housed individually (i.e., socially isolated), while two were housed together in three groups of ten each.

APPARATUS. In an underwater test, developed to evaluate dominance behavior, a pair of rats swam through an underwater tube and competed to escape from it. The testing apparatus consisted of a clear plastic tube, 4 ft. long and 2 in. in diameter. Small holes on the top of the tube allowed air bubbles to escape. The tube was placed horizontally in a $5\frac{1}{2}$ ft. by 12 ft. by 6 in. tank and submerged in $3\frac{1}{2}$ in. of 23°C water. Two detachable start tubes, one at each end of the swim tube, were used. Small holes all around the start tubes permitted easy submersion. Rats weighing 180–230 gm could swim through it, but could not turn around in it or pass another rat swimming in the opposite direction.

PROCEDURE. During the 4 weeks from the time of weaning until the beginning of pretraining, Ss were fed ad lib. During pretraining and training they were maintained on a paired feeding schedule so that the weights of the isolated Ss would be approximately equal to those of the nonisolated litter-mates.

In the first pretraining trial Ss were placed one at a time in a start tube, and the tube was partly submerged at one end of the half-submerged 4-ft. tube. As the start door was raised Ss traversed the long tube and escaped at the other end. They were never allowed to escape from the starting point. On all trials after the first pretraining trial, Ss were fully submerged and forced to swim through a fully submerged tube to escape. On the second pretraining trial they swam through a 27-in. tube, because it was too difficult for them to swim through the fully submerged 4-ft. tube if they had not previously swum through a fully submerged 27-in. tube. On the third and fourth pretraining trials they swam through the long tube. In these four pretraining trials, the time taken to swim the full length of the tube was measured in seconds with a stopwatch.

On the basis of swimming times of the third and fourth pretraining trials, the faster swimmer of the two isolates was matched with the faster of their two nonisolated littermates. During the 12 training trials, 14 isolated Ss swam from the right to the left end of the 4-ft. tube. The other 14 swam from left to right. The group-living Ss swam in the direction opposite to that of their matched isolated littermates. Three trials per day were given, allowing 2 hr. between trials. In the training trials, the time taken to swim to the midpoint of the tube was recorded. The half-length swimming time was considered to be comparable in general to the time S would take to swim from the entrance of the tube to its approaching opponent in a competition situation.

In the survival competition situation, an isolated S was placed in its start tube, and its matched group-living littermate was placed in the start tube at the opposite end of the 4-ft. tube. Both were submerged simultaneously. The start doors were lifted to allow them to swim toward their "escape ends." As they approached each other, the S that forced the opponent to move backward to the opponent's starting point was scored as "dominant."

RESULTS

Out of 27 pairs that competed, 21 isolates were scored as dominant and only 6 as submissive. This clearly indicates that socially isolated Ss under survival motivation were more dominant than their nonisolated littermates. According to the binomial test for a large sample (Siegel, 1956, pp. 40–41), the percentage of dominant animals in the isolated group was significant ($Z = 2.88$, $p < 0.01$). The average duration of the 27 dominance tests (i.e., from the time the start door was lifted until one of the pair was forced out) was 11.7 sec.

On the first pretraining trial, when the isolated Ss were placed in the start tube and partly submerged, many of them withdrew to the back of the tube and remained immobile for a relatively long time, revealing "freezing behavior." On this trial the two groups differed significantly in the time taken to escape ($t = 4.5$; $N = 28$ pairs; $p < 0.01$). However, this difference did not seem to be related to swimming ability. In the following pretraining and training trials, when both groups were fully submerged, the swimming times of the two groups were not significantly different.

The two groups did not differ significantly in body weight at the time of weaning, during pretraining and training, and at the time of dominance testing.

DISCUSSION

The data, indicating that socially isolated animals were more dominant than nonisolated ones, are consistent with those of Ginsburg and Allee (1942), Janssen et al. (1960), Kuo (1960), Uyeno and Benson (1965), and Yen et al. (1958). However, the results appear to disagree with those of Hutchinson et al. (1965), King and Gurney (1954), Rosen and Hart (1963), and Thompson and Melzack (1956), who all found relatively slow, inactive, and "submissive" responses in isolated animals. These responses were probably due to typical naive unadaptive behavior such as timidity or freezing behavior, attributable to a novel test environment.

Also in the present study, on the first pretraining trial the isolated animals revealed freezing behavior in the partly submerged start tube. However, on the following pretraining and training trials, when they were fully submerged, freezing behavior disappeared and their times to escape from the underwater tube were as fast as the nonisolated ones. Therefore, freezing behavior of the isolates dissipated before they were given the dominance test.

The relatively inactive and submissive behavior of the socially isolated animals of Hutchinson et al. (1965), King and Gurney (1954), Rosen and Hart (1963), and Thompson and Melzack (1956) might also be due to their methods of testing. In these studies animals were motivated, but maximally demanding conditions necessary to arouse survival motivation appeared to be absent. In the foot shock experiment of Hutchinson et al. (1965), the behavior of the shocked rats that stood on their hind legs facing, pawing, and shrieking at each other was probably not "aggressive," but defensive. Moreover, the apparent discrepancies between the results of the present study and those of previous works may be due to species differences.

REFERENCES

ANDERVONT, H. B. Influence of environment on mammary cancer in mice. *J. Nat. Cancer Inst.*, 1944, **4**, 579–581.

GINSBURG, B. and ALLEE, W. C. Some effects of conditioning on social dominance and subordination in inbred strains of mice. *Physiol. Zoöl.*, 1942, **15**, 485–506.

HARLOW, H. F. and HARLOW, M. K. Social deprivation in monkeys. *Sci. American*, 1962, **207**, 136–147.

HUTCHINSON, R. R., ULRICH, R. E. and AZRIN, N. H. Effects of age and related factors on the pain-aggression reaction. *J. comp. physiol. Psychol.*, 1965, **59**, 365–369.

JANSSEN, P. A. J., JAGENEAU, A. H. and NIEMEGEERS, C. J. E. Effects of various drugs on isolation-induced fighting behavior of male mice. *J. Pharmacol. exp. Ther.*, 1960, **129**, 471–475.

KING, J. A. and GURNEY, N. L. Effect of early social experience on adult aggressive behavior in C57BL/10 mice. *J. comp. physiol. Psychol.*, 1954, **47**, 326–330.

KUO, Z. Y. Studies on the basic factors in animal fighting. IV. Developmental and environmental factors affecting fighting in quails. *J. genel. Psychol.*, 1960, **96**, 225–239.

MASON, W. A. The effects of social restriction on the behavior of rhesus monkeys: I. Free social behavior. *J. comp. physiol. Psychol.*, 1960, **53**, 582–589.

MUHLBOCK, O. F. E. Invloed van het millieu af de kunkerontwikkeling. Onderzalkingen by de melkklierkanker van de muis. *Ned. Tijdschr. Geneesk.*, 1950, **94**, 3747–3752.

REYNOLDS, H. H. Effect of rearing and habitation in social isolation on performance of an escape task. *J. comp. physiol. Psychol.*, 1963, **56**, 520–525.

ROSEN, J. and HART, F. M. Effects of early social isolation upon adult timidity and dominance in peromyscus. *Psychol. Rep.*, 1963, **13**, 47–50.

SIEGEL, S. *Nonparametric statistics for the behavioral sciences.* New York: McGraw-Hill, 1956.

THOMPSON, W. R. and MELZACK, R. Early environment. *Scient. American*, 1956, **194**(1), 38–42.

UYENO, E. T. and BENSON, W. M. Effects of lysergic acid diethylamide on attack behavior of male albino mice. *Psychopharmacologia*, 1965, **7**, 20–26.

YEN, H. C. Y., STANGER, R. L. and MILLMAN, N. Isolation-induced aggressive behavior in ataractic test. *J. Pharmacol. exp. Ther.*, 1958, **122**, 85A.

SOCIALLY FACILITATED REDUCTION OF THE FEAR RESPONSE IN RATS RAISED IN GROUPS OR IN ISOLATION*

Bruce J. Morrison and Winfred F. Hill

SEVERAL experimenters have investigated the effects of group vs. individual testing of a fear response on the magnitude of the fear response, testing whether rats which have been conditioned to fear a particular stimulus situation show greater fear when placed in the situation alone than when placed in the situation along with other rats.

Anderson (1939) found no significant difference in emotional behavior, as measured by defecation, in rats in pairs as opposed to rats individually. However, he had to change the emotion-provoking situation many times as the animals tended to adapt to each one. Davitz and Mason (1955), under the assumption that the more movement in a testing field the less fear, tested fearful rats alone, with another fearful rat, and with a non-fearful rat. Fearful Ss were those which in a previous experiment had been negatively conditioned to a flickering light which was presented while they were in the testing situation. The presence of a nonfearful S reduced the strength of the fear response exhibited by a fearful S. The presence of another fearful S influenced behavior in the same direction, but to a degree just short of significance. Davitz and Mason (1955) hypothesized that the introduction of another S into the situation increased the number of stimuli to be explored or decreased the amount of generalization from the fear-provoking situation. Rasmussen (1939) found that it took longer for groups of three thirsty Ss to stop drinking in a situation where there was a 3-sec. delayed shock than for the individually tested Ss. This was interpreted as showing that groups of Ss were less fearful than individual Ss. This, on the other hand, may have been due to less learning of fear in the group

situation. The two experiments in the present study further investigate social reduction of fear and the variables involved.

EXPERIMENT 1

Experiment 1 presents a more definite answer as to the reality of social reduction of fear and determines some of the variables involved. The experiment was a 2×2 factorial design where Ss learned that the situation was dangerous either in groups or individually and were tested for amount of fear in groups or individually. The measure of fear was the amount of approach behavior in an approach-avoidance conflict situation.

Method

Subjects. The Ss, 72 experimentally naive, female albino rats of the Sprague-Dawley strain, were housed in cages of six. They were 60 days old at the beginning of the study, which lasted about 3 weeks. During the experiment three Ss died; one S was in the group learning, individual testing condition and two Ss were in the individual learning, group testing condition.

Apparatus. The Ss were tested in a 7.5 by 9 in. runway which was 5 ft. long and painted gray, with a vertical black line every 10 in. along its length. The goal box, 7.5 by 10 in., was painted white and had a removable feeding dish 7.5 by 3 in. The floor of the goal box was an electric grid. To adjust the grid's electrical circuit so that the amount of shock received by one S standing on the grid alone could be made comparable to the amount of shock received by each of three Ss

* From the *Journal of Comparative and Physiological Psychology*, 1967, **63** 71–76. Copyright, 1967 by the American Psychological Association. The first author conducted this research in partial fulfilment of the requirements for the M.A. degree at Northwestern University, with the second author as adviser. Appre-

ciation is expressed to Carl P. Duncan and Albert Erlebacher for serving on the thesis committee.

The second experiment was conducted by the first author when he was on a summer fellowship from the National Science Foundation.

standing on the grid together, the average resistance of one S on the grid was determined and then two resistors of this size were added in parallel with the grid when one S was shocked. Average resistance was determined from 10 60-day-old Ss used in a different experiment. The size of the start box, 8.25 by 7.5 in., was large enough so that three Ss could move around in it without bumping each other.

Procedure. Each S received two 5-min. pre-handling sessions each day for 5 days. During each session S was allowed to explore a large unpainted wooden box and was picked up and replaced five times by E. Since Ss were rewarded with 15 sec. of feeding on Purina lab chow mash, rather than the pellets with which they were usually fed, they became familiar with the mash during this period. On the first day of prehandling, mash was presented both in the home cages and in the prehandling box. Thereafter, mash was only presented in the prehandling box. The Ss were on a 23-hr. food deprivation schedule which started the day before prehandling began and continued until the third day of testing, when Ss were placed on a 47-hr. deprivation schedule for the duration of the experiment. The Ss were fed within 1 hr. of running their trials for the day.

All of the Ss first learned to eat at the end of the runway. During this preexperimental stage each S learned to run the straight maze both individually and in a group of three. Groups always consisted of the same three cage mates. The sequence of the group or individual trials was random and continued until S ran the length of the runway in just under 15 sec. in both conditions. As the individual S approached criterion, future individual trials were eliminated. If one S in the group situation reached criterion before the other Ss, it continued to run with the group. If, by the time the others reached the 15-sec. criterion, the leader was traversing the maze in 14 sec., it was brought to a 14-sec. criterion in the individual situation. However, because of what appeared to be almost perfect transfer from one situation to the other, E was able to manipulate the number of individual trials for each S in such a way that all Ss in a group reached criterion in the group situation within one or two trials of each other.

After Ss reached criterion, one of the two preexperimental groups in each cage was assigned randomly to the group learning condition and the other to the individual learning condition. The groups of three and the individual Ss were then run on one trial in which, instead of food, they received a 2-sec., 3.4-ma., 250-V. shock in the goal box. The day before they were shocked, half the Ss ran in the condition (group or individual) in which they were to be shocked and the other half ran in the other condition.

For the testing condition, half the cages were randomly assigned to group testing and the other half to individual testing. The Ss were given two 15-sec. test trials each day. These trials were about 7 min. apart. The number of trials before Ss would again run to the goal box within 15 sec. was recorded. In addition to this criterion, the amount of approach behavior on each trial was recorded by noting the farthest line that each S crossed going toward the goal.

During the experiment, Ss were taken from the colony room to the experimental room in group cages when they ran in groups and in individual cages when they ran individually. When three Ss were run together, they were all placed in the start box at the same time. The start gate opened 1 sec. after S or Ss were placed in the start box. When the trial was over, Ss were taken out of the straight maze as quickly as possible. To facilitate removal of Ss, the goal-box door was closed after 15 sec. or whenever all three Ss entered the goal box, whichever was first. When groups were run, this usually took about 3 sec. Half the time the S closest to the goal box was removed first, and half the time the S farthest from the goal box was removed first. The individually run Ss were removed so that they averaged the same amount of time before removal as did those Ss run in

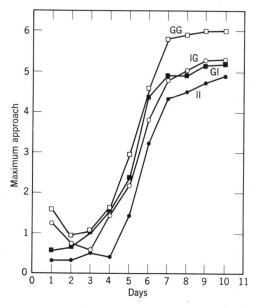

FIGURE 1. Average maximum approach per trial in 10-in. units for Experiment 1.

groups. On the day Ss were shocked, first a group of three Ss was shocked and then three individual Ss were shocked. The position of each S of the group in the goal box was noted and one individual S was shocked in approximately the same position as each S in the group. During the shock period the feeding dish was removed.

Results

The average amount of approach behavior per trial on each day of testing is shown in Figure 1 for each condition. The first of the two letters in the abbreviated condition label designates whether these Ss were shocked in a group (G) or individually (I). The second letter designates whether these Ss were tested in a group or individually. The mean of approach behavior for GG Ss was greater than for any other condition on every day of testing. The mean of the II Ss was less than that of any other condition on every day. Since three Ss were lost during the experiment, the data of three other Ss had to be randomly eliminated to make the cells of the factorial design proportional for an analysis of variance. Those Ss tested in groups approached significantly closer than those tested individually ($F=11.4$, $df=1/62$, $p<0.05$). Whether or not Ss were in groups when shocked was not significant; neither was the interaction between learning and testing (Fs $= 6.44$ and 1.53, respectively, $df=1/62$). The effect of trials was significant even with the conservative df recommended by Geisser and Greenhouse (1958) ($F=284.56$, $df=1/68$, $p<0.001$). Trials × Testing, Trials × Learning, and Trials × Testing × Learning interactions were not significant (Fs $= 1.44$, 0.51, and 0.81, respectively, $df=1/68$). The mean square for Ss within groups for this analysis was 26.43.

The average number of test trials before S ran to the goal box showed the same results as did the amount of approach behavior on each trial. (Any S failing to reach criterion received a score of 20.) The Ss in the GG condition took an average of 11.94 trials to reach criterion, while the Ss in the II condition took an average of 15.39 trials. Between these two were the IG and the GI conditions with averages of 12.37 and 13.41 trials, respectively. An analysis of variance confirmed that Ss in the group testing condition reached criterion sooner than those tested individually ($F=10.05$, $df=1/65$, $p<0.01$). Tests of the learning variable and the interaction between learning and testing conditions were not significant (Fs $= 2.90$ and 1.20, respectively). The mean square within groups for this analysis was 8.63.

In order to test whether the lower average number of trials to criterion of the group testing condition was due to timid Ss following the boldest S in each group, the interaction between timidness-boldness and testing condition was tested. The number of trials to criterion for the most timid S and the boldest S from each group or dummy group was used. Dummy groups consisted of three Ss tested individually and were made up before the experiment began. Timidness-boldness was determined by the number of trials to criterion. The means were: bold Ss in groups $=$ 10.6, timid Ss in groups $= 13.7$, bold Ss individually $= 11.5$, and timid Ss individually $= 18.4$. The greater effect of testing condition on timid than on bold Ss yields a significant interaction ($F=9.28$, $df=1/47$, $p<0.01$). As expected, the timidness vs. boldness and group vs. individual testing main effects were significant (Fs $= 63.16$ and 20.88, respectively, with $df=1/47$, $p<0.01$). The mean square within groups for this analysis was 4.75.

During the first day of testing there was a total of three feces dropped in the maze for all 72 Ss. This extremely low value made this commonly used measure of emotion useless.

EXPERIMENT 2

Experiment 2 attempts to determine whether living in groups vs. in isolation has any effect upon socially facilitated reduction of the fear response. If socially facilitated reduction of the fear response is caused by the increased number of distracting stimuli, as was suggested by Davitz and Mason (1955), then Ss raised in isolation should show greater socially facilitated fear-response reduction, since they have not been so well habituated to the stimulation from the presence of another animal as have those raised in groups. On the other hand, if the reason for the reduction of the fear response is learned assurance when with another rat, then rats raised in isolation should not show this fear reduction. Finally, if this type of social facilitation is an innate characteristic of the organism, then both groups should exhibit the phenomenon to the same degree.

Method

Subjects. The Ss were 48 experimentally naive, female albino rats of the Sprague-Dawley strain. Beginning at 25 days of age, half the Ss were housed in groups of six and half were housed individually. They were 60 days old at the beginning of training. The experiment lasted 4 weeks.

Apparatus. The group cages were 18 by 14.5 by 9 in. and the individual cages were 6.7 by 9.5 by 6.7 in. in size. Isolation was only partial since Ss in the individual cages could smell and hear each other, but could not see or make physical contact with one another. Light came in through the front of the cage. The testing situation was the same as that used in Experiment 1.

Design. This experiment was a 2×2 factorial design using the two most extreme groups from Experiment 1 (group-group and individual-individual) and the two different housing conditions (individual and group). Because of the possible effect of the difference in cage sizes, Ss in the group and individual rearing conditions were tested for activity 2 days after testing in the alley was over. Each S was placed in a revolving drum for 5 min. and the number of revolutions was recorded. The same revolving drum was used for all Ss tested in an ABBA sequence from the two rearing conditions on the same day.

Procedure. All Ss were trained in the pre-experimental stage to a 10-sec. criterion and were rewarded with Nabisco 100 per cent bran cereal. The Ss remained on a 22-hr. food deprivation schedule throughout the experiment. Otherwise, all other aspects of Experiment 2 were the same as Experiment 1.

Results

The average amount of approach behavior per trial on each day of testing is shown in Figure 2 for each condition. The convention for designing conditions is the same as experiment 1 except that a third letter is added at the end to tell whether the Ss were raised individually (I) or in groups (G). Both GGG and GGI conditions showed significantly more approach behavior on each day than did either the IIG or III conditions ($F = 51.56$, $df = 1/44$, $p < 0.01$). Those Ss reared in groups showed significantly more approach behavior than those reared in isolation ($F = 22.83$, $df = 1/44$, $p < 0.01$). The mean square for Ss within groups for this analysis was 95.50. The interaction between rearing and testing was significant ($F = 6.35$, $df = 1/44$, $p < 0.05$). The variable of trials was significant ($F = 30.16$, $df = 1/44$ for conservative test, $p < 0.001$), but the interaction between trials and rearing was not significant ($F = 0.06$, $df = 1/44$). The Trials × Testing and Trials × Testing × Rearing interactions were significant ($F = 3.62$ and 7.68, respectively, $df = 1/44$, $p < 0.01$).

Any S failing to reach criterion was given a trials-to-criterion score of 30. The mean trials to criterion were: GGG = 9.58, IIG = 28.92, GGI = 21.50, and III = 30.00. Mann-Whitney U tests had to be used because there was no variance between the Ss in the III condition. The Ss tested in groups

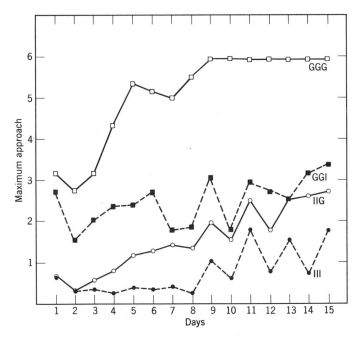

FIGURE 2. Average maximum approach per trial in 10-in. units for Experiment 2.

approached significantly closer ($Z = 4.04$, $p < 0.01$). The Ss raised in isolation approached significantly less ($Z = 2.35$, $p < 0.05$). The interaction was tested by combining the scores of the III and GGG Ss and comparing this with the scores of the GGI and IIG Ss combined. The interaction was not significant ($Z = 1.60$).

Figure 3 shows the percentage of Ss in the GGG and in the IIG conditions reaching criterion on or before each day of testing and the average amount of approach behavior per trial for those Ss not reaching criterion on or before each day. Since both of these measures increased in magnitude over the test period, it can be assumed that the increase in Figure 2 for GGG and IIG conditions over trials is due to a gradual increase in each S's approach behavior and not just to a few Ss reaching criterion in an all-or-none fashion.

Figure 4 shows for Ss raised in isolation what Figure 3 shows for Ss raised in groups. The amount of approach behavior for GGI groups did not increase over days, although the percentage of GGI Ss reaching criterion did increase over trials. The day-to-day fluctuations of amount of approach behavior for the GGI conditions seemed much larger than those of any other condition. Therefore, the gradual increase of average approach behavior in Figure 2 for the GGI group does seem to be due to those Ss reaching criterion in an all-or-none manner.

The difference between the two rearing conditions in the activity measure was not significant ($t = 0.46$, $df = 46$). Those Ss raised in isolation had

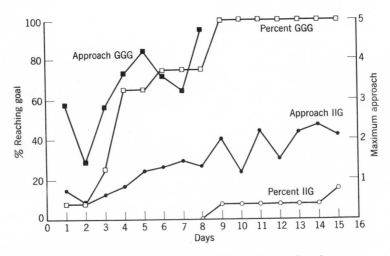

FIGURE 3. Accumulative percentage of Ss reaching goal, and average maximum approach for those Ss not reaching goal, for group-reared Ss.

a mean of 24.8 revolutions, while those raised in groups had a mean of 22.8 revolutions during the 5-min. test period.

DISCUSSION

Rats in groups approached more and took fewer trials to criterion in an approach-avoidance conflict than did Ss run individually. Although differences in learning have been suggested by other Es to account for differences between grouped and individual Ss, this variable was insignificant in Experiment 1. The group effect was not due to competition between members of the group because the feeding dish was too large

for any two Ss to prevent the third from eating. It also was not due to differences in generalization from learning to testing, as Davitz and Mason (1955) have suggested, since that factor was counterbalanced and also found to be insignificant in Experiment 1. The phenomenon seems to be relatively strong, since it was significant even with 47-hr. food deprivation, which is assumed to be close to maximum motivation possible. The effect seems to be due at least in part to timid Ss following the bold Ss in the group testing situation. It is as though the timid Ss were given some assurance by being in the presence of other Ss and therefore tried to remain as close as possible to the other Ss.

Experiment 2 tested to see if this reduced fear

was due to distraction, learned reassurance, or innate reassurance from the presence of other *S*s. Since those *S*s raised in groups tended to show more social facilitation than those raised in isolation, the learned-reassurance hypothesis received some support. This conclusion is consistent with the above interpretation of the boldness-timidness finding.

Figures 3 and 4 show that all groups steadily increased in approach behavior over days, except the GGI group. It is hypothesized from the data and from *E*'s observations that those *S*s in the GGI conditions who reached criterion did so in an all-or-none manner. By all-or-none manner is meant that instead of approaching the feed box closer and closer on each trial before finally entering it, *S* either went from a position close to the start box to the goal in one trial, or else remained

close to the start box. The *S*s raised in isolation became very excited when they first encountered another *S*. There was much mutual grooming, pushing, and tumbling, and little goal-oriented behavior. During the first few trials the *S*s in the GGI condition did not seem to be aware of either the fear cues or the goal cues of the situation. Thus, Davitz and Mason's (1955) distraction hypothesis appears to work for *S*s reared in isolation. However, the social facilitation of those *S*s reared in groups is greater than that of *S*s reared in isolation and does not appear to be of an all-or-none nature. Even if distraction is operating in the group situation, something else which has a stronger effect is also operating. This something else appears to be timid *S*s following bold *S*s due to learned assurance in the presence of other *S*s.

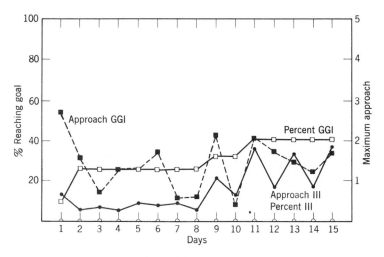

FIGURE 4. Accumulative percentage of *S*s reaching goal, and average maximum approach for those *S*s not reaching goal, for individually reared *S*s.

REFERENCES

ANDERSON, E. E. The effect of the presence of a second animal upon emotional behavior in the male albino rat. *J. soc. Psychol.*, 1939, **10**, 265–268.

DAVITZ, J. R. and MASON, D. J. Socially facilitated reduction of a fear response in rats. *J. comp. physiol. Psychol.*, 1955, **48**, 149–151.

GEISSER, S. and GREENHOUSE, S. W. An extension of Box's results on the use of the F distribution in multivariate analysis. *Ann. math. Stat.*, 1958, **29**, 885–891.

RASMUSSEN, E. E. Social facilitation in albino rats. *Acta psychol.*, 1939, **4**, 275–294.

RELATION BETWEEN ADRENAL WEIGHT, BRAIN CHOLINESTERASE ACTIVITY, AND HOLE-IN-WALL BEHAVIOR OF MICE UNDER DIFFERENT LIVING CONDITIONS*

D. D. Thiessen, James F. Zolman and David A. Rodgers

COMPARISON of results of several areas of investigation suggests a link between endocrine function, brain chemistry, and level of behavioral alertness, and therefore strongly suggests that naturally occurring variations in endocrine balance and brain chemistry will be associated with important variations in behavior.

Levine and others (see Levine, 1960, for a partial review) have demonstrated that differences in gross amount of stimulation of young rats result in differences in endocrine responsiveness to stress situations at a later age. Levine (1960), Bernstein (1952), and others also report differences in activity following early stimulation, the stimulated animals generally showing a higher activity level. Krech, Rosenzweig, and Bennett (1960) have demonstrated that differences in amount of exposure of young rats to environmental complexity result in differences in amount of cholinesterase (ChE) in the brain. It is conjectural whether or not Levine's procedures of gross stimulation would constitute exposure to environmental complexity in the Krech et al. (1960) sense. However, the experimental procedures and results are sufficiently parallel to suggest the possibility of interaction of, or interrelationship between, endocrine function and brain chemistry. Rosenzweig, Krech, and Bennett (1960) posit that their observed changes in brain chemistry reflect alteration in the general responsiveness of the brain. Supporting this hypothesis, they have found strain differences in maze learning ability to be associated with strain differences in brain ChE levels. These results taken in conjunction with the work of Levine and others

lend support to the possibility that differences in brain chemistry might be related to differences in endocrine level and that both might be related to important behavioral differences.

In another area of investigation, Magoun and co-workers (see Magoun, 1958, for a partial review of this work) have demonstrated with many species that the general alertness and responsiveness of the organism are closely related to activity within the reticular formation of the brainstem. Dell and co-workers (Dell, 1958) have shown with the cat that intravenous injection of the adrenal hormone epinephrine in as minute a dose as 5 μg/kg will stimulate the reticular system such as to enhance cortical activation. Conversely, Harris and others (see Harris, 1958, for a review) have shown that electrical stimulation of the hypothalamus can produce increased pituitary secretion of adrenocorticotrophic hormone (ACTH). Again, the implication is strong that an interaction exists between endocrine function, brain responsiveness, and behavior.

Recent studies have indicated that endocrine activity can be altered in many species by varying the number of animals per cage (see Thiessen and Rodgers, 1961, for a review). These studies grew out of the observation that population density is self-limiting in many mammalian species, even in the presence of abundant food and shelter. This density-limitation effect is apparently mediated by pituitary-adrenal-gonadal activity, presumably through an increase in ACTH secretion by the pituitary as size of population increases, a resulting increased secretion of adrenocorti costeroids, and related alteration of other endocrine levels

* From the *Journal of Comparative and Physiological Psychology*, 1962, **55**, 186–190. Copyright 1962 by the American Psychological Association. This work was supported in part by NSF Grant G9936 made to D. A. Rodgers and G. E. McClearn, by USPHS Fellowship MF-11, 174, by USPHS Fellowship MF-12, 827, by NIH Grant M-1292 made to D. Krech, M. R. Rosenzweig and E. L. Bennett, and by the United States Atomic Energy Commission. The authors wish to express their appreciation for invaluable technical assistance provided by E. L. Bennett, H. Morimoto and Marie Hebert.

such as to decrease the reproductive capabilities of the population. These results suggest not only a possible important source of variance in behavioral studies, but also a convenient procedure for manipulating endocrine levels.

The present study explores the interrelation between endocrine function, brain chemistry, and behavior. Population density and environmental complexity are varied, and the effects on endocrine level, behavior, and brain chemistry are examined. The measure of endocrine activity used is adrenal weight per unit body weight. The behavioral measure used is running time on a hole-in-the-wall apparatus. The brain-chemistry measure used is total ChE activity per unit weight of brain tissue.

METHOD

SUBJECTS. Subjects were 50 male C3H/Crgl/2 mice. Following weaning at approximately 24 days of age, they were maintained under standard laboratory conditions in groups of 8 or 12 per cage. At 48 days of age they were randomly divided into five groups of 10 each and ear-punched for identification. The groups were then housed for 28 days in the following conditions:

Group A. Housed singly in solid-wall individual 3-in. by 5-in. cages 4 in. deep. No visual or tactual interaction between animals was possible.

Group B. Housed singly in solid-wall individual cages $8\frac{1}{2}$ in. by $11\frac{1}{2}$ in. by $3\frac{1}{2}$ in. deep. No visual or tactual interaction was possible.

Group C. Housed individually in $3\frac{1}{2}$-in. by $4\frac{1}{2}$-in. wire-mesh compartments within a single large social cage $3\frac{1}{2}$ in. deep. Visual, auditory, and olfactory, but not tactual, interaction was possible.

Group D. Housed as a group in a single 10-in. by $16\frac{1}{2}$-in. social cage $3\frac{1}{2}$-in. deep.

Group E. Housed as a group in a single 10-in. by $16\frac{1}{2}$-in. social cage $3\frac{1}{2}$-in. deep. The cage contained several small wooden "toys" that were changed periodically. These *S*s received 1 hr. of "free play" daily in an open-field arena described by McClearn (1959).

For all groups, standard food and water were available ad lib. Once a week, each animal was weighed and cages were changed. Note that living space per animal is approximately equated for all *S*s except those in Group B, where living space is approximately six times as great as for the other groups.

APPARATUS and PROCEDURE. After 4 weeks of the differential treatments, all *S*s were tested individually for speed of running in a hole-in-the-

wall apparatus described by McClearn (1959). This device consists of two 3-in. by 4-in. compartments 3 in. deep separated by a partition containing a $\frac{7}{8}$-in.-diameter hole that can be uncovered by a guillotine door. One compartment, the start box, has a clear plastic cover and gray wooden sides and floor. The other compartment, the goal box, has a gray opaque wooden cover as well as gray wooden sides and floor. Running time was recorded as the amount of time to the nearest second it took the *S* to pass through the hole with all four feet. Raising of the guillotine door automatically started the time. Runs not completed within 32 min. were terminated. The apparatus was housed in a lighted room so that the goal box, with an opaque cover, was appreciably darker than the start box, with a clear plastic cover. Lower scores in the apparatus should be associated with higher levels of activation.

On the following day, the animals were killed by decapitation. Their adrenals were immediately removed and weighed to the nearest 0.01 mg. A sample of tissue was removed from the somesthetic and visual areas of both hemispheres of the cerebral cortex, and designated the cortical sample. The rest of the dorsal cortex, the cerebellum, medulla, and pons were discarded. The remaining brain was kept as a second, subcortical, sample. The ChE determinations followed the procedures used by Krech et al. (1960). The brain samples were removed by gross dissection, weighed, quickly frozen on dry ice, and stored at $-20°C$ until they were analyzed. All tissues were removed and weighed and the brain samples frozen within 10 min. following decapitation of the *S*. Within a few weeks of their removal, the brain samples were assayed for total amount of ChE according to the analytical procedure, using an automatic titrator, reported by Rosenzweig, Krech, and Bennett (1958). The ChE was determined as moles acetylcholine $\times 10^{10}$ hydrolized per minute per milligram of tissue.

RESULTS

Hole-in-the-wall latencies, adrenal weights, and brain ChE activity are summarized by groups in Table 1. The intercorrelations of these measures are summarized in Table 2. In all comparisons, adrenal weights are expressed in milligrams per 100 grams of body weight, and ChE activity is computed per unit tissue weight. Rank-order correlations were used in all comparisons involving hole-in-wall latencies because the latency data were markedly skewed and because 3 of the

TABLE 1. *Averages and Variabilities of Behavioral and Physiological Measures*

Group	Hole-in-wall Latencies (Sec.)		Adrenal Wt. (mg/100 gm body wt.)		Specific ChE Activity (Moles Ach × 10^{10}/min/mg tissue)					
					Cortical (C)		Subcortical (S)		C/S Ratio	
	Mdn.	Range	M	s	M	s	M	s	M	s
A	507	1,238	11.96	1.75	67.7	12.7	252.0	12.1	.269	.069
B	882	1,882	11.92	1.95	66.3	12.7	240.3	14.0	.277	.062
C	187	1,877	12.82	1.77	66.5	8.2	249.7	6.8	.266	.028
D	40	302	12.04	0.78	68.0	8.7	249.5	10.0	.273	.041
E	16	45	13.92	1.98	67.9	11.8	264.9	8.3	.256	.050

50 Ss were terminated after 32 min. without having completed the task.

Differences in rate of weight gain have been reported for growing animals under different stress regimens (Bernstein, 1952, and others since. See Bernstein and Elrick, 1957, for a partial review). In the present study, there were no significant differences or suggestive trends in body weights of the five groups at the conclusion of the experiment. There were, of course, no significant differences at the beginning of the experiment. Consequently, no further analyses of the weight-gain data were made. Adrenal weights were expressed as a function of unit body weight, consistent with common practice in studies of effects of population density.

Hole-in-the-wall latencies, the behavioral measure, showed the greatest sensitivity of all the variables to the experimental procedures (Table 1,

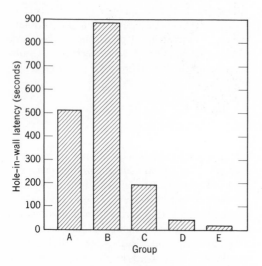

FIGURE 1. Behavioral effects of differential housing. (See text for description of groups.)

Fig. 1). Group differences as tested by Wilson's distribution-free test of analysis-of-variance hypothesis, were highly significant (chi square = 29.6, $df = 4$, $p < 0.001$). Groups D and E each differed significantly from all others (Mann-Whitney U Test, $p = 0.05$), showing both the predicted effect of population density and the added effect of extra stimulation. The individually housed groups, A, B, and C, did not differ significantly among themselves, although the "apartment dwellers," Group C, tended to fall, as expected, intermediate between the grouped animals and the isolates (Fig. 1). Consistent with previous population-density studies concerning endocrine response (Thiessen and Rodgers, 1961), living space *per se* had negligible effect on the behavioral or other measures (Group A vs. Group B).

The groups differed significantly in subcortical ChE activity as measured by analysis of variance ($F = 6.33$, $df = 4$ and 45, $p < 0.01$). Consistent with previous findings with rats (Krech *et al.*, 1960), the extra-stimulation group, E, was significantly higher in subcortical ChE than the apartment dwellers, C, the social-living group, D, and the isolates, A and B (Duncan's multiple-range test, $p < 0.01$). The subcortical differences among Groups A through D were not significant, indicating that Group E was the primary source of intergroup variability.

The groups did not differ significantly in cortical ChE activity, and no suggestive trends were present. In previous work with rats, cortical ChE has been variably reported to decrease (Krech *et al.*, 1960) and to increase or remain the same (Krech, Rosenzweig, Bennett, and Longueil, 1959) following early stimulation. Consequently, the present results cannot be considered inconsistent with previous findings. Although the basis for differences in cortical ChE response is not definitely known, there is suggestive evidence that it may be a function of age of Ss at time of stimulation and testing (Zolman, 1960).

There were no significant group differences in the ratio of cortical to subcortical ChE activity (C/S), a measure found in some previous work with rats (Krech *et al.*, 1960) to vary with amount of stimulation. Nevertheless, consistent with such results, the C/S ratio for the group receiving extra stimulation was noticeably lower. As with cortical ChE response, an age factor may operate to alter the relation of C/S to stimulation.

Group differences in adrenal weights, as measured by analysis of variance across the five groups, closely approached but did not quite reach the 0.05 confidence level ($F=2.55$, $df=4$ and 45). Comparison of adrenal weights of Group E, the extra-stimulation Ss—the ones showing a clear difference from all others in subcortical ChE—against all other groups combined, however, reveals a highly significant difference by analysis of variance ($F=8.49$, $df=1$ and 48, $p<0.01$). The conclusion therefore seems warranted that the same conditions that produced a significant change in subcortical ChE level also produced significant adrenal hypertrophy.

DISCUSSION

Results of the present study clearly confirm the interrelationship of several lines of current investigation. A single experimental procedure led to predicted changes in brain chemistry, endocrine response, and behavior. The results are consistent with the view that the same kinds of early stimulation that produce endocrine response also produce brain-chemistry changes, and vice versa. Not only the between-group data but also the within-group data indicate that brain chemistry, endocrine function, and behavior are closely interrelated.

The evidence that changes in amount of ChE in the mouse subcortical brain will occur after 28 days of exposure to a complex environment extends the Krech, Rosenzweig, and Bennett findings with rats to an inbred strain of mouse. Further support is thus lent to their important hypotheses linking changes in brain chemistry to behavior. It is interesting to note that animals just living together in a large social cage did not differ significantly from the isolated groups, thereby indicating that a certain amount of extra stimulation was necessary before reliable changes in brain ChE could be demonstrated. The fact that the behavioral measure showed detectable differences, however, suggests that more sensitive assay procedures or larger numbers of Ss might have revealed other significant but small differences in brain chemistry.

Several interesting relationships appear from the correlations in Table 2. In all groups, adrenal weight is positively related to both cortical and subcortical ChE activity. The correlations across all groups combined are significant. Of special interest is the fact that adrenal weight relates significantly to both cortical and subcortical ChE activity even though cortical and subcortical ChE levels are not themselves highly correlated ($r=0.19$ for all groups combined, $p>0.15$). The multiple correlations of adrenal weight compared with the two ChE levels were therefore computed. Within the five groups, they range from 0.37 to

TABLE 2. *Correlations Among Behavioral and Physiological Measures*

	Product-Moment or Rank-Order Correlation between				
Group	Hole-in-Wall and Adrenal Wt.	Hole-in-Wall and Cortical ChE	Hole-in-Wall and Subcortical ChE	Adrenal Wt. and Cortical ChE	Adrenal Wt. and Subcortical ChE
	R	R	R	r	r
A	−.31	−.13	−.33	.43	.47
B	−.54	−.59	.23	.28	.36
C	.21	−.24	−.20	.68*	.54
D	−.16	−.43	−.19	.39	.03
E	.02	−.34	−.04	.18	.36
Combined	−.31	−.27	−.42**	.33*	.42**

* $p<0.05$.
** $p<0.01$.

0.70. Across all groups combined, the multiple correlation is 0.56. The consistency and size of these obtained correlations strongly suggest that direct interaction or common factors link endocrine response and brain ChE function. Whether the relationship is one of adrenal hormones stimulating brain function, as is suggested by the work of Dell and colleagues (Dell, 1958), or one of brain activity stimulating the pituitary-adrenal system, as is suggested by the work of Harris and others (see Harris, 1958, for a review), or one of common factors underlying both endocrine and brain response is not indicated by the present data. It seems probable, in light of results of these other areas of investigation, that all three processes may be involved and may interact.

From Table 2 it can be seen that behavior is significantly related to the physiological measures. With 3 exceptions out of 15, all within-group correlations of hole-in-the-wall latencies with adrenal weights and with each of the two samples of brain ChE activity are in the predicted direction, as are the correlations for the combined groups. The most consistent within-group relationships are with cortical ChE level, whereas the highest combined-group relationship is with subcortical ChE (Table 2), a reflection of the differential effects of the experimental conditions on the various measures. To assess further the relationship between physiology and behavior, all measures were converted to rank scores for all groups combined, and the multiple regression of latencies (L) on adrenal weights (A), cortical ChE (C), and subcortical ChE (S) was computed, yielding the equation $L = -0.098A - 0.212C - 0.367S$. The negative signs indicate that short

running times are associated with large adrenal weights and high levels of ChE activity. The multiple rank-order correlation is 0.47. From the regression equation, it is apparent that the best predictors of the behavioral measure are brain ChE levels, especially subcortical ChE activity. The relative weights could shift markedly, of course, for relationships within a single experimental condition or for summation across other conditions than those used in the present study. Also, of course, other behavioral measures might relate differently to the physiological variables.

SUMMARY

C3H/Crgl/2 mice show marked differences in running time on a hole-in-the-wall test as a result of being housed individually or in groups for 4 weeks with different amounts of stimulation present from the environment. The group-housed extra-stimulation animals showed both the shortest running time in the test apparatus and an increase in both adrenal weight and subcortical cholinesterase activity. Cortical and subcortical cholinesterase levels do not correlate highly with each other and do not show the same differential response to the experimental conditions, but both correlate positively with adrenal weight under all stimulus conditions. All three physiological measures correlate as predicted with the behavioral measure under most stimulus conditions. The results indicate that housing conditions can markedly influence both behavior and physiological states and that a relationship exists linking endocrine response, brain chemistry, and behavior.

REFERENCES

BERNSTEIN, L. A note on Christie's "experimental naivete and experiential naivete." *Psychol. Bull.*, 1952, **49**, 38–40.

BERNSTEIN, L. and ELRICK, H. The handling of experimental animals as a control factor in animal research—a review. *Metabolism*, 1957, **6**, 479–482.

DELL, P. C. Humoral effects on the brain stem reticular formation. In H. H. Jasper, L. D. Proctor, R. S. Knighton, W. C. Noshay and R. T. Costello (Eds.), *Reticular formation of the brain*. Boston: Little, Brown, 1958. Pp. 365–379.

HARRIS, G. W. The reticular formation, stress, and endocrine activity. In H. H. Jasper, L. D. Proctor, R. S. Knighton, W. C. Noshay and R. T. Costello (Eds.), *Reticular formation of the brain*. Boston: Little, Brown, 1958. Pp. 207–221.

KRECH, D., ROSENZWEIG, M. R. and BENNETT, E. L. Effects of environmental complexity and training on brain chemistry. *J. comp. physiol. Psychol.*, 1960, **53**, 509–519.

KRECH, D., ROSENZWEIG, M. R., BENNETT, E. L. and LONGUEIL, C. L. Changes in the brain chemistry of the rat following experience. *Amer. Psychologist*, 159, **14**, 427. (Abstract.)

LEVINE, S. Stimulation in infancy. *Scient. American*, 1960, **202**, 80–86.

MCCLEARN, G. E. The genetics of mouse behavior in novel situations. *J. comp. physiol. Psychol.*, 1959, **52**, 62–67.

MAGOUN, H. W. *The waking brain*. Springfield, Ill.: Charles C. Thomas, 1958.

ROSENZWEIG, M. R., KRECH, D. and BENNETT, E. L.

Brain enzymes and adaptive behaviour. In. G. E. W. Wolstenholme (Ed.), *Neurological basis of behaviour*. London: J. & A. Churchill Ltd., 1958. Pp. 337–358.

ROSENZWEIG, M. R., KRECH, D. and BENNETT, E. L. A search for relations between brain chemistry and behavior. *Psychol. Bull.*, 1960, **57**, 476–492.

THIESSEN, D. D. and RODGERS, D. A. Population density and endocrine function. *Psychol. Bull.*, 1961, **58**, 441–451.

ZOLMAN, J. F. Changes in the level of cholinesterase activity in the rat brain following training at different ages. Paper read at Western Psychological Association, San Jose, April 1960.

MOTHER-INFANT SEPARATION IN MONKEYS*

Billy Seay, Ernst Hansen and Harry F. Harlow

THE problem of mother-child separation and the mechanisms underlying the disturbance it produces in the child have been discussed by Bowlby (1960, 1961), who has reviewed the literature and presented a theoretical position concerning separation anxiety, grief, and mourning in children and infants. In the later paper he outlined six theories of separation and concluded that separation anxiety results from activation of the component instinctual response systems, which are the base of the infant's attachments to the mother, in a situation in which the familiar object is not available. The effects on the child occur in three phases: protest, despair, and detachment, the first stemming from anxiety, the second from grief and mourning, and the third from defensive reactions.

If the mechanisms producing Bowlby's separation syndrome are as biologically basic as the activation of "basic component instinctual responses," a similar syndrome should be produced in infant monkeys following maternal separation. There is great similarity in the variables which Bowlby (1958) has described in his Component Instinctual Response theory and the variables producing the infant monkey's tie to its mother. Although Harlow would attach more importance to clinging and somewhat less importance to sucking than Bowlby does, there is little fundamental difference (Harlow, 1960; Harlow and Zimmermann, 1959). It is clear that clinging, which results in contact comfort, is a stronger and

more persisting variable than nursing in the monkey, and any such differences in the importance of the components for monkey and man may be a function of the macaque monkey's maturational acceleration. The variable, following (visual and auditory responsiveness to the mother), has also been identified in studies concerning infant attachments of both species to their mothers (Bowlby, 1958; Harlow, in press), the responses appearing earlier in monkey than in man and being far stronger in the monkey's early infancy than in man's, probably because of the monkey's maturational acceleration.

The number of quantitative investigations on maternal separation in non-human primates is limited. Spence (1937), in a study involving permanent separation of infant chimpanzees from their mothers at various ages, observed the mothers' emotional responses to the sight of their infants. Five of seven mothers showed continuing own-infant specific emotional responses. Violent attempts to regain contact with the infants dropped out very quickly, but differential recognition of their own as opposed to other infants was still evident after separations of 6 months or more.

Short-term effects of baby exchanges of less than an hour's duration in two pig-tailed macaque mother-infant pairs have been reported recently by Jensen and Tolman (1962). The mothers in this study exhibited infant specificity after separation, but the two infants, which were 5 and 7

* From the *Journal of Child Psychology and Psychiatry*, 1962, **3**, 123–132. Copyright 1962 by Pergamon Press, New York. This research was supported in part by funds supplied by Grant No. M-722, National Institutes of Health, and by a grant from the Ford Foundation. The senior author was a Woodrow Wilson Fellow, and the second author held a National Institute of Mental Health Predoctoral fellowship during the course of this research. The authors are indebted to Mr Robert Dodsworth for his assistance in the planning and execution of this research.

months old, initially showed positive responses to the strange mother. However, these responses dropped out very quickly in the face of rejection or disregard. In the process of being separated, both infants and mothers showed emotional responses; the babies screamed almost continuously, and the mothers attacked their cages and attempted to escape. The mothers were ferocious toward attendants, and the babies resisted attendants by clinging to the mother or to any available object. When the infants were returned to their mothers, there were striking initial increases in infant-mother and mother-infant responsiveness.

Over 200 infant rhesus monkeys have been separated from their mothers in the first 12 hours of life at the University of Wisconsin Primate Laboratory. In virtually all cases a team of men is required to restrain the mother while removing the infant, and the activity is not without some risk. The emotional reactions of these mothers are brief, and no increased animosity toward laboratory personnel has been observed consequent to separation.

During the last two years, a number of informal separation studies of an exploratory nature have been conducted at the Wisconsin Primate Laboratory on 30- to 90-day-old rhesus infants raised with their own mothers. The effects of separations ranging from a few hours to a few days in duration indicated acute disturbance of both mother and infant during separation, and a marked increase in maternal protectiveness after return of the infant. In one case the mother, after a 3-day separation, kept her infant literally within reach for a period of more than a month. Following these preliminary observations a more formal study of mother-infant separation was initiated.

METHOD

SUBJECTS. Two pairs of infant rhesus monkeys and their four mothers served as subjects. Pair 1,

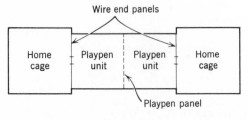

FIGURE 1. Deprivation apparatus. See text for dimensional details.

Male A58 and Female A59, were separated at 170 and 169 days of age, respectively, and Pair 2, Female A63 and Male A64, were separated at 207 and 206 days of age. All animals had previously been removed from their mothers within 12 hours of birth for weighing and identification tattooing, and then returned. Infant A63 had also been removed from its mother at 36 days of age for 3 days' treatment of an infected hand.

APPARATUS. The apparatus, shown in Fig. 1, included two woven-wire living-cages 36 in. by 36 in. by 36 in. adjoining two woven-wire playpen units 30 in. by 30 in. by 30 in., separated by a removable wire-mesh panel. The infants had access to the playpen units through 3.5- by 5.0-in. openings between home cages and playpen units.

Each pair of infants and each pair of mothers lived in such a test apparatus throughout the entire experiment, except for 1 hour each week when the apparatus was being cleaned. Pair 1 had lived in the apparatus since birth. Pair 2 had previously been housed in a slightly modified situation, and was placed in the standard separation apparatus 3 days prior to the first session of the pre-separation period.

PROCEDURE. The experiment proper was divided into three periods of 3 weeks each, as follows: (1) Pre-separation: the infants continued to live with their mothers and essential base-line data were taken; (2) separation: the infants were separated from their mothers and lived together in the playpen; and (3) post-separation: the infants were allowed to return to their mothers and to live with them throughout the remainder of the experiment.

A day before the first separation test session the wire end-panels were replaced by clear Plexiglas end-panels of the same size. To permit the infants access to the playpen units, the plastic panels were fixed in a position 5 in. above the floor until the moment of experimental separation. When each infant entered a playpen unit in the first separation session, the appropriate Plexiglas end-panel was lowered, thus depriving the infant of all but auditory and visual contact with its mother. Just prior to the first post-separation test session the Plexiglas end-panels were replaced by the original wire end-panels. During pre-separation and post-separation the removable panel between playpens was in place except for the daily 30-min. test sessions. During separation, however, this panel remained out at all times, allowing the infants free access to each other 24 hours a day.

Detailed observations were made of the behavior of the animals 30 min. a day, 5 days a week, using a symbol system and behavior inventory developed in this laboratory. In some cases two or more individual categories of behavior were combined for this presentation to provide a more general behavior index. The list that follows includes only those items which occurred with reasonable frequency:

1. *Mother-Infant Responses*
(a) Cradle: support of the infant with one or both arms or with the legs while the infant is in contact with the mother.
(b) Visual contact: oriented "looking at" or visual exploration of the infant.

2. *Infant-Mother Responses*
(a) Clinging: clasping the mother's fur with hands or feet and so effecting some degree of body-weight support.
(b) Ventral contact: gross body contact with the ventral surface of the mother.
(c) Support contact: contact in which the infant received body-weight support, including the ventral contact positions plus sitting or riding on the mother's back or head, or hanging from the central surface of the mother.
(d) Non-specific contact: bumping into, brushing or leaning against, or otherwise touching the mother in an accidental or casual manner.
(e) Nipple contact: sucking, mouthing, or orally manipulating the breast of the mother.
(f) Visual contact: as in 1(b), but oriented toward the mother.

3. *Infant-Infant Responses*
(a) Non-contact play: activity characterized by running toward or away from the partner, involving two distinct directional changes or a direct bee-line mock attack; this category also includes integrated play involving nipping and clasping and pulling, as well as non-contact play elements.
(b) Threats: facial expressions involving ears back, mouth open, brow wrinkled, or some combination of these response elements oriented toward the other infant.
(c) Approach: oriented movement of at least one body-length toward another monkey.
(d) Withdrawal: oriented movement of at least one body-length away from another monkey.
(e) Non-specific contact: as in 2(d), but involving another infant.
(f) Facilitation: imitative exploration or manipulation of objects immediately following similar manipulation by the other infant.

4. *Infant Non-social Responses*
(a) Crying: soft, high-pitched vocalization (ooh, ooh).

(b) Self-mouth: sucking or mouthing the self, usually involving a digit but scorable if directed to other parts of the body, e.g., the forearm.

The daily 30-min. test session was divided into 120 15-sec. periods. Three experienced observers participated, R.D. observing in each session and B.S. and E.H. alternating in a balanced order. Each observer recorded the behaviors of one infant-mother combination according to a predetermined order, so that subjects and testers were counterbalanced as far as possible. Behaviors were scored sequentially, but the maximum score for any one behaviour in any 15-sec. period was 1, thus preventing overweighting of repetitive behaviors. Reliability sessions were conducted in a slightly different situation between E.H. and R.D. about 3 months prior to this study. All inter-tester product-moment reliability coefficients were above 0.90 for individual behavior categories except for infant-mother non-specific contact ($r = 0.83$), infant-mother visual contact ($r = 0.64$), and the mouth-open component of infant-infant threats ($r = 0.78$). B.S. reached a training criterion of $r \geq 0.85$ with E.H. and R.D. on all categories prior to testing.

RESULTS

The initial reactions of all mothers and all infants after separation indicated a high degree of emotional disturbance. The infants' behavior generally included disoriented scampering, high-pitched screeching and crying, attempts to pass through the Plexiglas barriers, and huddling up against the barrier in close proximity to the mother. The mothers showed an increase in barking vocalization and in threats directed toward the experimenter, but the mothers' emotional response appeared to be both less intense and of shorter duration than that of the infants.

Over-all effects before, during, and following separation were tested by a repeated-measures analysis of variance on each behavior category listed except support contact. Results of the F and Duncan Multiple-Range tests (1955) show that two indices of infant behavior, infant-mother visual contact (Fig. 2) and crying vocalization were significantly higher ($P < 0.05$) during separation (see Table 1). Crying is consistently associated with disturbance in the infant monkey. Both responses showed a decrease as a function of weeks of separation, indicating either adaptation to the total situation or extinction of the specific responses. The mothers' data yielded no

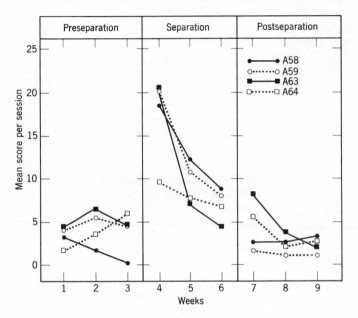

FIGURE 2. Infant–mother visual contact as a function of weeks.

such clear-cut indication of disturbance common to the mothers as a group, and informal observations indicated that the mothers adapted earlier and became more acceptant of the separation than the infants. Although we were reasonably certain that the mothers were disturbed for several days, no behavior category indicating maternal disturbance changed significantly.

TABLE 1. *Infant Crying Vocalization: Before, During, and After Separation*

Source	d.f.	MSS	F	P
Subjects	3	190.50		
Conditions	2	1335.62	8.23	< 0.025
Ss × C	6	162.17		
Weeks/C	6	56.22	1.39	—
Ss × W/C	18	40.56		

Infant-mother visual contact: before, during, and after separation

Subjects	3	3.99		
Conditions	2	241.97	17.79	< 0.05
Ss × C	6	13.60		
Weeks/C	6	39.42	7.56	< 0.01
Ss × W/C	18	5.21		

During the separation period, most complex infant-infant social behaviors exhibited a drastic decrease in frequency of occurrence. Non-contact play, illustrated in Figure 3, showed a significant

decrease (see Table 2); indeed, it was for all practical purposes obliterated. Threats, approaches, and withdrawals also decreased, but not significantly ($P > 0.05$). As shown in Table 3, infant-infant imitative exploration of toys and apparatus increased significantly ($P < 0.05$), but the increase in infant-infant non-specific contact was not significant ($P > 0.05$). An increase in infant self-mouthing and sucking during separation had been expected, but, instead, three of the four animals showed a decrease, A59 being the exception. A decrease in self-sucking with extreme emotional disturbance has been found previously by Bernstein and Mason (1962) and confirmed by Benjamin (1961). Intermediate levels of disturbance seem to bring about an increase in sucking. The experimenters had anticipated that separated infants might show the mutual ventral-ventral clinging seen in animals raised together from birth (Harlow, 1961), but no such increase was observed, despite the opportunities afforded by continuous association 24 hours a day.

TABLE 2. *Infant-Infant Non-Contact Play: Before, During, and After Separation*

Source	d.f.	MSS	F	P
Subjects	3	135.48		
Conditions	2	193.02	8.99	< 0.025
Ss × C	6	21.47		
Weeks/C	6	63.67	2.01	—
Ss × W/C	18	31.59		

Infant-directed behaviors by the mothers as a group were at a surprisingly low level throughout separation, but individual differences were large. The high variability is illustrated by the data on mother-infant visual contacts in Figure 4. A comparison with Figure 2, which shows infant-mother visual contact, further emphasizes the high variability of the mothers' behavior compared with that of the infants.

TABLE 3. *Infant-Infant Imitate: Before, During, and After Separation*

Source	d.f.	MSS	F	P
Subjects	3	0.41		
Conditions	2	5.03	11.03	< 0.01
Ss × C	6	0.46		
Weeks/C	6	0.07	0.74	—
Ss × W/C	18	0.10		

An assessment of the effects of separation on initial post-separation behavior for the group as a whole was made by comparing behavior on the first day of post-separation with that on the last day of pre-separation. The results of *t* tests applied to these data are given in Table 4. Infant-mother clinging and mother-infant cradling show a significant increase, and infant-mother non-specific contact shows a significant decrease. Ventral contact with the mother more than trebled, but the difference was not significant. Infant contact with the mothers' nipples showed a negligible average decrease.

Effects of separation throughout the 3-week post-separation period were assessed in terms of individual animals. The results of simple between-within analyses of variance, comparing the frequencies of mother-directed behaviors of clinging, support contact, nipple contact and non-specific contact before and after separation for each infant, are presented in Table 5. Three infants showed a pattern of increased clinging, support contact, and nipple contact, although the differences did not reach statistical significance in all cases, but A58 showed a significant decrease in support contact and nipple contact, a non-significant decrease in clinging, and a significant increase in non-specific contact for the 3-week period.

DISCUSSION

The results of this investigation appear to be in general accord with expectations based upon the human separation syndrome described by Bowlby. The original response of the infants was certainly one of protest—violent and prolonged protest. This was followed by a period which can be appropriately described as despair. In the apparatus just prior to separation, as in our playpen studies generally, the infants invariably took every possible opportunity to engage in mutual play. The drastic depression of play following separation (see Fig. 3) is best explained by the trauma of separation. It might be noted that frequent daily informal observations through a one-way mirror in the wall separating the test room from the darkened hallway also revealed almost no play

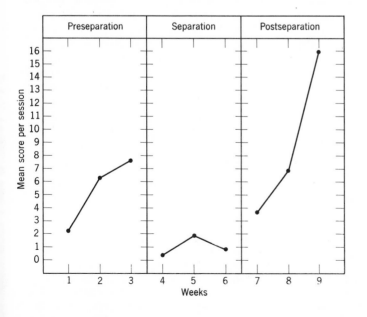

FIGURE 3. Approach–withdrawal play as a function of weeks.

TABLE 4. *Pre- and Post-Separation Infant-Mother and Mother-Infant Behavior*

Behavior	Pre-Separation Mean (Last Day)	Post-Separation Mean (First Day)	P
Mother-infant cradling	22.25	160.75	< 0.05
Infant-mother clinging	2.75	89.00	< 0.025
Infant-mother ventral contact	29.25	107.00	< 0.10
Infant-mother nipple contact	24.75	16.25	< 0.80
Infant-mother nonspecific contact	35.50	3.50	< 0.01

activity during the separation period. At least one monkey, No. A58, exhibited extreme disturbance during the first few separation days, showing definite signs of lack of sleep, and at times it seemed hardly able to stand or hold its eyes open. A58 also showed a decrease in thumb-sucking, as mentioned previously. Prior to separation, monkey A58 and its mother had shown significantly more contact scores than the other three pairs, suggesting a closer relationship for them than for the other pairs; A58's behavior may therefore represent an extreme example of separation trauma.

We have not seen in our monkeys the full pattern constituting Bowlby's detachment phase, and except for monkey A58, we have not even observed a partial pattern. Three of the infants immediately attached to their mothers when separation was terminated, and showed some increase in all categories of mother-directed behaviors except non-specific contact throughout post-separation testing (Table 5). On the other hand, A58 showed a decrease in most mother-directed behaviors except non-specific contact. Possibly this animal was showing some transient defence reactions against reattachment to the mother. We have not, after the reunion, observed signs of aggression against the mother such as those described by Robertson (1955) for children. Many factors could account for such differences: the period of maternal separation was relatively brief for the monkeys, the monkey mother had no part in the act of separation, strong aggressive

patterns had not yet matured in the infant monkeys (these patterns appear in the second year), and the monkey mother is probably less acceptant of infant aggression than the human mother. However, it is more than possible that the two species differ in the degree to which resentment of the mother is formed and, in turn, expressed as aggression during separation and reunion. We believe that the human being, even the human child, is capable of more subtle and more profound behavioral disturbances than any non-human animal.

Bowlby (1961) has summarized various analytical theories concerning separation anxiety, and it is interesting to view them in terms of our separation data. Freud's signal theory cannot apply, since bodily needs of food, water, and even companionship were always available. Sexual capacity was certainly not threatened, since they were at an age when sexual responses to the mother had ended and shifted to playmates. Klein's depressive anxiety theory is not applicable, for even if infant monkeys were capable of believing that the mother disappeared because the infant ate her up, the mother monkey was continually present visually (Klein, 1935). Furthermore, we do not believe that our infants' distress can be explained in terms of persecutory anxiety, and seriously doubt that the rhesus infant possesses symbolic capability essential for such mechanisms. The fact that such theories seem untenable to explain separation anxiety in the monkey does not mean that they do not apply to the human being, but

TABLE 5. *Individual Infant-Mother Responses: Three-Week Post-Separation v. Three-Week Pre-Separation Periods*

Behavior	A58 % Increase	P	A59 % Increase	P	A63 % Increase	P	A64 % Increase	P
Clinging	−30.2	< 0.10	65.2	< 0.20	58.8	> 0.20	128.0	< 0.05
Support contact	−57.1	< 0.001	72.8	< 0.05	29.5	> 0.20	74.2	< 0.05
Nipple contact	−58.1	< 0.001	241.1	< 0.001	45.7	> 0.20	13.2	> 0.20
Non-specific contact	28.8	< 0.001	5.2	> 0.20	−26.0	< 0.001	25.2	> 0.20

our data clearly show that separation anxiety can be produced in a primate infant by mechanisms as simple and direct as Bowlby subsumes in his concept of Primary Anxiety.

SUMMARY

The present study involved separation of four monkey mother-infant pairs for a 3-week period and measurement of the behavior of the subjects before separation, during separation, and after reunion. All mothers and all infants showed emotional disturbance in response to separation, but the infants' disturbances were more intense and more enduring than those of the mothers. The findings were interpreted as being generally in accord with Bowlby's theory of primary separation anxiety as an explanatory principle for the basic primate separation mechanisms.

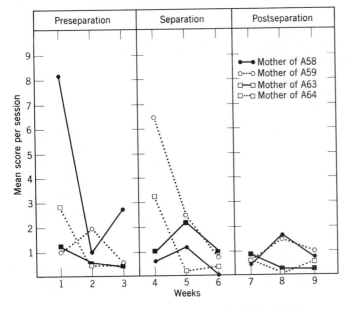

FIGURE 4. Mother–infant visual contact as a function of weeks.

REFERENCES

BENJAMIN, L. S. (1961). The effect of bottle and cup feeding on the non-nutritive sucking of the infant rhesus monkey. *J. comp. physiol. Psychol.*, **54**, 230–237.

BERNSTEIN, S. and MASON, W. A. (1962). The effects of age and stimulus conditions on the emotional responses of rhesus monkeys: Responses to complex stimuli. *J. genet. Psychol.*, **101**, 279–298.

BOWLBY, J. (1958). The nature of the child's tie to his mother. *Int. J. Psycho-anal.*, **39**, 1–24.

BOWLBY, J. (1960). Grief and mourning in infancy and early childhood. *Psycho-anal. Study Child*, **15**, 9–52.

BOWLBY, J. (1961). Separation anxiety: A critical review of the literature. *J. child Psychol. Psychiat.*, **1**, 251–269.

DUNCAN, D. B. (1955). Multiple range and multiple *F* tests. *Biometrics*, **11**, 1–42.

HARLOW, H. F. (1960). Primary affectional patterns in primates. *Amer. J. Orthopsychiat.*, **4**, 676–684.

HARLOW, H. F. (1961). The development of affectional patterns in infant monkeys. In *Deter-minants of Infant Behaviour*. Edited by Brian Foss. Methuen, London, 1961.

HARLOW, H. F. (1962). The maternal affectional system. In *Mother Infant Interactions*. Edited by Brian Foss. Methuen, London. (In press.)

HARLOW, H. F. and ZIMMERMANN, R. R. (1959). Affectional responses in the infant monkey. *Science*, **130**, 421–432.

JENSEN, G. D. and TOLMAN, C. W. (1962). Mother-infant relationship in the monkey, *Macaca nemestrina*: The effect of brief separation and mother and infant specificity. *J. comp. physiol. Psychol.*, **55**, 131–136.

KLEIN, M. (1935). A contribution to the psychogenesis of manic-depressive states. *Int. J. Psychoanal.*, **16**, 145–174.

ROBERTSON, J. (1955). Some responses of young children to loss of maternal care. *Nursing Times*, **49**, 382–386.

SPENCE, K. W. (1937). Réactions des mères chimpanzés a l'égard des enfants chimpanzés aprés séparation. *J. Psychol. nor. pathol.*, **34**, 445–493.

SOCIALIZATION AND IMPRINTING IN BROWN LEGHORN CHICKS*

Philip Guiton

INTRODUCTION

DETAILED investigations of wild Mallard duck-lings (Ramsay and Hess, 1957), and of New Hampshire Red chicks (Jaynes, 1956, 1957), have shown the critical period during which imprinting is possible in these two species to have upper limits of 20 and 54 hours, respectively. Following William James' early lead, several workers have suggested that the restriction of imprinting to a critical period is due to the development of fear or avoidance of moving objects, so that following is inhibited and imprinting therefore no longer possible (Verplanck, 1955; Hinde, Thorpe and Vince, 1956). This hypothesis has received support from Hess (1957) who has shown that the success of imprinting in Mallard ducklings depends on the amount of following; that the end of the critical period coincides with the appearance of "emotionality"; and that certain drugs which reduce emotionality (chlorpromazine, meprobamate), also extend the critical period.

Fear is probably not the only factor involved in terminating the critical period since some Mallard ducklings (Weidmann, 1956), and some domestic chicks (Jaynes, 1956; Guiton, 1958) which have exceeded the age limit show neither fright nor following-responses to a strange moving object. Moreover, studies with Brown Leghorn chicks (Guiton, 1958) have shown that the following-response to a strange object is lost at an earlier age in socially-reared chicks than it is in isolated ones; and that absence of a following-response in socially-reared birds is not invariably associated with any sign of fear or avoidance of the object. It was also found that chicks, reared together to an age when they no longer followed an unfamiliar object, and subsequently isolated for some days would then follow that object, whereas the non-isolated controls did not respond.

This reappearance of the following-response after isolation may seem, at first sight, to indicate that no irreversible process is involved in the original loss of the response. It can be supposed that the tendency to follow an unfamiliar moving object decreases when a chick receives certain visual, tactile or other types of stimulation from other chicks, but increases again after this stimulation has ceased for a certain length of time. According to this *drive-satiation hypothesis* the following-response to a strange object would persist for an indefinite period in chicks isolated from all social contact until the age of exposure and would be independent of any learning.

The results can be equally well explained, on the other hand, by supposing that the chicks learn to follow one another during the initial period of social-rearing; and that this response is subsequently generalized to other stimulus-objects—any moving object—but only after a prolonged period of isolation from other chicks. Without this period of isolation before exposure to the moving object, no following-response is obtained because the chick can still discriminate the moving object from its fellow chicks. It therefore ignores the object while searching for its companions. On the *generalization hypothesis*, the response to an unfamiliar object could still occur in chicks isolated after social contact, even at an age when it may have become difficult or impossible to evoke in totally isolated, socially-inexperienced chicks, due to the development of fear.

The following experiments were carried out to test these hypotheses.

METHODS AND MATERIALS

The chicks were F_1 hybrids of two inbred lines of the Brown Leghorn flock at the Poultry Research Centre. At 4 to 24 hours old each chick was transferred from its individual hatching box in the incubator, to an individual cage in a constant temperature room and supplied with

---

* From the *Journal of Animal Behavior*, 1959, 7, 26–34. Copyright 1959 by Baillière, Tindall & Cassell, London.

food and water. While in these cages, the birds had no visual or physical contact with any moving objects or with one another, so that social experience (other than that of an auditory kind) prior to experimental treatments, was confined to the very brief period of handling when they were transferred. The room temperature was gradually reduced from 37°C on the first day to 35.5°C on the fourth day.

Two models were used for imprinting the birds or for testing their following responses: Model M, a light green box 40 cm long by 17 cm high, by 7 cm wide, and having a head, tail and four legs; and Model B, a brick-red box 25 cm by 28 cm by 12 cm, with a tail, head and two legs. The head, tail and legs were dissimilar in the two models and were incorporated to make the models more easily distinguishable by the chicks.

These models could be suspended from the ends of an overhead arm rotated by an electric motor. By means of this apparatus they were propelled at a constant speed of 14 cm per second along a circular route 180 cm in diameter. Each model contained a loudspeaker connected by wiping-contacts to a tape recorder. This was adjusted to play a continuous-loop recording of a broody hen calling her chicks (clucking).

The response measurements made were: (1) the number of seconds per 5-min. period, during which the chick walked within 30 cm to the side of, or behind the model, or not more than 15 cm in front of it, and in the same direction; and (2) the time taken for the chick to make its first full response, by which is meant the bird making a direct approach to the model or walking with it, while making at the same time, contentment calls.

TRAINING SESSIONS. These took place when the birds were from 20 to 54 hours old. Each one was transferred in a small enclosed wooden box from the cage to the runway, and there placed 60 cm in front of the stationary model and about 20 cm to the side of its track. The model was then started up, and driven along its circular route without interruption for 30 min.: clucking for the first 25 min., silent for the last 5 min. The amount of following was measured during the last two 5-min. periods.

TEST SESSIONS. These also lasted 30 min., but differed from the training sessions in that the model was silent throughout, and following was measured during the second, fourth and sixth 5-min. periods. Test sessions took place about 72 hours after training sessions, when the birds were between 96 and 126 hours old.

All chicks were trained or tested individually.

The temperature of the runway was maintained at 25–27°C.

EXPERIMENTS

Experiment 1

Generalization of the Following-Response

The aims of this experiment were: (1) to compare the effect of an unfamiliar moving object on birds totally isolated from the day of hatching, with its effect on birds which were only isolated after some social contact, in order to find out whether the reappearance of the following-response after isolation is due to earlier social experience; (2) to determine if chicks imprinted on one model generalize their response to other models, when isolated; and (3) to see whether the inhibitory effect of socialization on the following-response can be distinguished from the effect of fear.

Thirty-four chicks were divided into five groups each of which received one of, the following treatments:

1. *Socialization-isolation.* Seven chicks were transferred from their cages to a social brooder at 20 to 54 hours of age, left there for three and a half hours, and then placed, still as a group, in the experimental runway for 30 min. without a model. Each bird was then returned to its isolation cage. Approximately 72 hours later, each chick's following-response was tested with Model B.

2. *Imprinting-isolation.* Six chicks were imprinted—trained to follow Model M—at the age of 20 to 54 hours, replaced in their isolation cages, and then tested with Model B 72 hours later.

3. *Total isolation.* Nine chicks received no training or social experience until, at the age of 96 to 126 hours, they were tested with Model B. Six of these chicks were, however, placed singly in the runway for 30 min. each at the age of 20 to 54 hours—but without a model—to habituate them to the experimental situation.

4. *Imprinting-socialization.* Six chicks were trained to follow Model M, as in group 2, but were then reared for the next 72 hours in a social brooder, before being tested with Model B.

5. *Total socialization.* Six chicks were reared together in a social brooder from the age of 20 to 54 hours until they were tested with Model B at 96 to 126 hours old.

TABLE 1. *Mean Total Duration of Following in Each Group of Model B during Test Sessions*

Group and Treatment	Mean Total Duration of Following, in Seconds	Number of Birds in Group	Standard Error of the Mean
1. Socialization—isolation	136	7	32.14
2. Imprinting—isolation	325	6	74.35
3. Total isolation	41	9	11.64
4. Imprinting—socialization	16	6	12.36
5. Total socialization	10	6	4.76

The whole experiment was spread over three consecutive weeks and involved three separate broods.

RESULTS. The results of test sessions are summarized in Table 1. The differences between the scores of the following groups are statistically significant:

Groups	t	Number of degrees of freedom	$P <$
1 and 2	2.52	11	0.05
1 and 3	2.68	14	0.02
3 and 5	2.47	13	0.05
1 and 4	3.49	11	0.01

It follows from these data that the differences between the scores of Groups 2 and 3, 2 and 4, 1 and 5, and 2 and 5 are also statistically significant. There is, however, no significant difference between the scores of Groups 3 and 4, or between those of 4 and 5.

These differences were due in part to differences in the number of birds responding to the model; differences in the amount of following in the responding birds; and lastly, in the time taken for the birds to start following (Table 2).

There was no apparent difference in behavior between the three birds of Group 3 which had received no habituation experience before the test, and the six that had: their respective mean scores were 44 seconds (three birds) and 39 seconds. Moreover, of the two groups in which there was virtually no following, one (Group 4) consisted of birds with prior experience of the runway, while the other (Group 5) consisted of

TABLE 2

Group and Treatment	Mean Score of Responding Birds: Duration of Following in the 2nd, 4th and 5th 6-Min. Period (Seconds)		
	5–10 Mins.	15–20 Mins.	25–30 Mins.
1. Socialization—isolation	30	33	72
2. Imprinting—isolation	88	126	111
3. Total isolation	5	27	33
4. Imprinting—socialization	0	9	45
5. Total socialization	0	28*	2*
	Number of Birds Per Group Which had made a Full Response by the 2nd, 4th and 6th Period		Number of Birds in Group
1. Socialization—isolation	6	7	7	7
2. Imprinting—isolation	6	6	6	6
3. Total isolation	1	5	7	9
4. Imprinting—socialization	0	1	2	6
5. Total socialization	0	0	0	6

* Following, but no full response.

inexperienced birds. Limited experience of the experimental situation at an earlier age can therefore have had little effect on the results compared with the effect of social experience and imprinting.

Since chicks of Groups 1 and 2—previously exposed to other chicks or to Model M—responded to Model B much better than chicks with no such treatment (Group 3), it is evident that the following-response was facilitated by the previous experience. This effect, however, was only evinced in those chicks where the social contact was restricted to the second day of life, for Groups 4 and 5—in both of which the chicks remained together until they were tested—had lower following scores than the totally isolated birds (Group 3).

Emotional Behavior

Four models of behavior are included in this category: avoidance of the model; crouching; trying to escape from the runway; and distress calls. As a rule, when avoidance of the model was strong the bird was rather silent; when crouching it was always silent except for a few seconds before standing up, when it started to give distress calls. Loud and persistent distress calls chiefly occurred as the chick walked about the runway, apparently searching for something; or they occurred in chicks trying to get out of the runway, though such birds were occasionally silent.

TABLE 3. *Incidence of Strong Avoidance Behavior During Test Sessions*

Group	Number of Chicks Showing Strong Avoidance	Number of Chicks in Group
1. Socialization–isolation	2	7
2. Imprinting–isolation	1	6
3. Total isolation	6	9
4. Imprinting–socialization	2	6
5. Total socialization	1	6

Avoidance of the moving model was most marked in the totally isolated birds (Group 3), (Table 3). In spite of this, 7 out of the 9 chicks in this group eventually followed, whereas, of the

totally socialized chicks (Group 5) none gave a full following-response although showing less fear of the model. The essential difference between these two groups was this: in Group 3 there was avoidance of the model at the start of the test, but at some moment before the end of it, fear rapidly fell and the bird suddenly switched over to following; in Group 5, however, where avoidance (or crouching) ceased completely in 5 out of 6 chicks, this was succeeded not by following but by "searching" with distress calls, the model being almost altogether ignored. Group 4 chicks (imprinting-socialization) displayed much the same pattern of behavior as Group 5, though here, 2 out of the 5 chicks which had ceased to avoid the model before the end of the test, then started to follow it.

Although the chicks of Groups 1 (socialization-isolation) and 2 (imprinting-isolation), like those of the three other groups, either crouched or avoided the model at the beginning of the test, three of them very strongly, all but one of them had switched to following before the end of 10 min., and thereafter showed few or no signs of being emotional; and the last one had switched by the fifteenth minute.

Conclusions

A comparison of the behavior of the totally isolated chicks with that of the chicks reared socially until the test, confirms that, as previously indicated (Guiton, 1958), socialization leads to an earlier loss of the following-response. This effect does not appear to be the consequence of increased fear of the model in socialized chicks, although non-following in the isolated birds may have been primarily due to fear of the model.

However, since better following was obtained in the socialization-isolation group than in either of the above groups, it would seem that previous social experience is necessary if the response is to persist, and that subsequent isolation causes the response to be generalized to strange objects. A comparison of the test performance of Group 2 chicks (imprinting-isolation) and Group 4 chicks (imprinting-socialization) supports this interpretation, since the response established during training is only generalized to a strange model if the first experience is not interfered with by subsequent socialization. The superiority of Model M-trained chicks over socialized-isolated ones (Group 1) with respect to the response to Model B, may well be due to the much greater similarity of Model M to Model B than of a chick to Model B; or to the circumstance that

TABLE 4. *The Mean Time (in Seconds) Taken to Respond to Each Model*

Group	Model M	Model B	Difference
M-trained chicks	154	259	105
B-trained chicks	223	143	80

social contact involved much more actual following in the one case—chicks trained with model M —than in the other.

If these interpretations are correct then it is probable that continuously socialized chicks become sufficiently imprinted or conditioned to one another to be able to discriminate between a chick and a strange moving object such as a box, and that in the test situation it is the absence of its companions rather than fear of the strange moving object that inhibits the socialized chick from following that object.

Experiment 2
Discrimination Between Models

The response of M-trained chicks to Model B in Experiment 1, could conceivably mean that the birds were inherently incapable of discriminating between the two models and not, as suggested, that the response was generalized because the birds were isolated after their training with Model M. An experiment was therefore carried out to test the chick's ability to discriminate between the two models presented simultaneously after a previous training session with one of the models only.

The chicks were divided into two groups, the birds of one receiving training with Model M (Group M), and those of the other with Model B (Group B), both at the age of 20 to 54 hours. The training sessions and the method of scoring were the same as those in Experiment 1. In the test sessions, carried out 72 hours later, the chicks were transferred from their isolation cages to the runway where they were placed centrally, midway between Models M and B which were situated diametrically opposite to one another. As soon as the chick was in place and released from its box, the two models were set in motion in the same direction round the runway. The time taken by each chick to start following each model, and the number of seconds which it spent following each model during the second, fourth and sixth 5-min. period, were both scored.

A preliminary experiment indicated that the effect of the previous training was partly counteracted by an inherent preference for Model M. Physical contact between chick and model appears to have a strong stimulating effect on following, and inequalities in this respect were thought to account for the above difference. In the definitive experiment, therefore, the legs of the models were removed, and the models themselves were sufficiently raised off the ground to give the chick complete head-clearance.

RESULTS. The first following-response was correct—that is, to the training model—in 11 out of the 12 chicks. The exception was a B-trained chick (Table 5). In spite of this high initial degree of discrimination, all but one chick—an M-trained bird—subsequently responded to the other model as well (Table 4).

The relative amounts (group means) by which Model M was followed during the test by Group M and Group B chicks, is given in Table 5. The statistically significant difference between the scores of the two groups ($t = 2.6$, arc sin transformation; $P < 0.05$ for 10 degrees of freedom) shows that the nature of the training continued

TABLE 5. *Number of Chicks in Each Group Making a Correct First Choice Between Models M and B; and the Mean Relative Amount by Which Model M was Followed, in Each Group*

Group	Number of Birds Making Correct 1st Choice	Number of Birds in Group	Mean Following Score for Model M, as Percentage of Total Following Score for Both Models
M-trained	6	6	88
B-trained	5	6	50

to exert an influence. The fact that the mean score of Group B chicks was not significantly different from 50 per cent, and only 1 out of the 6 birds in this group showed a pronounced preference for Model B, suggests however, that an inherent preference for Model M still existed after physical contact between the chicks and the models had been eliminated.

This preference might be explained by supposing that Model B was a less stimulating object than Model M. However, although the mean training score for Group M was greater than that of the other group, the difference was not significant at the 5 per cent level of confidence (t = 2.12, for 10 degrees of freedom).

Whereas the mean following score of M-trained birds tested with Model B alone (Experiment 1, Group 2) was 325 sec., the similarly trained chicks in this experiment followed the same model for an average of only 38 sec. out of a total average amount of following of 261 sec. for both models. In Experiment 1 the high level of response to the strange model was, therefore, quite clearly due to the absence of the model used for imprinting.

Experiment 3
Group Behavior in Socialized Chicks

The effect of socialization on chicks is similar in many respects to that of imprinting (Experiment 1). If it is assumed that with continuous socialization chicks become sufficiently imprinted or conditioned to one another to be able to discriminate between a fellow chick and a strange box, then their emotional behavior in the test situation: "searching" and giving distress calls (Experiment 1, Group 5), can be supposed to be the immediate consequence of their isolation from other chicks. That this is so, is well shown by the following observations.

1. Six chicks, socialized from the age of 20 to 54 hours onwards, were placed at the age of 96 to 126 hours around the perimeter of the runway about 50 cm apart, and observed for 30 min. For the first 6 min. all chicks crouched silently. In the seventh minute one chick started to give distress calls, softly and hesitantly at first, then boldly and loudly as it began to walk about; by the end of the eighth minute the other five chicks, one by one, followed suit. Soon after this the chicks began to join up: the first two coming together after 7 min. 45 sec., and a third one joining them a few seconds later. Distress calls ceased to be given by a chick within 5 sec. of joining up. By the beginning of the thirteenth

minute all six chicks had come together, and the distress calls had entirely ceased and been replaced by contentment calls.

From the thirteenth to the twenty-fifth minutes there were no more distress calls. During this time the chicks remained close together, scratching and pecking at the ground, and giving a continuous "twittering" of contentment calls. No chick ever ventured more than about 20 cm away from its companions.

After 25 min. they were forcibly scattered by waving a piece of hardboard overhead, and this at once produced a great volume of distress-calls which, however, died down again as they joined up, within about 30 sec. of being scattered.

2. (a) Six other chicks of the same age were transferred from their social brooder to the runway, in a group. They immediately all crouched silently, but at the end of 40 sec. one chick got up and gave distress-calls until the seventieth second. After 1 min. 35 sec. all six chicks had risen and begun pecking on the floor and giving contentment calls. Occasionally distress calls were heard after this for 2 min., but by the end of the fifth minute these had entirely ceased. The contentment calls, pecking and scratching continued with little interruption, the group of chicks moving very little and very slowly around the runway.

(b) After 20 min. six other (ringed) chicks were placed, one at a time, and in turn, in the runway about 140 cm away from this group of chicks. The behavior of each ringed chick was observed for 5 min. It was then removed and the next chick put in its place, and so on.

Each ringed chick behaved in the following manner: first, it crouched silently for some seconds; then it started giving distress-calls which increased in volume as it stood up and began to walk. Then, after a further interval of seconds the bird switched over to giving contentment calls and ran towards the group of six chicks; hesitated when about 30 cm away from it, giving more distress calls, and finally joined the group. With but one exception, the distress-calls then ceased. The one exception continued its calls, but intermittently, to the end of the 5 min. The times taken by each chick to complete each stage are given in Table 6.

(c) After the sixth ringed chick had been tested and removed, the original six making up the group were left in the runway, and the two models M and B, legless as in Experiment 2, were started up on their circular run. With these models continuously moving for the next 45 min., observations were made on the behavior of the chicks during the first and last 15 min. of the period.

TABLE 6

Chick	Time Taken to Start Distress Calls	Time Taken to Start Moving Towards Other Chicks	Time Taken to Join Up With Group
1	10 seconds	30 seconds	75 seconds
2	15 ,,	58 ,,	97 ,,
3	11 ,,	40 ,,	65 ,,
4	14 ,,	44 ,,	55 ,,
5	12 ,,	64 ,,	71 ,,
6*	55 ,,	100 ,,	115 ,,

As far as could be observed all the chicks behaved alike at any one moment. The group was on the periphery of the runway within about 15 cm from the track of the models for the whole period. During the first 3 min. the passage of one or the other model gave rise to an increase of movement and excited "twittering" as each chick appeared to be trying to move centrally into the group; but it gave rise to no crouching, scattering or distress calls. By 4 min. excitement appeared to have subsided, but some of the chicks gave occasional distress-calls. Although these calls then became very infrequent, the group continued to show slight signs of being disturbed: there was less pecking and scratching, and the chicks were quieter than before the models started moving. At the end of 15 min. the chicks were scattered by waving a piece of hardboard overhead. This provoked an outburst of distress-calls which ceased as soon as all chicks had grouped again: that is, after about 20 sec. During the time they were scattered however, one model passed by, and those chicks which were very close to it, or across its track, took slight evasive action: they "ducked" and ran a few paces away, running from the model rather than towards the other chicks. In the second 15-min. period of observation, the general behavior of the chicks had not noticeably changed. At no time, in either period, did any of the birds show strong fear of the model, or, on the other hand, signs of following it. Finally, at the end of 45 min., the birds were scattered a second time, with exactly the same results as before.

These observations, like those of Collias (1952) all show that distress in the experimental situation is the outcome, in the main, of the isolation of the chick from its companions. Once they have come together the chicks show none of the behavior patterns which have been called "emotional," and which are characteristic of socialized chicks tested individually, in which they persist for the whole period of the test. And since the individuals of a group stick very closely together and rapidly reform when scattered, it is difficult to suppose the social response in these socialized chicks of being "satiated." Furthermore, exposure of a group of chicks to a strange model, while producing only a very few signs of distress or fear, nevertheless does not stimulate the chicks to follow that model, the tendency being, presumably completely opposed by the attraction which the chicks have for one another. Finally, it may be noted that a single chick is more quickly attracted to a group of several other chicks than to a single one: compare the speed with which chicks responded in tests 1 and 2b. The greater volume of "twitterings," the greater mass and greater amount of movement of a group presumably provides stronger stimulation than does a single chick.

DISCUSSION

It is evident that communally reared chicks behave as if they were imprinted on one another. Whether the learning involved is essentially similar to imprinting or not, the results, at any rate, would appear to be closely comparable. One consequence of this process is that the post-hatching period of responsiveness to a strange moving object is shortened, so that the "critical period" itself must be reduced. Although strange objects may then be avoided rather than followed, this is a secondary consequence of the effect of the socialization and not the principle or primary factor limiting responsiveness. By inference one may then suppose that imprinting on the parent bird would also influence the duration of the critical period. Fear would seem to exert a primary effect only on those birds with no prior social experience and, as has been shown, only when they are older.

The stages in the argument are as follows:

1. Completely socialized chicks cease to follow an unfamiliar moving object at an earlier age than do totally isolated ones.

2. The response of the isolated birds does, however, decline with age, later becoming very difficult to evoke as a result, probably, of increasing fear. In contrast to this, the following-response reappears in socialized chicks which have been isolated for several days, and it is much more readily elicited in these birds than in the totally inexperienced birds, and depends therefore, on previous social experience.

3. This effect seems to be due to the stimulus-generalization of a response acquired during the earlier period of socialization, a conclusion supported by the fact that a similar generalization occurs in chicks initially exposed to one model, then isolated, and later tested with a second and unfamiliar model.

4. Whereas non-following in totally isolated chicks seems to be mainly associated with the development of strong fear of the moving object, that of socialized chicks is not. Although the latter may avoid the model, sometimes persistently, this behavior is generally less strong than in the former. On the other hand, these socialized birds are characterized by more persistent searching, distress-calls and other forms of emotional behavior which are the direct consequence of isolation from their companions. There is no indication that the following-response is reduced in strength by socialization, for the attraction the chicks have for one another remains very strong. It is relevant to note here that the following-response of young imprinted coots is virtually inexhaustible except by physical fatigue (Hinde, Thorpe and Vince, 1956).

The effects of socialization are not, therefore, easily explained by the drive-satiation hypothesis. On the other hand it seems that the following-response ceases to be elicited by a strange moving object as soon as the chick has already learnt to follow one object—which may be another chick—sufficiently well to be able to discriminate between this and other objects. This effective stage of socialization is already reached by the age of 72 hours (Guiton, 1958). At this age, however, fear is not yet sufficiently well developed to prevent a totally inexperienced chick from following a strange object. Fear only begins to interfere seriously with the response 24 hours later. The inhibitory effects of socialization on the following-response to unfamiliar objects thus take place

considerably earlier, so that the development of fear does not seem to be the primary factor which determines the age at which the response is lost except in completely isolated birds.

Previous studies have shown that imprinting of ducklings to one object later restricts the response to that object (Fabricius, 1951; Ramsay and Hess, 1954); and that ducklings become imprinted to the ducklings they are reared with (Fabricius, 1951; Collias and Collias, 1956). These workers do not, however, appear to have found any evidence that the age at which the following-response to a strange moving object is lost is affected by previous imprinting or socialization. The termination of this period of responsiveness has been attributed either to increasing fear of unfamiliar objects and situations (Hinde, Thorpe and Vince, 1956) or to emotionality unrelated to previous imprinting (Hess, 1957), or to "an endogenously conditioned decrease of the internal motivation of the following reaction" (Fabricius, quoted from Hinde, Thorpe and Vince, 1956).

The results, especially of experiment 1, raise a number of questions which are at present being studied. In particular, it would be interesting to know:

1. To what extent following under the effect of stimulus-generalization can lead to further imprinting; and

2. How far the critical period can be extended by reducing fear in isolated birds; or by increasing the inherent stimulating value of the model.

The answer to these questions will probably reveal to what extent and under what conditions imprinting is irreversible as suggested by Lorenz (1937).

SUMMARY

1. In a previous study of Brown Leghorn chicks it was shown that the following-response to a strange moving object was lost at an earlier age in socially-reared chicks than in isolated ones, but that the response reappeared in the former if they were subsequently isolated for several days. This could mean that the loss of the response was due to a short-term, reversible process—the drive satiation hypothesis; or it could mean that socially-reared chicks learned to respond to one another, as opposed to following other moving objects, but that this response was then generalized under conditions of isolation. Experiments, described in this paper, were designed to examine these two hypotheses.

2. The effect of previous experience on the following-response of 4–5-day-old chicks was studied. The chicks were tested singly. Those with no previous visual or physical experience of moving objects or of other chicks followed a moving object (Model B) better than those reared socially from their second day onwards. On the other hand, if chicks were allowed only 4½ hours of social contact with one another on their second day and then isolated until tested, their response to Model B was stronger than that of either of the two above groups. Restricted experience of other chicks thus potentiated the later response to a strange moving object. In chicks trained on the second day to follow one object (Model M) for 30 min., the response was generalized to a different object (Model B) 3 days later but only if kept in isolation from one another for these intervening days.

3. Chicks trained to Model M on the second day, isolated for 3 days and then tested with Model M and Model B simultaneously, responded well to Model M but very little to Model B. It is clear therefore, that the models could be discriminated, and that the strange model was only followed in the absence of the one on which the chicks had been imprinted.

4. Non-following by chicks in the first experiment (see 2 above) was associated in most instances with avoidance of the model, or with "searching" accompanied by distress calls, or with attempts to escape from the test-situation. It is noticeable that although initial avoidance of the model was stronger in isolated chicks than it was in socialized ones, the following-response scores of the former were higher than those of the latter. The socialized chicks were, on the other hand, characterized by more persistent "searching" and "distress." This suggests that the absence of following in socialized chicks is not primarily due to fear of the moving object.

5. Observation of 4–5-day-old chicks in the experimental situation, in groups of six instead of singly, showed that distress and fear only occur because a chick is suddenly isolated from its fellows. The grouped chicks showed no signs of "emotionality"; but nevertheless showed no tendency to follow a model. Their social responses to one another, however, were very strong: as soon as a chick was forcibly separated from the other chicks it started making distress calls until it saw the others and ran towards them. The effect of socialization is not therefore, to reduce the strength of the following-response but to increase its selectivity.

6. The results supported the hypothesis that chicks become imprinted or otherwise conditioned to one another, and as a result are henceforth inhibited from following unfamiliar objects, though this may again become possible after a subsequent period of isolation. Since the initial, unselective response is lost in socialized chicks before fear is sufficiently well developed to interfere with it, it is suggested that, under normal conditions, it is this effect of socialization which determines, primarily, the age at which the chicks cease to follow a strange object, and the age at which, therefore, imprinting ceases to be possible.

REFERENCES

COLLIAS, N. E. and COLIAS, E. C. (1956). Some mechanisms of family integration in ducks. *Auk.*, **73**, 378–400.

FABRICIUS, E. (1951). Zur Ethologie junger Anatiden. *Acta. Zool. Fenn.*, **68**, 1–178.

GUITON, P. (1958). The effect of isolation on the following response of Brown Leghorn chicks. *Proc. roy. phys.*, *Soc. Edinb.*, **27**, 9–14.

HESS, E. (1957). Effects of meprobamate on imprinting in waterfowl. *Ann. N.Y. Acad. Sci.*, **67**, 724–732.

HINDE, R. A., THORPE, W. H. and VINCE, M. A. (1956). The following-response of young Coots and Moorhens. *Behaviour*, **11**, 214–242.

JAYNES, J. (1956). Imprinting: the interaction of learned and innate behaviour: I. Development and generalization. *J. comp. physiol. Psychol.*, **49**, 201–206.

JAYNES, J. (1957). Imprinting: the interaction of learned and innate behaviour: II. The critical period. *J. comp. physiol. Psychol.*, **50**, 6–10.

LORENZ, K. (1937). The companion in the bird's world. *Auk.*, **54**, 245–273.

RAMSAY, R. O. and HESS, E. (1954). A laboratory approach to the study of imprinting. *Wilson Bull*, **66**, 196–206.

VERPLANK, W. S. (1955). An hypothesis on imprinting. *Brit. J. anim. Behav.*, **3**, 123.

WEIDMANN, U. (1956). Some experiments on the following and flocking reaction of Mallard ducklings. *Brit. J. anim. Behav.*, **4**, 78–79.

CHAPTER FOUR

AFFILIATION

No doubt, the most significant aspect of the animal's environment are other animals around it. The animal seems to depend upon them in a variety of ways and for a variety of ends, and restriction of social contact may have disastrous effects on the animal's adaptive behavior. If contact with others is indeed so important, we would expect that it constitutes a source of significant incentives for the animal. It should be relatively easy, therefore, to place the animal's behavior under control of the incentives deriving from social contact. Such data are indeed found. Butler (1954), for instance, investigated exploratory behavior of monkeys that was under the control of social and nonsocial incentives. Monkeys were kept in restraining cages equipped with doors which, when opened, exposed to the subject either another monkey, a toy train, food, or empty space. Butler's monkeys opened their doors considerably more often to look at another monkey than to look at the other stimuli.

Since social contact is a source of significant incentives for the animal, we may conjecture that deprivation of social contact acts as a state of need or drive, and that the animal will engage in instrumental behavior leading to the removal of the state of deprivation. As a matter of fact, affiliation has been thought of as a need whose goal state is social contact. Shipley and Veroff (1952) and Atkinson, Heyns and Veroff (1954) working with human subjects, have found negative correlations between popularity and scores of need for affiliation taken from the TAT. These results must be interpreted with caution, however, because the lack of popularity may not entail the lack of social contact, and we may have here not an instance of social deprivation but of social *rejection*.

If we think of need for affiliation in the same terms as, let us say, we think of hunger, we will be prone to expect that social deprivation has similar consequences to food deprivation, and that when we deprive the animal of contact with others, like depriving it of food, its behavior will become oriented toward securing contact with others. The socially deprived animal will be more likely to seek contact with others than an animal whose social interaction has not been restricted (e.g. Gewirtz & Baer, 1958). The data on the effects of social restriction on affiliative behavior, however, are in almost complete contradiction to such an expectation. On the whole, socially restricted animals tend to approach others less than animals allowed normal social interaction. Affiliative attempts of restricted rhesus monkeys towards others of the same species are considerably depressed, as we note from Mason's experiments in this chapter. In fact, these animals show relatively little social interest. The Pratt–Sackett research confirms this result, and Ashida's experiment shows the same effect for rats. Pratt and Sackett also demonstrate that social choice of these monkeys is influenced to some extent by familiarity; that is, animals favor interaction with another that was reared under the same conditions than with one reared under different conditions.

But, as is the case in all areas of research, the data on affiliative behavior are not completely consistent. In their two experiments, Shelley and Hoyenga find that rats deprived of social contact at weaning and kept in restriction for a number of weeks show stronger interest in a caged rat than rats reared and kept under conditions of fairly free social interaction. The restricted rats maintain this interest for a number of consecutive days, while socially reared animals' interest quickly disappears. While the Shelley–Hoyenga data may be interpreted as giving support to the contention that affiliative behavior responds primarily to social deprivation, these authors seem rather reluctant to explain their results in those terms.

In research on affiliation with animals, an orienting response directed toward another animal is sometimes regarded as an indicator of affiliative tendencies. But there is a danger in equating the two. In the Shelley–Hoyenga experiment, for example, the caged rat was in fact both a social object of which the restricted animals were deprived, *and a novel stimulus*. It could not be unequivocally determined therefore, if the restricted rats attended to the caged animal because of its novelty or because of the arousal of their own social motives.

REFERENCES

ATKINSON, J. W., HEYNS, R. W. and VEROFF, J. The effect of experimental arousal of the affiliation motive on thematic apperception. *Journal of Abnormal and Social Psychology*, 1954, **49**, 405–410.

BUTLER, R. A. Incentive conditions which influence visual exploration. *Journal of Experimental Psychology*, 1954, **48**, 19–23.

GEWIRTZ, J. L. and BAER, D. M. Deprivation and satiation of social reinforcers as drive conditions. *Journal of Abnormal and Social Psychology*, 1958, **57**, 165–172.

SHIPLEY, T. E., and VEROFF, J. A projective measure of need affiliation. *J exp. Psychol.*, 1952, **43**, 349-356.

THE EFFECTS OF SOCIAL RESTRICTION ON THE BEHAVIOR OF RHESUS MONKEYS: II. TESTS OF GREGARIOUSNESS*

William A. Mason

DIFFERENCES in the spontaneous social behavior of feral monkeys and laboratory-reared monkeys whose social experience was restricted from infancy were described in a previous communication (Mason, 1960). Monkeys reared in the laboratory showed deficiencies in the organization of the sexual act, briefer and less frequent grooming episodes, and more frequent and prolonged fighting.

In view of these differences one might expect socially restricted monkeys to be less gregarious, i.e., less highly motivated to engage in social interaction, and to be less attractive as social incentives.

The present research was designed to test these possibilities by investigating the performance of feral and laboratory-reared rhesus monkeys in several situations presenting a choice between social and nonsocial alternatives.

EXPERIMENT 1
INTRAGROUP COMPARISONS

This experiment was designed to assess the nature and strength of gregarious motivation in feral and socially restricted monkeys when tested with individuals of similar social history.

Method

SUBJECTS. The two groups of six rhesus monkeys are described in detail in a previous paper (Mason, 1960). The Restricted Ss, three males and three females, were born in the laboratory and separated from their mothers at birth or before the end of the first month of life. They were housed from infancy in individual

cages and were 3 years old at the start of the present experiments. The second group of three males and three females, the Feral group, was the same age as the Restricted group as estimated from body weight and dentition records. They arrived at the laboratory at an estimated age of 20 months and had been housed in individual cages for at least 5 months when the present research began. Prior to the present experiment each of the 15 pairs obtainable from each group had been observed in 16 test sessions, each 3 min. long, to provide information on the form and frequency of free social interactions (Mason, 1960).

APPARATUS. The apparatus consisted of detention cages, 18 in. high, 12 in. wide, and 12 in. deep in which the incentive animals were confined. These cages were constructed of transparent plastic framed with angle iron. The door of each cage was counterweighed to raise automatically when an 8-in. chain at the front of the cage was pulled. It was possible for E to limit S to just one opening when two cages were present. The detention cages were placed within a 6-ft. by 6-ft. by 6-ft. test chamber equipped with one-way-vision panels and a sliding door through which S entered and left the test chamber (Mason, 1960).

PROCEDURE. Prior to social testing, each S was individually trained until it was thoroughly proficient at opening the doors of the detention cages to secure food. To insure that the cage-opening response was maintained throughout the experimental series, food tests preceded and followed each social test session. All trials were 2 min. long. A trial was started by raising the sliding door and introducing S into the test chamber from an adjacent carrying cage. In

* From the *Journal of Comparative and Physiological Psychology*, 1961, **54**, 287–290. Copyright 1961 by the American Psychological Association. This research was conducted at the University of Wisconsin. Support was provided through funds received from the Graduate School of the University of Wisconsin, Grant G-6194 from the National Science Foundation, and Grant M-722 from the National Institutes of Health.

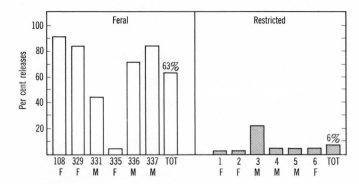

FIGURE 1. Releases in dual-cage tests. *M*, *F* designate male and female *S*s.

addition to data on cage-opening, all aggressive interactions following a social choice were recorded. The experiment consisted of three phases:

Dual-Cage Tests. Two detention cages were present at opposite corners of the chamber, each containing one member of *S*'s group, and *S* was allowed to release one of these incentive animals during the 2-min. trial period. The testing schedule was arranged so that each of the 15 pairs obtainable from each group was presented as an incentive pair once in a 4-day period (3 or 4 pairs a day), with the restriction that the same animal did not serve in two successive pairs on the same day. All *S*s were tested once with a given incentive pair before the next pair was introduced. The monkeys were tested in balanced order for 20 days, providing each *S* with 5 trials with each of 10 possible incentive pairs. Inasmuch as every incentive animal was presented to the *S* in 4 different incentive pairs, each *S* was given a total of 20 opportunities to release each of the other monkeys in its group.

Single-Cage Tests. Observations during the dual-cage tests suggested that the presence of an incentive animal which was feared by the chooser tended to inhibit releases of the second incentive animal. In the present phase, therefore, only a single cage was used. The monkeys were tested for five days, in balanced order. Each *S* received one trial daily with each of the other animals in its

group serving as an incentive, making a total of five trials with each incentive *S*.

Inside-Release Tests. The *S*'s behavior in the detention cage during the preceding tests indicated that confinement in the cage was aversive. The detention cage was modified, therefore, so that the release chain was inside the cage and the monkey confined within could release itself. This was permitted, however, only when another monkey was in the test chamber. Thus, the present test situation placed in opposition the tendencies to avoid social contact and to escape from confinement. The monkeys were tested for 20 days in balanced order. Each *S* was given one trial a day with each of the other members of its group present in the test chamber, making a total of 20 trials for each *S* with each incentive animal.

Results

DUAL-CAGE TESTS. The most important outcome of the dual-cage tests was the extremely low level of releases in the Restricted group. Total releases for this group, shown in Figure 1, were 6 per cent compared with 63 per cent for the Feral group. Data on individual performance, also presented in Figure 1, show that all but one monkey in the Feral group made more releases than the highest-scoring Restricted *S*. The difference between groups is significant at the 0.02

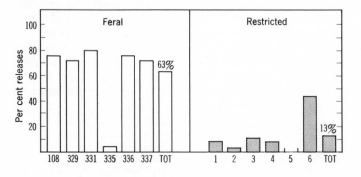

FIGURE 2. Releases in single-cage tests.

level. Unless otherwise indicated, in this and all subsequent comparisons between groups, statistical evaluations were based on the Mann-Whitney test (Siegel, 1956). It is noteworthy that four monkeys in the Feral group showed strong and consistent preferences for one or more members of their group. There was no evidence of preferential responses among Restricted Ss.

SINGLE-CAGE TESTS. Feral Ss again made substantially more releases than the Restricted Ss, and the difference between groups was significant at the 0.04 level. Individual performance is shown in Figure 2. There was no evidence of differential preferences or aversions for specific individuals in the Restricted group. In the Feral group, however, S 329 was generally nonpreferred, being released only once, and S 335 was generally unresponsive, making only 1 release in 100 trials.

INSIDE RELEASE TESTS. As shown in Figure 3, the percentage of releases in both groups increased markedly in this test situation, but the Restricted Ss released themselves less frequently than did the Feral Ss ($p = 0.01$). With one exception, all trials resulted in releases for the Feral group. Every member of the Restricted group, however, failed to release itself on at least 12 trials. In contrast with the results of the dual- and single-cage tests, the present experiment revealed consistent differential responses to individual incentive animals by Restricted Ss.

AGGRESSION. Even though releases were fewer for the Restricted group in each of the experiments described in this series, thus reducing the opportunity for fighting to occur, the total incidence of aggression was substantially higher in this group. The mean total frequency of aggression was 22.2 and 5.5 for Restricted and Feral groups, respectively ($p = 0.03$). Rank-order correlation coefficients were computed, based on the total frequency of aggression and total number of

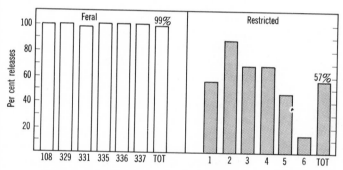

FIGURE 3. Performance on inside-release tests.

cage openings for each pair. The values obtained were -0.09 and -0.58, for Feral and Restricted pairs, respectively. Only the latter coefficient was statistically significant ($p = 0.05$).

EXPERIMENT 2

RESPONSES OF FERAL AND RESTRICTED MALES TO SOCIALLY EXPERIENCED FEMALES

The purpose of this experiment was to compare the reactions of Feral and Restricted males to socially experienced female incentive animals.

Method

SUBJECTS. The Ss were the three Feral and three Restricted males described in Experiment 1. Three field-reared adolescent females served as incentive animals. The females had been paired equally often with each of the males in a previous experiment (Mason, 1960), and had had no contact with any of these males prior to that experiment.

APPARATUS AND PROCEDURE. A single detention cage with the release chain outside the cage, was used in the test chamber previously described. The females were used as incentive animals only. The males were tested in balanced order for 15 days and each day received one 2-min. trial in the morning and in the afternoon with a different female, for a total of ten trials with each incentive S.

Results

The total percentage of releases for Restricted and Feral males was 87 and 68, respectively. In each group, one S released a female on every trial, one S failed to release on one trial only, and one S failed to release on more than one trial. These results suggest that the Experiment 1 differences between Restricted and Feral *S*s were related to the specific incentive animals used and to the *S*s' previous experiences with them.

EXPERIMENT 3

RESPONSES OF SOCIALLY EXPERIENCED FEMALES TO FERAL AND RESTRICTED MALES

The purpose of this experiment was to determine the reactions of socially experienced females to Feral and Restricted males presented as social incentives.

Method

SUBJECTS. The *S*s were described in experiment 2. The males served as incentive animals and the females, as choosers.

APPARATUS AND PROCEDURE. Testing was conducted in the situation previously described. The experiment was divided into two phases:

Dual-Cage Tests. Females were given one 2-min. trial a day for 10 days with each of the nine possible Feral-Restricted male pairs presented as incentive animals in the dual-cage situation previously described. Only one release was permitted on a trial. Information was obtained identifying the animal released and noting the amount of time prior to release which the female spent within 6 in. of the detention cages.

Inside-Release Tests. The inside-release cage, previously described, was used. The females were placed inside the cage and each female was given ten 2-min. trials with each male in the test chamber. The males were presented in balanced order, and each female received two trials in the morning and afternoon, each trial with a different incentive animal.

Results

DUAL-CAGE TESTS. The females released on 67 trials (25 per cent), the percentages for individual *S*s ranging from 9 to 55. The feral male was selected on all trials in which a release occurred. The difference between incentive groups is significant at the 0.01 level as determined by sign tests. Duration of proximity measures on those trials when a release did not occur also consistently favored the Feral *S*s. Mean duration of proximity to the detention cage was 4.06 sec. and 32.06 sec. with Restricted and Feral males, respectively ($p = 0.01$, sign test). Despite this evidence of overwhelming preference for Feral males, the females were observed early in testing to approach and present the elevated hindquarters

(Carpenter, 1942a) to the Restricted males, frequently before releasing a Feral monkey. These presentations before a release was made were recorded, and the percentage of trials in which presentation occurred to the caged Feral and Restricted males was 11 and 94, respectively ($p = 0.01$, sign test).

INSIDE-RELEASE TESTS. The frequency of openings in the inside-release tests was 97 per cent and 99 per cent with Restricted and Feral males, respectively, providing no evidence of differential responsiveness under these test conditions.

DISCUSSION

These experiments indicate that monkeys whose social experience was restricted from infancy were not highly motivated to interact with individuals of the same social history with which they had been previously paired in brief test periods. The low level of social choices by Restricted *S*s was probably influenced by the high incidence of fighting observed in this group. This interpretation receives support from the significant negative correlation found for Restricted pairs between number of releases and frequency of aggression. The present results are in agreement with Nowlis' (1941) finding with chimpanzees that aggressive behavior is more frequent with nonpreferred partners.

It cannot be concluded that the performance of the Restricted group in Experiment 1 represents a generalized and persistent reduction in gregariousness inasmuch as Restricted males when subsequently tested with socially experienced female incentive animals showed a sharp increase in the frequency of social choices. It is noteworthy, however, that the incidence of fighting was low in this test series and in the preceding investigation of spontaneous social behavior in which the same females were used (Mason, 1960).

The attraction of Restricted males to socially experienced females was clearly not mutual, as evidenced by more frequent female withdrawals from Restricted males during previous tests (Mason, 1960), by the failure of these females to release Restricted males in the research described here, and by the higher frequency of presentation which they directed toward Restricted males in the dual-cage tests. Although presentation is an essential component of the female mating pattern, several workers (Carpenter, 1942a; Chance, 1956; Hamilton, 1914; Maslow, 1936) have indicated that this response is elicited in rhesus monkeys of both sexes by an actual or potential aggressor.

The results are generally consistent with the

thesis that orderly and harmonious intraspecies social relations in rhesus monkeys are highly dependent upon previous socialization experience. Among the specific factors which are responsible for orderly social interactions, species-specific gestures appear to be of particular importance. On the basis of his field observations, Carpenter (1942b, p. 202) states that every species of non-human primate which he has observed "has its own repertory of condensed, stereotyped patterns of overt responses or gestures which characteristically occur in definable situations and which stimulate a definable range of responses in associates." The behavior of Restricted monkeys suggests that the effective development of these elementary forms of social coordination and communication is dependent upon learning. It is probable that the absence of these social skills contributed to the turbulent relations between socially restricted monkeys and to the apparent aversion which socially experienced females displayed toward interaction with Restricted males.

SUMMARY

1. The gregarious behavior of monkeys raised in the laboratory (Restricted group) and monkeys captured in the field (Feral group) was investigated in several situations all presenting a choice between social and non-social alternatives.

2. Pairs of Restricted monkeys made fewer social choices and fought more frequently following a social choice than did pairs of Feral monkeys.

3. Restricted and Feral males were subsequently tested with the same socially experienced female incentive animals, and the number of social choices by Restricted males increased sharply as compared with their performance in the first experiment.

4. When given an opportunity to choose between Feral males and Restricted males, socially experienced females uniformly preferred the Feral Ss.

5. The results support the conclusion that orderly and harmonious intraspecies social relations in rhesus monkeys are dependent upon previous socialization experience. Socially restricted monkeys are apparently not preferred either by individuals of similar social history or by monkeys born in the field.

REFERENCES

CARPENTER, C. R. Sexual behavior of free ranging rhesus monkeys (*Macaca mulatta*): I. Specimens, procedures and behavioral characteristics of estrus. *J. comp. Psychol.*, 1942, **33**, 113–142. (a)

CARPENTER, C. R. Societies of monkeys and apes. *Biol. Symp.*, 1942, **8**, 177–204. (b)

CHANCE, M. R. A. Social structure of a colony of *Macaca mulatta*. *Brit. J. anim. Behav.*, 1956, **4**, 1–13.

HAMILTON, G. V. Sexual tendencies of monkeys and baboons. *J. anim. Behav.*, 1914, **4**, 295–318.

MASLOW, A. H. The role of dominance in the social and sexual behavior of infra-human primates: I. Observations at Vilas Park Zoo. *J. genet. Psychol.*, 1936, **48**, 261–277.

MASON, W. A. The effects of social restriction on the behavior of rhesus monkeys: I. Free social behavior. *J. comp. physiol. Psychol.*, 1960, **53**, 582–589.

NOWLIS, V. Companionship preference and dominance in the social interaction of young chimpanzees. *Comp. psychol. Monogr.*, 1941, **17**, 1–56.

SIEGEL, S. *Nonparametric statistics*. New York: McGraw-Hill, 1956.

SELECTION OF SOCIAL PARTNERS AS A FUNCTION OF PEER CONTACT DURING REARING*

Charles L. Pratt and Gene P. Sackett

THE early experiences of primates often have profound consequences on later behavior. In rhesus monkeys exploratory, maternal, sexual, and social behaviors appear extremely vulnerable

* From *Science*, 1967, **155**, 1133–1135. Copyright 1967 by the American Association for the Advancement of Science.

to early social and sensory restriction (1). Monkeys reared in isolation tend to withdraw from other animals and huddle by themselves. The fact that socially normal monkeys may avoid contact with monkeys reared in isolation further retards rehabilitation. We varied the amount of peer contact during rearing and investigated its effect on physical approach to a social partner, in order to determine whether monkeys reared under identical conditions prefer each other to monkeys reared under different conditions.

Three groups of rhesus monkeys were reared from birth in the laboratory without mothers. Each group contained four males and four females. Sets of three animals were matched across groups for age, sex, and test experiences after rearing was complete. The first group (A) was reared from birth to 9 months in individual closed cages. On the first 5 to 7 days they experienced physical, but minimal visual, contact with a human during feeding. No other physical or visual contact with humans or live monkeys occurred during rearing. Changing visual experiences throughout rearing were limited to presentation of pictures of monkeys engaged in various behaviors and pictures of people and inanimate objects (2). From months 9 through 18 the monkeys in group A were housed individually in bare wire cages from which they could see and

hear other isolates and humans, but physical contacts were unavailable.

Subjects in the second group (B) were reared individually in a large nursery room in bare wire cages from birth to 9 months. Other monkeys and humans could be seen and heard, but physical contact was not available. From month 9 through 18 the monkeys in group B were housed in the same room as the monkeys in group A; they were in wire cages where they could see and hear, but not touch, one another.

The third group (C) lived in wire cages in peer groups of varying sizes during the first 18 months of life. Rearing conditions and social behavior tests provided physical peer contact during this period. In summary, group A had no early contact with live peers, group B had visual and auditory but no physical contact with peers, and group C had complete peer contact during the rearing period.

When they were 18 months old, sets of monkeys from all groups interacted during social behavior tests in a large playroom (3). Each animal was tested weekly for 12 weeks in three 30-min. sessions. In one weekly session a constant set of one group A, one group B, and one group C monkey of the same sex interacted together; the same animals were always tested together. On the two other weekly sessions constant pairs of groups A and B, A and C, and B and C subjects interacted in groups of four monkeys. After social testing, each subject had received equal playroom exposure to one monkey from its own rearing condition and to two monkeys from each of the other rearing conditions. After playroom testing was completed, the monkeys were tested for their preference for other monkeys reared under the same conditions or for those reared under different conditions.

Testing was done in the "selection circus" (Fig. 1), which consists of a central start compartment that bounds the entrances to six adjoining choice compartments. Wire-mesh cages for the stimulus animals were attached to the outside of appropriate choice compartments. The front walls of the stimulus cages, the outside walls of the choice compartments, and the guillotine doors separating choice compartments from the start compartment were all made of clear Plexiglas.

For the testing, the subject was placed in the center start compartment with the Plexiglas guillotine doors down for a 5-min. exposure period. The subject could see and hear the stimulus animals, but could not enter the choice compartments near them. Unused choice compartments were blocked off by plywood walls

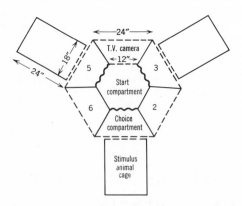

FIGURE 1. Scheme of the "circus" which is constructed of aluminium channels containing Plexiglas walls (dotted line), plywood walls (solid line), and Plexiglas guillotine doors (wavy line). Wire-mesh stimulus cages with a single Plexiglas wall are attached outside choice compartments. In testing, the subject is first placed in the start compartment. It can look into and through the choice compartments, but cannot enter them until the Plexiglas guillotine doors are raised by a vacuum lift. Plywood walls block physical and visual access to choice compartments that are not used in the experiment.

inserted in place of the Plexiglas guillotine doors. After the exposure period, a 10-min. choice trial was given. The Plexiglas guillotine doors were raised by a vacuum system; this procedure allowed the subject to enter and reenter choice compartments or to remain in the start compartment. The total time spent in each choice compartment during the test trial was recorded over a closed-circuit TV system.

The monkey's entry into different choice compartments served as our index of social preference. This measure of preference involves visual orientation, but, more importantly, it also involves locomotion toward a specific social object. It may be argued that a measure of viewing time, such as that used by Butler (4) in which monkeys inspected various objects through a small window, is not a proper index of social preference. Although actual physical contact was not available to our subjects, a great deal of nontactile social interaction was possible. Thus, our measure of preference based on physical approach toward a social object seems to be more analogous to an actual social situation than would be a simple viewing response.

Two types of trials were given. In the first, the stranger trial, one stimulus animal from each of the rearing groups was randomly positioned in a stimulus animal cage outside choice compartments 1, 3, or 5. These stimulus animals had received no previous social contact with the test subject but they were the same age and the same sex. A second test was identical with the stranger trial except that the three stimulus animals had received extensive social experience with the test subject during the playroom tests. Before the start of these tests, all 24 subjects had been adapted to the circus during nonsocial exploration tests. The order of serving first as a stimulus animal or as a test subject was randomized across groups.

Analysis of variance of the total time spent in the choice compartment had rearing condition as an uncorrelated variable, and type of stimulus animal and degree of familiarity as correlated variables. Familiarity did not have a significant main effect, and it did not interact with the other variables (all $P > 0.20$). Rearing condition had a significant effect ($P < 0.001$), which indicated that total choice time in all compartments differed as a function of early peer contact. Group A subjects spent half as much time (average = 220 sec.) in choice compartments as either group B (average = 422 sec.) or group C (average = 468 sec.) monkeys.

The interaction of rearing condition with type of stimulus animal was also significant ($P < 0.001$). Table 1 shows this effect, with choice times averaged over the trials with strange and familiar stimuli. These data show that like prefers like—each rearing condition produced maximum choice time for the type of stimulus animal reared under that condition. The data for individual subjects supports this averaged effect. In the group A, 2 of the 8 monkeys did not enter choice compartments. Of the 6 remaining monkeys, 5 spent more time in the group A choice compartment than in the other two compartments (two-tailed binomial, $P = 0.038$, with $p = \frac{1}{3}, q = \frac{2}{3}$). In the groups B and C all subjects entered choice compartments, and 7 out of 8 in each group spent more time with the animal reared like themselves than with the other animals (both $P = 0.0038$, two-tailed binomial).

TABLE 1. *Mean Number of Seconds Spent With Each Type of Stimulus Animal for Each Rearing Condition, Averaged Over the Two Test Trials.*

Rearing Condition of Experimental Animal	Rearing Condition of Stimulus Animal		
	A (Totally Deprived)	B (Partially Deprived)	C (Peer-Raised)
A (totally deprived)	156	35	29
B (partially deprived)	104	214	103
C (peer-raised)	94	114	260

The data indicate that social preferences are influenced by rearing conditions. In playroom testing the group C monkeys were the most active and socially advanced groups studied. Therefore, it was not surprising that they discriminated and showed large preferences for both strange and familiar group C animals. The group A monkeys, however, were highly retarded in their playroom behavior, and they did not show much progress over the 12 weeks of social interaction. As expected, these animals did exhibit a low degree of choice time in this study. We also thought that group A monkeys would be least likely to show preferences for a particular type of animal. It was, therefore, surprising to find that they did prefer each other to animals reared under other conditions. The group B animals, which were intermediate in social adequacy in playroom testing, also preferred each other. This result seems to strengthen the idea that animals of equal social capability, whether or not they are familiar with each other, can discriminate themselves from others, and not only discriminate but approach each other.

These results have important implications for studies designed to rehabilitate primates from the devastating effects of social isolation. The fact that socially abnormal monkeys prefer each other poses difficulties in the design of social environments which contain experiences appropriate for the development of normal social responses. Further, the finding that socially normal monkeys do not choose to approach more abnormal ones compounds the problem of providing therapy for abnormal animals.

These data also have implications for attachment behavior in mammals. Cairnes (5) suggests a learning theory approach to the formation of attachments in which the subject will approach a social object as a function of having made many previous responses while the social object was part of the general stimulus situation. Thus, indices of social attachment toward an object are expected to be higher with increases in the probability that this object occurs as part of the stimulus field in the subject's overall repertoire of responses. Although this seems a reasonable approach, the present data present some difficulties for this view. During rearing, the monkeys in group A did not have the same opportunity to learn the characteristics of other monkeys as did the monkeys in groups B and C. Yet, the monkeys

in group A did prefer each other to the alternative choices available. Thus, it is possible that the preference shown by group A monkeys was not based on the conditioning of approach behavior to specific social cues, as is suggested by the stimulus-sampling theory of attachment. It is possible that the behavior of group A was motivated by avoidance of cues contained in the social behavior or countenance of the other two types of monkeys. Thus, there may be at least two distinct kinds of processes in the choice of a social stimulus. The conditioning of specific social cues to the response systems of an animal may be one factor, and the avoidance of nonconditioned cues may be a second important factor in the formation of social attachments.

The specific cues used by the monkeys studied here are not known. Neither do we yet know how our animals differentiated between the stimuli. The discrimination may be based solely on differences in the gross activity of the stimulus animals, or on more subtle and specific social cues. Analysis of the specific stimulus components operating in this situation may clarify the nature of the social cues involved. The important question to be answered is whether the types of cues used in selecting a partner are qualitatively different for different rearing conditions, or whether the same aspects of stimulation are simply weighted differently as a function of an animal's rearing history.

REFERENCES AND NOTES

1. HARLOW, H. F. and HARLOW, M. K. *Sci. Amer.*, **207**, 136 (1962); SACKETT, G. P., *Child Develop.*, **36**, 855 (1965).
2. The rearing conditions are described fully by G. P. Sackett, *Science*, **154**, 1468 (1966).
3. The playroom situation is described by H. F. Harlow, G. L. Rowland and G. A. Griffin, *Psychiat. Res. Rep.*, **19**, 116 (1964).
4. BUTLER, R. A., *J. Comp. physiol. Psychol.*, **50**, 177 (1957).
5. CAIRNS, R. B., *Psychol. Rev.*, **73**, 409 (1966).
6. Supported by NIMH grant MH-11894.

MODIFICATION BY EARLY EXPERIENCE OF THE TENDENCY TOWARD GREGARIOUSNESS IN RATS*

Sachio Ashida[1]

INTRODUCTION

SOME research evidence (Denenberg, 1963; McClelland, 1956) has shown that the body weight and emotional behavior of rats are significantly influenced by their early experience. Other research evidence (Denenberg et al., 1964) has shown that the rat's behavior is affected by the postweaning social interaction with other organisms of the same species. One logical extension of these studies is to investigate the effect of postweaning social interaction of rats upon their tendency to approach another animal. The purpose of this study is to test the hypothesis that the tendency toward gregariousness is a function of rat's early experience in which the strength of such a tendency is positively related to the number of other rats housed together with the subjects in the homecage.

METHOD

The subjects were 30 male and 30 female Sprague-Dawley albino rats. One male and one female rat were used as the goal object of these subjects. The apparatus used was a commercial Jenkins-Warden Obstruction Box.

After being weaned at the age of 21 days, 10 male and 10 female rats were placed alone in individual steel cages 7 in. by 8 in. by 11 in. (Group 1). Ten male and 10 female rats were placed in pairs of the same sex in steel cages 9 in. by 9 in. by 15 in. (Group 2). Ten male and 10 female rats were placed in groups of five of the same sex in steel cages 18 in. by 12 in. by 20 in. (Group 3). During a period of 18 weeks after all rats were weaned, body weights and the amounts of food and water intakes were recorded twice a day.

For the experimental trials, each group was further divided into two subgroups of five rats of the same sex. One of these subgroups was designated as the experimental group and the other as the control group. Each subject was placed at the starting compartment of the obstruction box; then the plastic door of the compartment was opened in order to allow the subject to cross the narrow passway to the goal compartment via the transparent oneway-swing door. If the subject entered the goal compartment, it was allowed to stay there for 15 sec., and then was returned to the starting compartment by the experimenter. This procedure was repeated during a 10 min. period. For the experimental group, a rat of the same sex was placed at the goal compartment. For the control group, no rat was placed at the goal compartment.

RESULTS AND DISCUSSION

The mean number of crossings to the goal compartment is summarized in Table 1. An analysis of variance yields $F(1,48) = 26.389$ for between Sex, $F(1,48) = 7.003$ for between experimental and control Condition, and $F(2,48) = 12.281$ for between Groups. These differences are significant beyond the 0.001, 0.05, and 0.001 level of confidence respectively. Of particular interest is the Sex by Condition interaction, $F(1,48) = 5.948$, and the Groups by Condition interaction, $F(2,48) = 4.869$. Both interactions are significant beyond 0.05 level, but the Sex by Groups interaction, $F(2,48) = 2.640$, is negligible.

Interpretation of the obtained mean number of crossings is, however, complicated by the fact that each group has had different environmental conditions during the preceding 18 weeks. In order to compute the net strength of the tendency

---

* From *Psychonomic Science*, 1964, 1, 343–344. Copyright 1966 by editors of Psychonomic Journals.

[1] The author wishes to thank Mr. Keene Hueftle, University of Nebraska, for his laboratory assistance.

TABLE 1. *The Mean Number of Crossings to Achieve Goal Compartment*

Sex	Conditions	Groups		
		Group 1	Group 2	Group 3
Male	Experimental	1.0	6.0	3.8
	Control	3.4	4.4	2.6
Female	Experimental	4.0	9.8	11.6
	Control	3.6	6.8	5.2

toward greagariousness, the logarithmic transformations of the number of crossings will be applied. The rational basis is as follows: It is assumed that all three groups of rats have developed different sets of response tendencies during the 18 weeks of housing. Let T_1, T_2 and T_3 be the set of response tendencies of the rats in Group 1, Group 2, and Group 3 respectively. The components of T_1, for example, may be a tendency to move toward the corner of the cage, to move away from a certain object in the cage, to stay with another rat, to explore the inside of the cage, and so forth. In the experimental situation, some of these tendencies may be elicited by the environmental conditions, and some may not be elicited. Let T_{1g} be the approach tendency toward the other rat in Group 1, T_{1a} the exploratory tendency to a new situation in Group 1, and (T_{1x}) a set of unspecifiable tendencies of the rats in Group 1. Then, any possible response evocation, R, of the rats in Group 1, is assumed to be a function of these variables, i.e.:

$$R = F[T_{1a}, T_{1g}, (T_{1x})].$$

Let $m_1(R_{Ei})$ represent the number of crossings of ith experimental rat in Group 1 such that R_{Ei} is a specified component of R, $i = 1, 2 \ldots$,

5, and $m_1(R_{Ci})$ the number of crossings of ith control rat in Group 1. And assume each component of T_1, i.e., T_{1a}, T_{1g}, and (T_{1x}) is independent of each other. Then from the above assumptions,

$$m_1(R_{Ei}) = f(T_{1ai}) \cdot f(T_{1gi}) \cdot f[(T_{1xi})] \quad (1)$$

However, in the experimental situation, the environmental condition for the control group is quite different from the condition for the experimental group, i.e., a rat was placed in the goal compartment for the experimental group, but no rat was placed for the control group. Thus, for the control group, we have

$$m_1(R_{Ci}) = f(T_{1ai}) \cdot f[(T_{1xi})] \quad (2)$$

In order to factorize the net strength of the tendency toward gregariousness, i.e., $f(T_{1gi})$, take logarithm of (1) and (2), and subtract (2) from (1). We have

$$\mathrm{Log}\, m_1(R_{Ei}) - \mathrm{Log}\, m_1(R_{Ci}) = \mathrm{Log}\, f(T_{1gi}) \quad (3)$$

The mean strength of the tendency toward gregariousness,

$$1/n \cdot \sum_{i}^{n} \mathrm{Log}\, f(T_{1gi}),$$

can be computed easily from the equation (3).

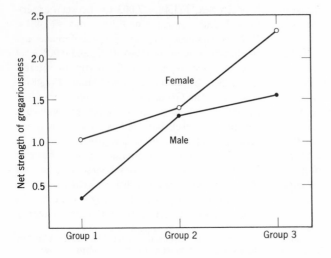

FIGURE 1. The net strength of the tendency toward gregariousness as a function of the number of rats housed together in their home-cage. The net strength is shown in terms of antilogarithm of the equation (4).

The data analysis of this type seems to have a certain advantage in which the main effect of the experiment in question could be evaluated (see Fig. 1).

The results show that the relative strength of the tendency toward gregariousness is clearly a function of the number of members housed in the homecage. It appears that at least some of the necessary conditions fostering such a functional relationship are effective early in the living conditions of the rats and that the tendency toward gregariousness can be acquired. The results also show that the differences between groups of the same sex in the body weight, the rate of weight gain, and the mean amounts of food and water intake during the 18 weeks of housing were not statistically significant.

Effects of early experience on the tendency toward gregariousness seem to be much more complex and extensive in rats than in birds or fishes. Collias (1952), for example, found that hand-reared chicks respond very slowly to the food-call chucking of the hen. Qualitative measure of approach tendency of isolated chick to another chick revealed that the approach was much slower than normal. When they were placed with the flock, such chicks tended to keep apart from the group (e.g., Collias, 1950). The results of the present study, however, clearly confirm our general hypothesis that the strength of the tendency toward gregariousness is a function of early experience. In summary, it is possible to estimate the net strength of such a tendency by use of a logarithmic transformation of the obtained data and by a simple computation of the following equation:

$$1/n \cdot \sum_{i}^{n} \mathrm{Log} f(T_{jgi}) =$$

$$= 1/n \cdot \sum_{i}^{n} \mathrm{Log}\, m_j(R_{Ei}) - 1/n \cdot \sum_{i}^{n} \mathrm{Log}\, m_j(R_{Ci}) \quad (4)$$

where j is the code number of a specified group of rats.

REFERENCES

Collias, N. E. Social life and the individual among vertebrate animals. *Ann. N.Y. Acad. Sci.*, 1950, **51**, 1074–1092.

Collias, N. E. The development of social behavior in birds. *Auk.*, 1952, **69**, 127–159.

Denenberg, V. H. Early experience and emotional development. *Scient. Amer.*, 1963, **208**, 138–146.

Denenberg, V. H., Hudgens, G. A. and Zarrow, M. X. Mice reared with rats: Modification of behavior by early experience with another species. *Science*, 1964, **143**, 380–1.

McClelland, W. J. Differential handling and weight gain in the albino rat. *Canad. J. Psychol.*, 1956, **10**, 19–22.

REARING AND DISPLAY VARIABLES IN SOCIABILITY*

Harry P. Shelley and Kermit T. Hoyenga

Lindzey *et al.* (1965) have identified genetic differences in the sociability of mice. Sociability is defined as the time spent next to caged display animals in opposite corners of an open field. Two experiments were conducted. Experiment 1 tested the hypothesis that post-weaning group living will result in higher sociability scores than post-weaning social isolation. When the results of Experiment 1 were significant but in the opposite direction from prediction, Experiment 2 was conducted to (1) verify the findings of Experiment 1 and (2) study the effect of variation in the display animal, since the effects of variation will provide evidence as to the nature of the sociability response. A chick was selected as a second display animal since it had appropriate auditory and visual properties, was of a size for the apparatus, and yet differed from the rat in significant phylogenetic ways.

* From *Psychonomic Science*, 1966, **5**, 11–12. Copyright 1966 by editors of Psychonomic Journals. This research was supported by NIMH Research Grant MH-0703102.

METHOD

SUBJECTS AND APPARATUS. Holtzman Sprague-Dawley rats were placed at weaning either singly or in groups of 8 or 9 in 38 in. by 12 in. by 12 in. cages. Food and water consumption was ad lib. The Ss were 107 days (Experiment 1) or 79 days old (Experiment 2) at testing. To accommodate rats rather than mice the open field was enlarged to 48 in. by 48 in., the wall height increased to 12 in., and the four display cages enlarged to 8 in. by 8 in. by 8 in.

PROCEDURE. Following Lindzey *et al.* a line was drawn on the floor 6 in. away from the two exposed sides of the display cages. When a rat placed two feet inside the line he was judged adjacent to the cage. An observer with stop watches recorded the time spent adjacent to the display and empty cages. The observer also noted any freezing (lack of motion for more than 3 sec. when not adjacent to a cage). The only light source was a 40-watt bulb 48 in. above the center of the floor.

The display cages were placed in diagonally opposite corners and systematically rotated for each successive trial. S was placed in the center of the open field and allowed to run freely for 5 min. Each S was given a trial on each of 5 (Experiment 1) or 6 (Experiment 2) consecutive days. In Experiment 2 a rat was used as the display for half of the Ss in each rearing condition; for the other half a 42-day-old Hyline chick was used.

RESULTS

EXPERIMENT 1. Analysis of variance of the data

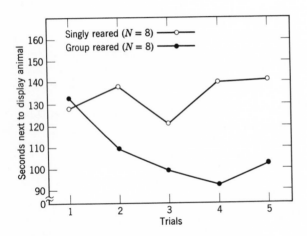

FIGURE 1. Experiment 1. Sociability as a function of social rearing conditions.

plotted in Figure 1 indicates a significant rearing by trials interaction ($F=4.36$, $df=4/56$, $p<0.01$) as well as a difference between rearing conditions ($F=25.06$, $df=1/14$, $p<0.01$). Only trivial chance differences are found when the data on time next to the empty cages are analyzed.

EXPERIMENT 2. Although rearing effects across all trials and all display conditions are not significant ($F=2.12$, $df=1/32$, $p>0.05$), the rearing by trials interaction is replicated (Fig. 2) and is of particular significance ($F=5.03$, $df=2/64$, $p<0.01$). The failure to replicate rearing effects is not particularly significant since inspection of Figure 2 indicates that a significant rearing effect would result with additional trials. Initially time spent by the display rat does not differ as a function of rearing conditions, but over the trials the curves separate. Although the time values are significantly less than with a rat display [over all

trials ($F=14.77$, $df=1/32$, $p<0.01$) the curves for the chick display condition are essentially parallel to those for the rat display, i.e., $F<1.00$ for the display by rearing by trials interaction]. Again only trivial chance differences are found when the data on time next to the display cages are analyzed.

When the time next to the empty cages is summed over all trials and subtracted from the time next to the chicken display cages summed over all trials, 14 of the 18 remainders are positive (sign test, $n=18$, $p<0.02$), i.e., the time spent next to the chick display is significantly greater than the time spent next to the empty cages.

EMOTIONALITY. In neither experiment was there a single instance of freezing; neither were there enough fecal boli to provide a measure of emotionality. Fear as it is usually measured in the open field was not a factor in Ss' behavior.

DISCUSSION

Starting at approximately the same point, the socially isolated Ss spend increasing amounts of time adjacent to the display animal while the group reared Ss spend less and less time there. Although the absolute times are less, the curves for the chick display are essentially similar to those for the rat display. Chick or rat more time is spent by the display cages than by the empty cages.

Accounting for the findings requires consideration of the following, not necessarily exclusive, conditions:

1. Novel stimulus and habituation effects (Berlyne, 1960). Novel stimuli or complex stimuli have the property of receiving attention and directing investigative behavior. Berlyne's habitu-

ation hypothesis holds that repeated exposure to a stimulus reduces or eliminates the tendency of an organism to respond selectively to that stimulus. For the group reared Ss the display rats become identified quickly and attention is directed to other parts of the open field. For the socially isolated Ss, not accustomed to other animals in the environment, the open field is more quickly explored and the more novel elements, the display rats, receive more attention in subsequent trials. However, if novelty and habituation are the sole factors operating, the chick displays should receive more attention than the rat displays. Such is not the case.

2. Social motivation-deprivation effects. During the preweaning period social motives are acquired which are then deprived under social isolation rearing conditions. The effects of depri-

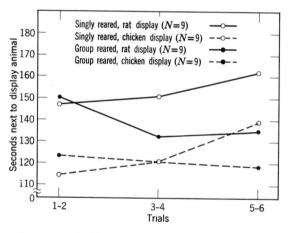

FIGURE 2. Experiment 2. Sociability as a function of social rearing conditions and display animal.

vation are masked in the early trials by exploration of the open field. As exploration is accomplished, more time is available for "socializing." The group-reared Ss, whose social motives have not been deprived, quickly identify the display animal and then spend more time in exploring the more novel open field. Generalization effects may be used to account for the time spent next to the chick display. However, the assumption of social motives for the laboratory rat rest, at best, on shaky and meager grounds.

3. Interaction effects. Not only are the stimuli from an animate display varied and irregular but the display animal may also react to S's behavior as well as S reacting to the display's behavior. Such interaction does occur and resembles play-fighting including tactile contact since it is possible for one animal to bite at the other animal's toes through the wire-mesh. To some

extent similar interaction occurred with the chick display. However, it was more limited and occasional pecks by the chick would briefly drive S away. Such differences in interaction may partially account for the lower sociability scores for the chick display.

4. Genetic effects. Although there are obvious problems in going from one species to another, the evidence that variations in sociability can be produced by variations in post-weaning rearing and the fact that sociability responses are made to the chick display indicate that whatever the genetic determinant it operates in a more general way. The problem is analogous to the imprinting situation. The following response is not a social response unless conditions are such as to make it social (Hess, 1959). Sociability behavior may be social only because the situation makes it so.

REFERENCES

BERLYNE, D. E. *Conflict, arousal, and curiosity.* New York: McGraw-Hill, 1960.

HESS, E. The relationship between imprinting and motivation. In M. R. Jones (Ed.), *Nebraska symposiums on motivation.* Lincoln: University Nebraska Press, 1959. Pp. 44–77.

LINDZEY, G., WINSTON, H. D. and ROBERTS, L. E. Sociability, fearfulness, and genetic variation in the mouse. *J. pers. soc. Psychol.*, 1965, **1**, 642–645.

CHAPTER FIVE

AGGRESSIVE BEHAVIOR

OUR language predisposes us to treat affection and aggression as opposites. The affectionate or the affiliative response eventually results in some benefit to its target, while the aggressive response brings about harm. And because it is possible to treat them as opposites, we are prone to assume that they lie on the same dimension, and hence share some underlying psychological processes. Actually, the experimental literature on affiliation and on aggression shows that the only aspects of significance that these two response patterns have in common are (a) that they both contain a strong instinctive component that manifests itself with increasing clarity as we go down the phylogenetic scale; (b) that they are both affected by early social experience; and (c) that both show inter- as well as intraspecific forms. Since these aspects apply to an entire host of other social behaviors, there probably isn't very much that we can bring from the study of affiliative behavior to help us understand aggression.

This chapter deals only with intraspecific aggression, leaving entirely aside the questions of fights between species and of predation. The experiments reported by King show that, like affiliative behavior, aggression responds to socialization. Just as it was necessary for the animal to have normal social contact in order to develop normal affiliative responses, so is it necessary for the development of a normal aggressive responses. And, for mice, the presence of another member of the same species may, from King's study, seem to be a sufficient condition for fighting. This result is somewhat paradoxical because the mere sight of members of the same species seems also sufficient for releasing the affiliative response, as we noted in the previous chapter from the studies of Mason and of Pratt and Sackett. Since it is not a matter of chance whether an aggressive or an affiliative response will be made, there must be situational stimuli which select between them.

A distinction among some forms of aggressive behavior should be made, which is probably also valid for affiliative behavior. Aggression occurs when there is a question of territorial occupancy; aggression occurs in the context of gaining mating privileges; and it is also present in the establishment of dominance hierarchies (see Chapter Nine). These forms of aggression are instrumental since, to a large extent, they are under the control of the consequences which ensue. Thus, the amount of fighting among pigeons, for instance, can be dramatically affected by the size of their living cage (Willis, 1966), and we note from the study by Ulrich and Azrin in this chapter, that the same is true for rats. Once the territory is secured or the position in the hierarchy established, fighting ceases. Similarly, those forms of affiliative behavior that lead to the establishment of social groupings and of hierarchies seem to have the same features. But there exists also a form of aggression which apparently is entirely under the control of the releasing stimulii, i.e. reflexive aggression. And, much of affiliation that occurs in sexual behavior and in caring for offspring is also reflexive in character. The reflexive form of aggression in rats and in other animals has been extensively investigated by Ulrich, Azrin, and their associates, and two of the studies from their laboratories are reported in this chapter. The first study shows the specific situational antecedents of reflexive aggression in rats, and Vernon and Ulrich demonstrate the noninstrumental aspect of this behavior by succeeding in classically conditioning it. The experiments by Tedeschi and his associates, on the other hand, show how reflexive fighting in mice can be inhibited by the administration of drugs.

REFERENCES

WILLIS, F. N. Fighting in pigeons relative to available space. *Psychonomic Science*, 1966, **4**, 315–316.

RELATIONSHIPS BETWEEN EARLY SOCIAL EXPERIENCE AND ADULT AGGRESSIVE BEHAVIOR IN INBRED MICE*

John A. King[1]

A. INTRODUCTION

A RELATIONSHIP between early social experience and adult aggressive behavior has been found in one inbred strain of mice (9). Male mice raised in isolation after weaning tend to be less aggressive adults than males raised in social groups. The effect of early social experience on adult aggressive behavior in mice may be used to test the critical period hypothesis (2, 12) and to discover the relationship between early experience and adult behavior.

According to the critical period hypothesis

* From the *Journal of Genetic Psychology*, 1957, **90**, 151–166. Copyright 1957 by The Journal Press, Provincetown, Mass. This investigation was supported by research grant MH-123 from the National Institute of Mental Health, of the National Institutes of Health, Public Health Service.

[1] The author wishes to acknowledge the assistance of Helen Connon in collecting the data and the staff of the Division of Behavior Studies, R. B. Jackson Memorial Laboratory for their suggestions during the course of the study: J. P. Scott, J. L. Fuller, R. Chambers.

there are "specific stages in ontogeny during which certain types of behavior normally are shaped and molded for life" (2). The purpose of this study is to identify (a) the period or age when the social experience is most effective in producing adult aggression in mice, (b) the duration of the experience, and (c) the nature of the experience. Further objectives were to learn if the effects of the early experience (d) perseverate throughout life, (e) generalize into different adult activities, and (f) are genetically specific.

B. SUBJECTS

There were 288 male mice of the C57BL/10

and BALB/c inbred strains used in this study. All mice were housed in 12-in. by 6-in. by 6-in. compartments of a wooden box containing two compartments. The tops of the compartments were covered by wire screen from which a food hopper and water bottle were suspended. The procedure describes the way the mice were reared and divided into groups.

C. APPARATUS

Mice were fought in boxes which were similar to the living cage except that the partition separating the two compartments was removable. During the test, the partition was removed,

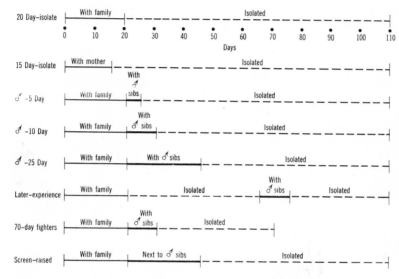

FIGURE 1. Diagrammatic representation of the treatments given to eight groups of mice prior to the aggression test given about 110 days of age.

permitting the mice in each compartment to come together. No handling of the mice prior to the test was required.

D. PROCEDURE

1. Constitution of Groups

Groups I through VIII consist of male C57-BL/10 mice. Group IX and X, which duplicate Groups I and V, consist of male BALB/c mice.

(a) Group I (20-day-isolates). Forty-two males were weaned at 20 days of age and isolated from all other mice until placed in the testing situation at 100 to 120 days of age. This group is a com-

bination of 24 mice from a previous study (9) and 18 additional mice which were given the same treatment. A graphic representation of the early experience of this group compared with seven other groups is presented in Figure 1.

(b) Group II (15-day-isolates). Twenty-two males were raised singly (other litter mates were discarded) by their mothers until 15 days of age, at which time they were separated from their mother and kept in isolation until placed in the testing situation at 100 to 120 days of age. This treatment eliminated sibling or paternal associations before weaning and maternal influences at the earliest possible age. The 22 mice in this group are the survivors of an initial 60 mice which were weaned at 15 days.

(c) *Group III* (♂-5-days). Twenty-four males were separated from their parents and all female siblings at 20 days of age and raised with only their male siblings until 25 days of age, at which time they were isolated until placed in the test situation at 100 to 120 days of age.

(d) *Group IV* (♂-10-days). Twenty-eight males were treated like Group III mice except they remained with the male siblings from 20 to 30 days of age.

(e) *Group V* (♂-25-days). Fifty-two males were treated as the preceding two groups, except they were left with their male siblings and father from 20 to 45 days of age. Data from a preceding study using 34 mice were added to the data from 18 mice in the present study. The addition of the fathers as a variable in this group made the young no more aggressive than those raised only with male siblings, as shown in the results.

(f) *Group VI* (later-experience). Thirty-four C57BL/10 males were isolated from other mice at 20 days of age until they were 65 days old, when they were placed with their male siblings, which had also been isolated. They remained together for 10 days and were isolated again at 75 days of age until placed in the test situation at 100 to 120 days of age.

(g) *Group VII* (70-day-fighters). Sixteen males were treated like Group IV mice, except that the interval of isolation between the social experience and the test situation was reduced to 40 days. This reduction in the duration of isolation necessitated fighting the mice at 70 days of age.

(h) *Group VIII* (screen-raised). Twenty male C5BL/10 mice were weaned at 20 days of age and were placed alone in one compartment of a living cage with a double ¼-in. wire-mesh screen instead of a solid wooden partition separating the isolates. The screen prevented physical contacts between the mice. After 25 days (45 days old) of this semi-isolated treatment, the mice were totally isolated until placed in the test situation at 100 to 110 days of age. This treatment corresponded to Group V mice except actual contact was prohibited.

(i) *Group IX* (20-day-isolate BALB/c). Twenty-four male BALB/c mice were given the same treatment as Group I in the C57BL/10 mice.

(j) *Group X* (♂-25-day BALB/c). Twenty-six male BALB/c mice were given the same treatment as Group V in the C57BL/10 mice.

2. Procedure of Aggression Tests

Two members of the same group were placed individually in each compartment of a fighting box 3 days prior to being tested. Pairs of combatants were formed from mice that had never lived together. The tests began by removing the partition between the two compartments and the mice could then come together. The frequency of social interactions (running away, nosing, smelling genitals, grooming, attacking) from the time the mice were exposed to each other until the first fight began was recorded. Fights were distinguished from attacks by the participation of both mice in a fight. The principal measure of aggressive behavior used in this study was the fighting response latency, which is the elapsed time from the moment when the partition was removed until the first vigorous fight began (4). The pair was allowed to fight approximately 5 sec. before the partition was lowered between them. Often the combatants had to be separated with forceps, although most battles were brief and the mice separated temporarily before returning to fight. Dominance was prevented by separating the mice after 5 sec. of fighting. A maximum of 5 min. was allowed for each trial if the mice failed to fight. Each pair was given ten trials at the rate of one each day. Since all mice in each group were not born on the same day, they were tested between 100 and 120 days of age with a median of approximately 110 days of age.

The response latency seemed the most satisfactory measure of aggression because it is reliable (a product moment correlation between even and odd trials gives $r = 0.99$, $N = 72$), it can be used without permitting dominance to form, it indicates the duration of a fight (3), and it is not affected by chance factors, such as a random bite which may terminate the fight by causing the injured mouse to submit.

The failure of the response latency data to meet the assumptions of normality required by parametric statistics necessitated the use of a non-parametric statistic. For this purpose the Mann-Whitney U test (11) was applied.

3. Procedure of Sexual Tests

Ten mice from Groups I, II, IV, V, and VIII, were given sexual tests approximately 3 months after the aggression tests, during which time they were kept in isolation. The mice were tested by placing one estrus female into the living cage of each male. Females were brought into estrus by 0.01 mg. injections of estradiol (Progyron-B, Schering Co.) and tested for receptivity with a sexually active male until intromission was achieved. The females were then placed with the male mice of each group for 10 min. each day for 10 days. The frequency of intromissions are the

TABLE 1. *Statistical Differences in the Second† to Tenth Trials of Eight Groups of Mice Raised With Different Early Social Experiences**

Group	VI	III	II	IV	VII	VIII	V	Mean Latency in Min.
I	NS	NS	NS	*2 3 4* *5 6 7* *8 9 10*	*2 3 4* *5 6 7* *8 9 10*	*2 3 4* *5 6 7* *8 9 10*	*2 3 4* *5 6 7* *8 9 10*	3.32
VI	—	NS	NS	*2 3 4* *5 6 7* *8 9 10*	*2 3 4* *5 6 7* *8 9 10*	*2 3 4* *5 6 7* *8 9 10*	*2 3 4* *5 6 7* *8 9 10*	3.16
III		—	NS	2 6 9	*2 3 4* 5 10	*4* 5 6 *8 9 10*	*2 3 4* 5 6 7 *8 9 10*	3.03
II			—	NS	10	*3* *9 10*	*2 3* 5 6 7 *8 9 10*	2.91
IV				—	NS	NS	a a 7	1.96
VII					—	NS	NS	1.72
VIII						—	NS	1.73
V							—	1.19

† No significant differences in the first trial.
* Figures in italics indicate trials which reached the 1 per cent probability level, others indicate the 5 per cent level as determined by the Mann-Whitney *U* test.

only measure of sexuality activity used in this experiment because ejaculations were too infrequent and mountings do not represent as high a level of sexual activity as intromission.

E. RESULTS

1. *Effect of Isolation*

The effect of isolation may reveal the influence early social relationships have upon behavior. Mice isolated at 20 days of age until tested for fighting at 110 days of age were inhibited in their aggressive behavior (9).

When the aggressive behavior of mice isolated at 15 days of age (Group II) is compared with the behavior of mice isolated at 20 days (Group I), it can be seen from Table 1 that there is no significant difference between them. It was expected that earlier isolation would reduce aggressiveness even more. However, the mean response latency of Group II was less than that of Group I, which indicates a tendency for the earlier isolated mice to be even more aggressive than the later isolated mice. It is possible that this tendency of the Group II mice to be more aggressive is the result of a selection of the strongest and most aggressive mice, since about two-thirds of the initial group died during the treatment.

The inhibition of aggression by isolation, whether at 15 or 20 days of age, is apparent in the slower response latencies of Groups I and II than in Group V mice (Table 1).

2. *Effect of Duration of Social Experiences*

The duration or quantity of an early experience which an organism must receive before it is permanently engraved upon its behavioral pattern may be limited to one traumatic experience, may extend throughout certain developmental periods, or may recur at widely separated intervals throughout its development. The purpose of treating Groups III, IV, V, and VI with varying lengths of early social contacts was to ascertain the minimum number of days necessary to result in low fighting response latencies in adult mice.

The treatment of Group III differed from Group I by delaying isolation 5 days until the mice were 25 days old and by restricting their social contacts between 20 and 25 days of age to their male siblings. This difference in treatment was not sufficient to distinguish Group III mice ($\bar{x} = 3.03$ min.) from Group I mice ($\bar{x} = 3.32$ min.), although it did tend to reduce Group III fighting response latencies somewhat as compared to Group I (Fig. 2).

When mice were given 10 days of social contacts with their male siblings (Group IV, $\bar{x}=$ 1.96 min.) after weaning at 20 days and then were isolated, they were significantly (Table 1) faster fighters than mice isolated at 20 days of age (Group I). On the other hand, their response latencies did not differ significantly from Group V ($\bar{x}=1.19$ min.) mice, which experienced 25 days of sibling and paternal contacts after weaning. These results indicate that social contacts for a period of 5 days immediately after weaning was not long enough to make the mice more aggressive than the completely isolated mice (Group I), but that a period of 10 days of social contacts produced

about the maximum amount of aggressive behavior. Longer periods of social contacts were not necessary.

3. *Effects of Age of Social Experience*

If 10 days of social interaction immediately after weaning is sufficient to produce low fighting response latencies in adult mice, we may ask if the same amount of social interaction during adulthood has the same effect. According to the critical period hypothesis, there are certain in 9 of the 10 trials as shown in Table 1. Since the "periods in development when the same experience may have a more profound effect than at

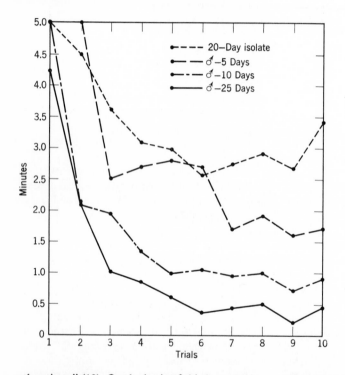

FIGURE 2. Median curves of the fighting response latencies of Groups I (20-day isolate), III (♂ 5-day), IV (♂ 10-day), and V (♂ 25-day) of C57BL/10 mice.

other times" (12). On the basis of this hypothesis, the 10 days of social exposure after weaning should produce different effects than 10 days of the same exposure after the mice have become sexually mature. Only one age, other than the 10 days after weaning, was tested, with Group VI mice. The age selected was 65 days, when most development has ceased, although the mice do continue to grow in size.

Group VI mice had response latencies closer to Group I mice than any other group and had significantly slower latencies than Group IV mice experiences of the VI mice were similar to those of Group I mice immediately after weaning when both groups were isolated, we may conclude that

this period is more critical in the development of aggressive behavior than at maturity when the mice experienced social contacts similar to those of Group IV mice.

4. *Effect of Duration of Isolation*

In order to keep the age at testing as constant as possible, it was necessary to vary the length of the period of isolation in the different groups. It is possible that this uncontrolled variable was contributing to some of the differences among the groups. One method of testing the effects of the duration of the isolation period is to change it radically and fight the mice at a different age. If the duration of isolation is contributing to the effects produced on adult behavior, two groups of mice treated alike, except for a different duration of isolation, should behave differently as adults.

Group VII mice, which were given the same social exposure as Group IV mice, but tested after a 40-day isolation period instead of an 80-day isolation period, did not differ from Group IV in any of the trials (Table 1). Within the limits tested, this variation in the isolation period cannot be considered to influence the adult aggressive behavior.

5. *Nature of the Social Experience*

In the preceding experiments, no attempts were made to understand the nature or the quality of the social experience obtained by the mice given periods of social contacts. Since no special training was given and no direct observations were made of the mice during the period they were living together, the results can only offer circumstantial evidence as to what actually occurred in the social groups. A preceding paper (9) suggested two alternative types of experience which may be influential in determining the adult behavior. One type of experience may be that of competition for food, water, or space, which acts as a type of latent learning for adult aggression. Williams and Scott have observed that conflict

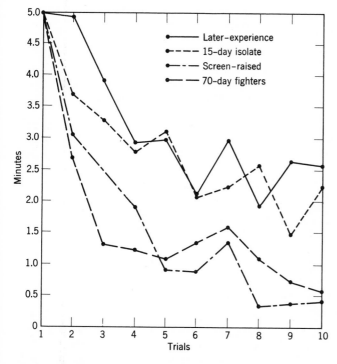

FIGURE 3. Median curves of the fighting response latencies of Groups VI (later-experience), II (15-day isolate), VIII (screen-raised), and VII (70-day fighters) of C57BL/10 mice.

behavior over food does occur often between 25 and 40 days of age (15). The other type of experience may simply be one of familiarization with other mice, so when they come together in a fighting situation they are not inhibited by the strangeness of another mouse.

In order to ascertain which of these experiences is affecting the adult behavior, a group of animals was given the opportunity to become acquainted with other mice through a screen for 25 days after weaning, but not allowed to compete. If a competitive situation was necessary for the mice to learn aggressiveness, one would expect that the Group VIII mice would be slow to fight. On the other hand, if the experience needed was only some familiarity with another mouse, one would expect the Group VIII mice to fight rapidly when tested.

A comparison between Group VIII and Groups IV and V as shown in Table 1 and the respective curves of the three groups in Figures 2 and 3 indicates that there was no significant difference among them. The unrestricted contacts among the siblings of Groups IV and V mice during the period of social living failed to distinguish them from Group VIII mice, which could only perceive each other through a screen during the same period. This failure of Group VIII mice to differ from the socially raised mice indicates that competition or any learning of aggression is not essential for developing a mouse eager to fight at maturity. Apparently the perceptual cues obtained through two wire screens was sufficient for establishing the pattern of behavior pursued at adulthood. These results suggest that the kind of experience obtained by the mice is that of familiarization with another mouse. It might be expressed in terms of an investigative drive. Once the investigative drive of a mouse is satisfied, it no longer serves to inhibit the tendency to fight.

6. Perseverance of the Effects of Early Experience

Since all of the groups were tested for aggression only once in their lifetime, it seemed desirable to learn if the effects would persevere until a later age. In order to test how long the effects of the early experience persists after their first encounter and an intervening period of isolation, the Group III mice were retested at 200 days of age. The mice were isolated for the 90-day interval between the tests.

In all of the nine trials used for comparison, the mice fought significantly faster ($P < 0.01$ on trials 3, 4, 6, and 8; $P < 0.05$ on trials 2, 5, 7, 9, and 10) in the retest at 200 days of age than on the first test at 110 days of age. This significant change in the behavior of the mice after a period of 3 months indicates that the effects did not persist in the mice as they became older. A product moment correlation of the mean response latencies for each pair of mice in both the 110-day and 200-day tests showed a positive correlation ($r = 0.73$; $P < 0.01$) between the two tests. The mice which had short latencies at 110 days tended to have shorter latencies at 200 days. It is likely that the experiences of the first test influenced the behavior of the mice at a later age. Age differences are probably not alone responsible for the increased aggression because Group VII mice are just as aggressive at 70 days as Groups IV and V at 110 days of age.

7. Generalization of Effects

If the early social environment of the mice can affect their adult aggressive behavior, will the same experiences generalize into other types of behavior? From the work of Stone (13), Beach (1), and Valenstein (14), we know that different social environments of developing rats and guinea pigs can affect the adult patterns of sexual behavior. It seems likely, therefore, that the sexual behavior of mice raised with different social experiences after weaning may also be affected. The further question may be asked: if the social experiences in infancy tend to make the mouse

aggressive, will he be more or less likely to exhibit high sexual activity?

In order to learn if the effects of early social experience generalize into types of behavior other than aggression, ten mice of the following groups were given sexual tests approximately 3 months after the fighting tests: Group I, II, IV, V, and VIII. During the 3-month period between the fighting and sexual tests, the mice were kept in isolation. This long interval between tests would tend to eliminate the carry-over of one test into the other. The use of the same mice for both tests permitted a correlation between the sexual and aggressive behavior of each individual to be made.

Group VIII mice had significantly fewer ($P < 0.005$) intromissions than any of the four other groups when the total number of intromissions over ten trials were tested by the Mann-Whitney U test. The Group II mice had fewer intromissions ($P < 0.05$) than Group V mice. These data demonstrate that Group VIII mice were less active sexually than the other groups and Group II mice were intermediate in their sexual activity. These differences cannot be attributed exclusively to early experiences because the mice also had the later experiences of the fighting tests. The experiences which produced an aggressive mouse did not necessarily enhance or inhibit its sexual behavior. For example, Group VIII mice, which were as aggressive as Group V mice, were less active sexually than Group V mice. The experiences which affect adult aggressive behavior apparently are specific and do not generalize into sexual behavior (10).

8. Genetic Effects

The genetic characteristics of the mice in Groups I through VIII were held constant by the use of a homozygous strain of inbred mice. Any differences in their behavior are consequently the result of their different environmental treatments. Before concluding that these treatments would affect animals of different genetic constitution in the same manner, it was necessary to try the same treatments on a different inbred strain. For this purpose, mice known to behave differently in aggressive situations (5), BALB/c, were given the same treatments as the two most divergent groups among the C57BL/10 mice. Group IX in the BALB/c mice corresponds in treatment to Group I of the C57BL/10 mice and Group X BALB/c mice corresponds to Group V C57BL/10 mice.

The slow fighting response latencies of the isolated Group I failed to appear in the Group IX mice, which were given the same early experiences as Group I. On the other hand, Groups V and X were similar in the adult aggressive behavior. The two BALB/c groups showed no significant differences in any of the trials (Fig. 4). Both Groups IX and X tended to be very aggressive, reaching an asymptote in their response latency curves below that of any group in the C57BL/10 strain. The faster fighting exhibited by Group IX than by Group I mice may be attributed to genetic differences, since both groups were treated alike.

F. DISCUSSION

The purpose of this study was to investigate the relationship between early social experience and adult aggressive behavior in mice in order to test the critical period hypothesis.

Since the age at which the mouse undergoes the experience is crucial to the hypothesis, it was necessary to learn if the same social exposure at different ages of the mouse have the same effect on its adult behavior. The results indicated that the age of the mouse was important because 10 days of social contact with their siblings following weaning had more effect on their adult aggressive behavior than 10 days of social contact with their siblings after the mice had become sexually mature. Although the same opportunities for social contacts were provided the mice at the different ages, it is impossible to conclude that the

nature of the social contacts was similar. Mice 20 to 30 days old have different types of social interactions than mice 65 to 75 days old. Social exposure during maturity probably induces the siblings to fight and establish a dominance hierarchy. Further fighting among the defeated mice in the test situation may be inhibited. These defeated mice, which are not aggressive, may account for the long latencies in Group VI. The fact that mammals are capable of exhibiting different behavioral patterns throughout their maturation does not invalidate the critical period hypothesis.

The quantity and quality of the experience also plays a role in determining the effect of the early experience. One purpose of this study was to learn the minimum number of days of early social contacts necessary to establish a high level of aggression in adult mice. From the three lengths of social exposure examined in this study, namely, 5, 10, and 25 days following weaning, we may conclude that longer periods have greater effects with diminishing returns. Mice exposed to siblings the first 5 days after weaning fight slower than those exposed the first 10 days, and mice exposed 10 days are slightly, but not significantly, slower to fight than mice with 25 days of exposure. Since the differences in the fighting latencies are greatest between 5 and 10 days, we may conclude that 10 days of social contacts immediately after

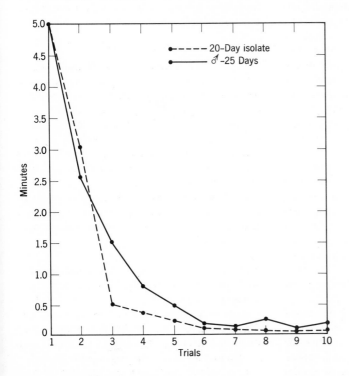

FIGURE 4. Median curves of the fighting response latencies of Groups IX (20-day isolate) and X (♂ 25-day) of BALB/c mice.

weaning is the minimum length of time needed to produce adult patterns of aggressive behavior.

The quality of the early experience was not directly observed in the present study in an effort to allow the social associations to be as natural as possible. In an experiment of Fredericson (5), one group of mice trained to compete for food between 29 and 36 days of age and another group for just one day at 33 days of age were more competitive as adults than another group without the training. Kagan and Beach (7) observed that the type of behavior of young male rats toward receptive females influenced the adult pattern of male sexual behavior. Kahn (8) subjected groups of young mice to 30 days of severe defeats by another mouse starting at 21, 35, and 60 days of age. He found the greatest differences in adult aggressive responses between mice which had the experience at 21 and 60 days of age. In the present study the nature of the social experiences of young mice living together was analyzed by restricting their contacts with a screen partition. This prohibition of overt fighting and competition for food, water, and space did not result in low fighting latencies in the adults. The contacts made through the screen were sufficient to produce maximum fighting responses. It appears that the slow fighting response latency of isolated mice results from an inhibiting mechanism. In a preceding paper (9), social investigation was suggested as the mechanism which inhibited fighting. Before mice fight, they must first investigate each other. The duration of this investigatory period in the test is prolonged among the isolated mice because they have not seen another mouse since weaning at 20 days. In contrast, those mice which have had the opportunity to investigate other mice through a wire screen for only 10 days following weaning have acquired sufficient experience with each other to react aggressively in a fighting situation.

Although the effects of early social experience may perseverate into adulthood when the mice are kept in isolation, the results from retesting Group III at 200 days of age suggest that intervening social experiences may eliminate the early effects. Since the acquisition of social experiences is a continuous process in the normal development of most mammals, the effects of the early experience may be altered by subsequent experiences.

Early experiences may affect adult aggressive and sexual behavior in mice (5, 8), rats (7), and guinea pigs (14), but there is no evidence that the effects of the early experience generalize from one type of adult behavior to another type. It has not been proven possible to predict the effect of an early experience on adult sexual behavior from the effects produced on adult aggressive behavior (10). An early experience which induces aggressiveness in adult mice may either inhibit or enhance their adult sexual behavior.

Genetic factors appear to be important in assessing the effect of early experience on adult behavior. A mouse of one genotype may be affected by an early experience differently than a mouse of another genotype. These results illustrate the need for caution in accepting the effect of early experiences as a general principle. In any heterogeneous society of mammals, some individuals may be more predisposed to modification by a given early experience than are others.

Although the results of this study confirm the critical period hypothesis in general, they also indicate that there are several qualifications in its application to mice. (a) The effects of the experiences gained during the critical period may be altered, reduced, or eliminated by subsequent experiences; (b) they may be specific for each type of adult behavior; and (c) there may be different effects on individuals of unlike genotypes. These qualifications suggest a modification of the critical period hypothesis previously quoted (2). There are specific stages in ontogeny during which certain capacities for behavior appear; the manner in which these capacities are utilized determines subsequent behavior; if these capacities are not utilized they tend to be inactivated.

G. SUMMARY

A total 288 mice of C57BL/10 and BALB/c strains were given different social experiences or the same experience at different ages in order to learn the effect of these experiences on adult aggressive behavior. Adult aggression was measured by the fighting response latency of mice initially exposed to each other. The following statements pertain only to the strains of mice used in this study.

1. Male mice with 10 or 25 days of social associations immediately after weaning are faster fighters than males with 5 days or without postweaning social associations.

2. Male mice which have social contacts immediately after weaning are faster fighters when adults than mice which have social contacts during maturity.

3. Male mice separated by a screen mesh from their siblings for a period of 25 days after weaning show no difference in aggression from those raised in contact with each other for the same period.

4. A second aggression test repeated 90 days

after the first test indicates that male mice are faster fighters in the second test. The effect of the early social associations did not perseverate.

5. Male mice which were given both sexual and aggressive tests failed to demonstrate that their sexual and aggressive behavior were similarly affected by their early experiences.

6. Male C57BL/10 mice isolated at 20 days of age were less aggressive adults than male BALB/c mice isolated at the same age.

REFERENCES

1. BEACH, F. A. Comparison of copulatory behavior of male rats raised in isolation, cohabitation, and segregation. *J. Genet. Psychol.*, 1942, **60**, 121–136.
2. BEACH, F. A. and JAYNES, J. Effects of early experience upon the behavior of animals. *Psychol. Bull.*, 1954, **51**, 239–263.
3. COLLIAS, N. Statistical analysis of factors which make for success in initial encounters between hens. *Amer. Nat.*, 1943, **77**, 519–538.
4. FREDERICSON, E. Response latency and habit strength in relationship to spontaneous fighting in C57 black mice. *Anat. Rec.*, 1949, **105**, 29 (abstract).
5. ———. Competition: The effects of infantile experience upon adult behavior. *J. Abn. & Soc. Psychol.*, 1951, **46**, 406–409.
6. GINSBURG, B. and ALLEE, W. C. Some effects of conditioning on social dominance and subordination in inbred strains of mice. *Physiol. Zoöl.*, 1942, **15**, 485–506.
7. KAGAN, J. and BEACH, F. A. Effects of early experience on mating behavior in male rats. *J. Comp. & Physiol. Psychol.*, 1953, **46**, 204–208.
8. KAHN, M. W. The effect of severe defeat at various age levels on the aggressive behavior of mice. *J. Genet. Psychol.*, 1951, **79**, 117–130.

9. KING, J. A. and GURNEY, N. L. Effect of early social experience on adult aggressive behavior in C57BL/10 mice. *J. Comp. & Physiol. Psychol.*, 1954, **47**, 326–330.
10. KING, J. A. Sexual behavior of C57BL/10 mice and its relation to early social experience. *J. Genet. Psychol.*, 1956, **88**, 223–229.
11. MANN, H. B. and WHITNEY, D. R. On a test of whether one of two random variables is stochastically larger than the other. *Ann. Math. Statist.*, 1947, **18**, 50–60.
12. SCOTT, J. P. and MARSTON, M. Criticial periods affecting the development of normal and maladjustive social behavior of puppies. *J. Genet. Psychol.*, 1950, **77**, 25–60.
13. STONE, C. P. The congenital sexual behavior of the young male albino rat. *J. Comp. Psychol.*, 1922, **2**, 95–153.
14. VALENSTEIN, E. S. The rôle of learning and androgen in the organization and display of sexual behavior in the male guinea pig. *Amer. Psychol.*, 1954, **9**, 486 (Abstract).
15. WILLIAMS, E. and SCOTT, J. P. The development of social behavior patterns in the mouse, in relation to natural periods. *Behaviour*, 1953, **6**, 35–64.

REFLEXIVE FIGHTING IN RESPONSE TO AVERSIVE STIMULATION*

R. E. Ulrich and N. H. Azrin

WHEN electric foot-shock is delivered to paired rats, a stereotyped fighting reaction results (O'Kelly and Steckle, 1939; Daniel, 1943;

Richter, 1950). The present investigation studies several possible determinants of this fighting reaction.

* From the *Journal of the Experimental Analysis of Behavior*, 1962, **5**, 511–520. Copyright 1962 by the Society for the Experimental Analysis of Behavior, Ann Arbor, Mich. This paper is based in part on a thesis submitted by the first-named author in partial fulfilment of the requirements for the Ph.D. degree at Southern Illinois University. This investigation was conducted at the Anna State Hospital and was supported by grants from NSF, NIMH, and the Psychiatric Training and Research Fund of the Illinois Department of Public Welfare. The assistance of W. Holz, R. Hutchinson, and Mrs K. Oliver is gratefully acknowledged.

FIGURE 1. Example of the stereotyped fighting posture.

METHOD

SUBJECTS. Male Sprague-Dawley rats of the Holtzman strain were used because rats of this strain were found to be very docile and non-aggressive in the absence of electric shock. At the beginning of the experiment and subjects were approximately 100 days old and weighed between 295–335 gm. None of the rats had prior experience with the apparatus.

APPARATUS. The experimental compartment measured 12 in. by 9 in. by 8 in., two sides of which were constructed of sheet metal and the other two of clear plastic. The floor consisted of steel rods, $\frac{3}{32}$ in. in diameter and spaced 0.5 in. apart. An open chest contained the experimental chamber, thereby permitting a clear view through the transparent door of the chamber. A shielded, 10-watt bulb at the top provided illumination, and a speaker produced a "white" masking noise. An exhaust fan provided additional masking noise as well as ventilation. The temperature was maintained at about 75°F. The various stimulus conditions used were programmed by electrical apparatus located in a room separate from the experimental chamber. A cumulative recorder, counters, and timers provided a record of the responses. Shock was delivered to the subjects through the grid floor for 0.5 sec. duration from an Applegate constant current stimulator. A

shock scrambler provided a changing pattern of polarities so that any two of the floor grids would be opposite polarity during a major part of each presentation of shock.

PROCEDURE AND RESULTS

DEFINITION OF THE FIGHTING RESPONSE. When two Sprague-Dawley rats were first placed in the experimental chamber, they moved about slowly, sniffing the walls, the grid, and occasionally each other. At no time did any fighting behavior appear in the absence of shock. Soon after shock was delivered, a drastic change in the rats' behavior took place. They would suddenly face each other in an upright position, and with the head thrust forward and the mouth open, they would strike vigorously at each other assuming the stereotyped posture shown in Figure 1.

This behavior has typically been referred to as fighting (Scott and Fredericson, 1951), and it was found to be readily identifiable provided that the topography of the response was well specified. For this experiment, a fighting response was recorded by an observer who depressed a microswitch for any striking or biting movement of either or both animals toward the other while in the stereotyped fighting posture. Once a shock was delivered, the subjects would typically assume and maintain this posture for brief periods during which several striking movements might be made. A new response was recorded only for those striking movements which were separated from previous striking movements by approximately 1 sec. Typically, rats struck at each other for only a brief duration (less than 1 sec.) following a delivery of shock; therefore, the number of fighting episodes was more easily recorded than

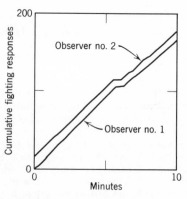

FIGURE 2. Agreement between observers in the simultaneous recording of fighting responses.

the duration of fighting. The duration for which the rats maintained the stereotyped fighting posture could not be reliably measured since this posture often blended imperceptibly in time into a more normal posture.

A measure of the reliability of recording was obtained by having two observers simultaneously score the fighting behavior. Figure 2 shows the cumulative records of the fighting responses which occurred during a 10-min. period in which shock was presented at a frequency of 20 shocks per min. The number of fighting responses recorded by each observer agreed within 5 per cent. The parallel slopes of the two lines indicate that there was close agreement between the two observers on both the total number of responses and also on the momentary changes in the rate of fighting.

FREQUENCY OF SHOCK PRESENTATION. Six rats were divided into three pairs, and each pair was exposed to electric foot-shock (2 mA) delivered at frequencies of 0.1, 0.6, 2, 20, and 38 shocks per min. Each of these frequencies was administered during each of three different sessions (10 min. per session) with a 24-hour interval usually allowed after each session. The order of presentation of frequencies was irregular. Figure 3 is the rate of fighting for each of the three pairs of subjects as a function of the frequency of shock presentation. The frequency of fighting for each pair of subjects increased from zero responses in the absence of shock to 33 fighting responses per min. at a frequency of 38 shocks per min. Individual differences between the pairs of rats were largely absent; the frequency of fighting of the different pairs of subjects was almost identical at each of the shock frequencies.

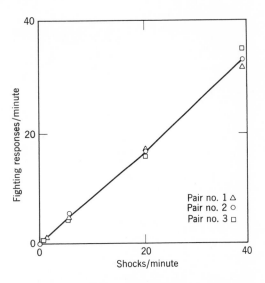

FIGURE 3. The elicitation of fighting responses as a function of the frequency of presentation of foot-shock for each of three pairs of rats.

If each delivery of shock produced a fighting response, the rate of fighting would be directly known from the frequency of shock presentation. Indeed, the higher frequencies of shock presentation did result in a relationship of this sort. Shock frequencies in excess of 6 per min. produced fighting in response to 82.93 per cent of the shocks (Table 1). Lower frequencies of shock (less than 1 per min.) produced fighting in response to no more than 66 per cent of the shocks. Visual observation of the rats revealed that shortly after a shock was presented, the subjects slipped out of the fighting posture and assumed other positions. It was also apparent that fighting in response to shock was more likely if the animals were facing each other at the moment of shock-delivery. Thus, the probability of fighting appeared to be lower at the lower frequencies of shock presentation because of the likelihood that the rats were at some distance from each other. This direct relationship between rate of shock presentation and rate of fighting reversed at very high frequencies. In an additional study with two pairs of rats, the shock was made so frequent as to be continuous. Although occasional fighting responses occurred, much of the behavior of the rats appeared directed toward escape from the experimental chamber. This "escape" behavior appeared to interfere somewhat with the usual reflexive fighting. Such behavior was also noted during the early part of the initial session when the subjects were first

TABLE 1. *Examples of the Consistency of Fighting Elicited by Shock from Three Pairs of Subjects During Two Sessions at Each of the Different Shock Frequencies. The Consistency of the Fighting Reflex is Expressed as the Percentage of Shocks that Resulted in a Fighting Response.*

Frequency of Shocks (Shocks/Min.)	Consistency of Fighting Reflex (Responses) (Shocks)		
	Pair No. 1	Pair No. 2	Pair No. 3
0.1	0.33	0.66	0.66
0.6	0.61	0.55	0.61
2.0	0.83	0.58	0.58
6.0	0.83	0.94	0.77
20.0	0.92	0.91	0.82
38.0	0.85	0.89	0.93

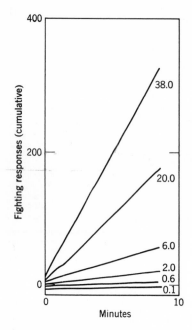

FIGURE 4. Typical curves for one pair of rats of the fighting responses at various frequencies of presentation of shock.

presented with shock. However, in this case the escape behavior did not persist.

Intrasession changes in fighting behavior were conspicuously absent (Fig. 4). The bottom curve is the cumulative record of the fighting for a 10-min. session in which only one shock was delivered at the middle of the session. This single shock produced an immediate fighting response. At a shock frequency of 0.6 shocks per min. (second curve from bottom) the rats did not fight after all of the six shock deliveries, but observation revealed that the four fighting responses which did occur were immediately preceded by the presentation of a shock. At no time did fighting occur during the interval between shock presentations although the stereotyped fighting was often maintained during that time. No warm-up period appeared at the beginning of the session; nor did the frequency of fighting decrease toward the end of the session.

SEQUENTIAL EFFECTS. Elicitation of the fighting reflex on a given day was virtually independent of the shock frequency used on preceding days or even on the same day. As a rule, the number of fighting responses at a given shock frequency varied less than 10 per cent, irrespective of the preceding shock frequency. On several occasions, the sessions followed within 10 min. of each other in order to determine the effects of a shorter

interval between sessions. At a frequency of 2 shocks per min., 68 per cent of the shocks were effective when 24 hours were allowed between sessions; 63 per cent of the shocks were effective when only 10 min. were allowed between sessions. This small difference in responding as a function of the interval between sessions was typical. The strength of the fighting reflex appears to be fairly independent of its history of elicitation.

REFLEX FATIGUE. Figure 2 revealed little change in the consistency with which the fighting reflex was elicited, even after 300 elicitations at the higher rates of shock presentation. In order to evaluate reflex fatigue, frequent shocks (every 1.5 sec.) were delivered to a pair of rats for an uninterrupted period of $7\frac{1}{2}$ hours. The fighting reflex proved extremely resistant to fatigue (Fig.5). During the first 2400 presentations (1 hour) of the shock, fighting was elicited after 82 per cent of the shocks. After 7200 presentations of shock (third hour), fighting still occurred after 70 per cent of the shocks. Only during the last 1.5 hours, after 6 hours and nearly 15,000 shocks, did the consistency of elicitation drop below 40 per cent. By this time the rats were damp with perspiration and appeared to be weakened physically. By the end of the 7.5 hours, approximately 10,000 fighting responses had been elicited. Several observers were required because of the extended observation period.

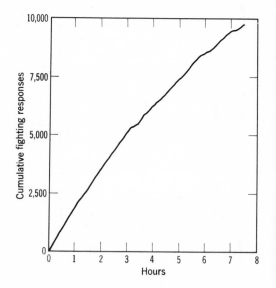

FIGURE 5. Cumulative record of the fighting responses that were elicited from a pair of rats during a long period (7.5 hr.) of frequent (every 1.5 sec.) shock presentation.

INTENSITY OF SHOCK PRESENTATION. Three pairs of rats were exposed to various intensities of shock at a fixed frequency of 20 shocks per min. Each intensity was presented for at least 10 min. The sequence of intensities was varied and several 10-min. periods were given at each intensity. The cumulative-response curves of Figure 6 for one pair of rats were typical of those obtained with all three pairs of rats. Increasing the shock intensity from 0–2 mA produced an increased frequency of fighting; at still higher intensities (3–5 ma.), the rate of fighting was somewhat reduced. Visual observations indicated that lower intensities produced a fighting response of less vigor and longer latency. Also, at the lower intensities, chance factors, such as the orientation of the rats relative to each other and to the grid floor, appeared to influence greatly the likelihood of a fighting response. If the rats were making good contact across several of the floor grids, and were also oriented toward each other, a fighting response was likely to result. Even so, this response was relatively short in duration, slow in onset, less vigorous, and less likely to result in a maintained fighting posture than the responses elicited by the higher current intensities. At these lower intensities, the definition of a movement as a fighting response often became arbitrary. At the higher intensities, the attack movement was unmistakable.

The slight decrease in fighting behavior at the highest intensity (5 mA) appeared to be partly a consequence of the debilitating effects of the shock. Prolonged exposure to this intensity often resulted in a complete loss of fighting because of the paralysis of one or both of the subjects. Even during the initial exposure to this very high intensity, fighting behavior appeared to be reduced by the strong tendency of the rats to engage in other shock-induced behavior, such as biting the grids, jumping, running, or pushing on the walls.

Thus, the optimal current intensity for eliciting fighting was approximately 2 mA. At lower intensities, the shock did not appear to be sufficiently aversive, while at higher intensities, the shock appeared to be debilitating and generated competing behavior. Tedeschi (1959) also found that 2- to 3-mA intensity is optimal for producing fighting between mice.

UNIFORMITY OF SHOCK PRESENTATION. All previous investigations of shock-produced fighting appear to have used the same type of shock circuit. Alternate bars of the floor grid have been wired in parallel so that adjacent bars were of opposite polarity, but many nonadjacent bars were of the same polarity. Such a design permits the rat to avoid the scheduled shocks by standing on bars of the same polarity. Skinner and Campbell (1947) found that this unauthorized avoidance could be eliminated by a scrambling circuit which insured that any two bars would be of opposite polarity during a major part of each shock delivery. A scrambling circuit of this sort was used throughout the present investigation. Three pairs of rats were now studied to determine the effects of omitting this scrambling circuit. An hour-long period of shock (2-mA intensity at a rate of 20 per min.) was given to each pair of rats on each of three successive days. On one or two of these days, the scrambler was omitted. For all three pairs of rats, the omission of the scrambler produced less than half as many fighting responses as were obtained with the scrambler. The curves in Figure 7 for one pair of rats reveal great variability in the frequency of fighting; periods of frequent fighting alternate with periods of little or no fighting. Visual observation revealed that one or both rats often avoided shocks by standing on bars of like-polarity. This safe posture was often maintained for several minutes during which no fighting was produced. When a part of the rat happened to contact a bar of different polarity, the resulting shock usually jolted the rat out of this safe posture. For the next few minutes, the rat was likely to receive the scheduled shocks and fighting resumed until once again a safe position was discovered. When the scrambler was in use, no safe position was possible and the rats typically fought immediately following each scheduled shock. The omission of a polarity scrambler in past studies may account for the frequent failure of shock to elicit fighting behavior (Miller, 1948; Richter, 1950).

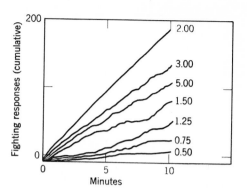

FIGURE 6. Typical cumulative records of the fighting responses that were elicited from one pair of rats at various intensities of footshock.

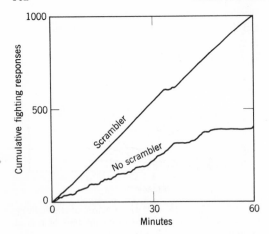

FIGURE 7. The elicitation of fighting responses by foot-shocks that were delivered with or without a polarity scrambler for the floor grids.

PREVIOUS EXPERIENCE. In this study, each rat had been housed individually and had no prior contact with his fighting-mate. This general unfamiliarity of the rats with each other might have been a factor in obtaining the fighting response to shock. This possibility was evaluated by housing two rats together in a single cage for several weeks. Subsequent exposure to foot-shock in the experimental chamber produced the same degree of fighting that had been obtained when the same rats had been housed separately. These results were replicated with 24 other animals. It appears, therefore, that previous familiarity of rats with each other does not appreciably effect the elicitation of fighting through foot-shock. On the other hand, nonreflexive fighting behavior has been found to be affected by previous familiarity (Seward, 1945).

SEX. Male rats are known to fight more often than female rats in a natural (no-shock) situation (Beeman, 1947; Scott and Fredericson, 1951). The relevance of sex for the elicitation of fighting by foot-shock was investigated by pairing a female rat with a second female, and a male rat with a female. Several such pairings revealed the same type of fighting in response to foot-shock (2 mA, 20 deliveries per min.) as had been obtained between the two male rats. Indeed, the sexual behavior between the male–female pair was completely displaced by the elicitation of fighting soon after the first few shocks were delivered. Unlike "natural" fighting behavior, reflexive fighting behavior does not appear to be appreciably affected by sexual differences.

NUMBER OF RATS. Reflexive fighting also resulted when more than two rats were shocked.

When 2, 3, 4, 6, or 8 rats were simultaneously given foot-shock, the same stereotyped fighting reaction occurred, two or more rats often aggressing against a single rat.

SIZE OF CHAMBER. Throughout the present study the size of the experimental chamber was 12 in. by 9 in. by 8 in. In this phase a pair of rats was given shock (2 mA) for 10 min. (20 shocks per min.) in a square chamber having an adjustable floor area. The height was held constant at 17 in. Figure 8 shows the number of fighting responses as a function of the floor area at each of the different floor sizes. With only a very small amount of floor space (6 in. by 6 in.) the fighting response was elicited by approximately 90 per cent of the shocks. At the larger floor areas, the number of fighting responses decreased; with the largest floor space (24 in. by 24 in.), only 2 per cent of the shocks elicited fighting. The amount of fighting between rats in response to shock appears to depend critically upon the amount of floor space in the fighting chamber. When the rats were only a few inches apart, the shock was likely to cause them to turn and lunge at each other. At the larger distances, the rats largely ignored each other.

STRAIN. As mentioned above, the Holtzman Sprague-Dawley rats are unusually docile in the absence of shock. Additional study revealed that other less docile strains of rats also exhibited this shock-elicited fighting. Two pairs of mature male rats from four other strains (Long-Evans hooded,

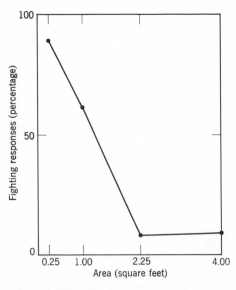

FIGURE 8. Elicitation of fighting responses from two rats by footshock in a square chamber of constant height and variable floor area.

Wistar, General Biological hooded, Charles River Sprague-Dawley)[1] were exposed to the optimal shock conditions (2 mA at 20 shocks per min.) in the same experimental chamber (12 in. by 9 in. by 8 in.) as had been used for the Holtzman strain. In all of the strains the same stereotyped fighting reaction occurred following the presentations of shock. However, less than 50 per cent of the shocks produced fighting between rats of the Wistar strain, whereas over 70 per cent of the shocks produced fighting between rats in each of the other strains. The Wistar rats appeared to be more sensitive to the shock since much competing behavior was generated by shocks of 2 mA intensity, and 2 out of the 4 Wistar rats died after exposure to these shocks. Apart from this seemingly greater sensitivity of the Wistar-strain rats, all of the strains showed the same stereotyped fighting response to foot-shock.

SPECIES. Mature guinea pigs and hamsters were studied under the same conditions of shock presentation and in the same experimental chamber to ascertain the existence of reflexive fighting in other species. Delivery of shock to a pair of hamsters produced a similar type of stereotyped fighting posture and attack as was seen with rats. These fighting responses could be consistently elicited at lower intensities of shock (0.75 mA) than was required with the rats. Also, the hamsters persisted longer in their fighting, often biting and rolling over each other. Tedeschi (1959) found that paired mice also fought vigorously in response to foot-shock. In contrast, the paired guinea pigs never showed the fighting posture or any attack movements in response to shock. Variations in the intensity and frequency of shock presentation, as well as food deprivation up to 72 hours, did not alter this failure to fight.

INTERSPECIES FIGHTING. When a Sprague-Dawley rat was paired with a hamster, shock produced the same fighting reaction by both animals. However, when a rat was paired with a guinea pig, all of the attacking was done by the rat. The guinea pig reacted only by withdrawing from the rat's biting attacks following the shock delivery. The rat attacked only the head of the guinea pig. During this attack, the rat assumed a semi-crouching position with the forepaws raised only slightly off the floor, a posture which differed from the upright position assumed by rats in fighting each other. Since the guinea pig never stood upright, the crouching position of the rat brought its head to the level of the guinea pig's head. The otherwise inflexible and stereotyped fighting posture of the rat appeared to be modified by the position of the guinea pig. No fighting occurred in the absence of shock.

INANIMATE OBJECTS. When an insulated doll was placed into the experimental chamber while a rat was being shocked, no attack was attempted. Similarly, no attack movements were made toward either a conducting doll or a recently deceased rat. Dolls moved rapidly about the cage also failed to produce fighting. Fighting responses were elicited only when the dead rat was moved about the cage on a stick.

ELECTRODE SHOCK. In using foot-shock, both rats are shocked simultaneously since they are standing on the same grid floor. Does the elicitation of fighting require that both rats be shocked? This question might be investigated by electrifying only that section of the grid under one of the rats. However, the rat quickly learns to stand on a nonelectrified section. A second solution is to shock the rats through implanted electrodes. The two rats were placed in an experimental chamber, and electrodes were implanted beneath a fold of skin on the back of one rat. A harness and swivel arrangement allowed the rat complete freedom in moving about. When a 0.5 sec. shock was delivered at an intensity of 2 mA, only a spasmodic movement of the rat resulted if no other rat were present. When the shock was delivered in the presence of a second rat, the stimulated rat usually assumed the stereotyped fighting posture and attacked the unstimulated rat. Upon being attacked, the unstimulated rat in turn often assumed the stereotyped posture and returned the attack. Once the attack was initiated by the shock, the continuance of the fighting appeared to be partly under social control. Fighting was elicited, then, even when only one member of the pair of rats was stimulated. Somewhat the same result was seen above when foot-shock elicited fighting in a rat paired with a guinea pig, in spite of the failure of the guinea pig to reciprocate. Similarly, in the course of delivering foot-shock to a pair of rats, occasionally a rat

---

[1] The different strains of rats were obtained from the following suppliers:

Long-Evans hooded: Small Animal Industry, Chamberland, Indiana;

Wistar: Albino Farms, Redbank, New Jersey;

Long-Evans hooded: Small Animal Industry, Chamberland, Indiana;

Wistar: Albino Farms, Redbank, New Jersey;

General Biological hooded: General Biological Supply House, Inc., Chicago 20, Illinois;

Charles River Sprague-Dawley: Charles River Laboratories, Inc., Brookline 46, Massachusetts;

Holtzman Sptrague-Dawley: Holtzman Company, Madison, Wisconsin.

would learn to eliminate the shock by lying motionless on its back, thereby producing a situation in which only one rat was being stimulated. Under these circumstances, the rat stimulated by foot-shock often attacked the supine rat in the same way that the rat stimulated by electrode shock attacked the unstimulated rat. It should be noted that in each of these situations where only one rat was being stimulated, the full-blown fighting response was elicited less frequently than when both rats were stimulated. Stimulation of a second rat is not a necessary condition for producing the fighting reaction but does, nevertheless, increase the likelihood of its occurrence.

INTENSE HEAT. The elicitation of the fighting reflex through electrode-shock as well as foot-shock suggested that other aversive stimulation also might elicit fighting. A pair of rats was placed in an experimental chamber with a thin metal floor that could be heated from below by a heating coil. After the heating coil was energized, the metal floor became progressively hotter and the two rats began jumping about and licking their feet. No fighting was produced in spite of the agitated movements of both rats. However, when the same pair later was placed on a preheated floor, fighting consistently resulted. The same results were obtained with additional rats. The rats scrambled about the chamber, interrupting their movements frequently to assume a fixed position and attack each other before resuming their running about. It is very likely that the rats received more painful heat stimulation during the fighting episodes than they would have received if they had jumped about. No more than 2 min. exposure to the heated floor was given because of the possibility of tissue damage. Nevertheless, the heated floor appeared to elicit fighting in much the same manner as a continuously electrified floor grid. It is probable that the gradual heating of the floor grid allowed the reinforcement of competing behavior, especially licking of the forepaws. This wetting the paws appeared to be effective in cooling the animal at the initially lower temperature of the gradually heated floor but not at the high temperature of the preheated floor. Once fighting was elicited by a preheated floor, subsequent exposure to a gradually heated floor did elicit some fighting, and the competing licking behaviors were reduced.

COLD AND INTENSE NOISE. In spite of the effectiveness of intense heat in eliciting fighting behavior, no fighting was elicited by placing rats on a sheet metal floor pre-cooled by dry ice. It is possible that the temperature induced by the dry ice was not sufficiently aversive; no pain was felt by a human observer upon touching the cooled floor for periods less than 2 sec. Since the rats were consistently moving about, it is quite likely that they did not allow a given paw to remain in contact with the cold floor for a sufficient period of time. Since the cool floor did not produce pain upon immediate contact, unlike electric shock and heat, the rat probably could eliminate pain completely in much the same manner as the rat lying upon its insulated back can completely eliminate painful foot-shock.

Intense noise was similarly ineffective in producing fighting behavior between paired rats. The noise was at an intensity of 135 dB (re 0.0002 dyne/cm^2) and enclosed a band from 200–1500 cps. The delivery of noise was varied from brief bursts of less than 1 sec. to periods of more than 1 min. No fighting resulted. A pair of guinea pigs was subjected to the same treatment in the expectation that guinea pigs might be more reactive to intense noise. No fighting resulted.

Fighting appears to be elicited by foot-shock, electrode shock, and intense heat, but not by intense noise or moderate cold.

DISCUSSION

The present investigation found that fighting behavior could be elicited from several paired species by several different types of aversive stimulation. The elicitation of this fighting occurred in almost a one-to-one relationship to the aversive stimulus when the optimal value of the aversive stimulus was used. When a response, such as salivation, is consistently made to a stimulus, such as meat powder, with no previous training, that response is referred to as an unconditioned response (Pavlov, 1927; Sherrington, 1947) or as a respondent (Skinner, 1938). Physiologists have supplied us with the term reflex to designate such specific stimulus-response relationships and in fact have extended the term to denote responses for which related stimuli are not always clearly observable (Keller and Schoenfeld, 1950). The consistent elicitation of the fighting response by aversive stimulation without prior conditioning appears to be best defined as an unconditional reflex. Miller (1948), however, has taken a different approach in the study of fighting behavior. He reports that he trained his subjects to fight by removing the shock each time the animals approximated the fighting position. In this case fighting is presumed to be an escape reaction that is reinforced by the termination of electric shock. In spite of the virtual one-to-one relationship between shock and fighting observed in the present study, it is possible that this apparently reflexive fighting was maintained by some unsuspected and perhaps subtle operant reinforcement. Several possible sources of operant reinforcement seem apparent. First, it is possible that the rats were simply attempting to stand on each other in order to eliminate the aversive stimulation. Several observations made during the course of these experiments bear upon this interpretation: (1) When one of a pair of rats was lying on its back and effectively avoiding all shock, the shocked rat, rather than attempting to climb upon the other rat, often directed an attack specifically at the other rat's head. (2) Fighting was maintained by electrode shock even though no escape was available to the rat stimulated through the electrodes. (3) Leaning against the other rat eliminated the shock no more than simply leaning against one of the insulating plastic walls of the experimental chamber. (4) On the heated floor, the fighting behavior served to increase rather than decrease the amount of aversive stimulation. (5) When an insulated doll was placed in the experimental chamber while a rat was given foot-shock, no attempt was made by the rat to jump upon the doll until several minutes of stimulation had elapsed.

A second possible source of operant reinforcement of fighting is that the fixed-duration shock delivery happened to terminate at the moment that the rats moved toward each other; thus, superstitious reinforcement of these movements would have resulted (Skinner, 1948). Again, several observations indicated that reinforcement of this sort was not operative in producing fighting: (1) Fighting often occurred with the onset of the first shock delivery when prior reinforcement through shock reduction was necessarily impossible. (2) Continuous and uninterrupted delivery of either foot shock or severe heat produced fighting. Of course, no reinforcement through the termination of the stimulus can result if the stimulus is not terminated.

A plausible interpretation of the fighting reflex is that a rat will attack any nearby object or organism upon being aversively stimulated. However, rats did not attack a nearby doll, either insulating or conducting, upon being shocked. Nor was the movement of an inanimate object in the presence of a shocked rat a sufficient condition for eliciting fighting. No fighting resulted when the dolls were moved about the cage at the end of a stick during and between shock presentations. Additional experiments revealed that even a recently deceased rat would not be attacked by a second rat that was given foot-shock, unless the dead rat was moved about the cage on a stick. It would seem, therefore, that a second moving animal either rat, guinea pig or hamster is a necessary condition for eliciting the fighting response from a rat stimulated by foot-shock.

REFERENCES

BEEMAN, E. A. The effect of male hormone on aggressive behavior in mice. *Physiol. Zool.*, 1947, **20**, 373–405.

DANIEL, W. J. An experimental note on the O'Kelly-Steckle reaction. *J. comp. Psychol.*, 1943, **35**, 267–268.

KELLER, F. S. and SCHOENFELD, W. N. *Principles of psychology.* New York: Appleton-Century-Crofts, Inc., 1950.

MILLER, N. E. Theory and experiment relating psychoanalytic displacement to stimulus-response generalization. *J. abn. & soc. Psychol.*, 1948, **43**, No. 2, 155–178.

O'KELLY, K. E. and STECKLE, L. C. A note on long-enduring emotional responses in the rat. *J. Psychol.*, 1939, **8**, 125–131.

PAVLOV, I. P. Conditioned Reflexes: *An investigation of the physiological activity of the cerebral Cortex.* London: Oxford University Press, 1927.

RICHTER, C. P. Domestication of the Norway rat and its implications for the problem of stress. *Assoc. Res. in Nerv. and ment. dis. Proc.*, 1950, **29**, 19.

SCOTT, J. P. and FREDERICSON, E. The causes of fighting in mice and rats. *Physiol. Zool.*, 1951, **24**, No. 4, 273–309.

SEWARD, J. P. Aggressive behavior in the rat. I. General characteristics: age and sex differences; II. An attempt to establish a dominance hierarchy; III. The role of frustration; IV. Submission as determined by conditioning, extinction, and disuse. *J. comp. Psychol.*, 1945, **38**, 175–197, 213–224, 225–238, **39**, 51–76.

SHERRINGTON, C. *The integrative action of the nervous system.* New Haven: Yale University Press, 1947.

SKINNER, B. F. *The behavior of organisms.* New York: D. Appleton Century Co., 1938.

SKINNER, B. F. "Superstition" in the pigeon. *J. exp. Psychol.*, 1948, **38**, 168–172.

SKINNER, B. F. and CAMPBELL, S. L. An automatic shocking grid apparatus for continuous use. *J. comp. physiol. Psychol.*, 1943, **40**, 305–307.

TEDESCHI, R. E. Effects of various centrally acting drugs on fighting behavior of mice. *J. pharmacol. exp. Therap.*, 1959, **125**, 28.

CLASSICAL CONDITIONING OF PAIN-ELICITED AGGRESSION*

Walter Vernon and Roger Ulrich

PAIRED animals fight when subjected to painful stimulation (1, 2). This behavior occurs within many species and between members of different species (3). Despite only minimum success in earlier attempts to develop conditioned fighting by pairing painful stimuli, such as electric shocks, with neutral stimuli (4), we have successfully conditioned the pain-aggression response in rats.

Eighteen young, adult, male Long-Evans hooded rats were subjects. A clear plastic experimental chamber measuring 38 cm by 39 cm by 36 cm had, as its floor, 23 rods, 3.18 mm in diameter and spaced 11.11 mm apart. The chamber was housed in a larger sound-attenuated chest and had a clear observation window. Electric shocks and electrically generated tones of 60 dB at 1320 cy/sec. were programmed by apparatus in a separate room. Initially it was established that the sound stimulus would not elicit fighting through 1000 repetitions, nor would a pseudoconditioned response occur to the sound stimulus when it was first presented immediately after 1000 electric shocks given at 10-sec. intervals. We paired the neutral auditory stimulus, hereafter referred to as the conditioned stimulus (CS), and the painful electrical unconditioned stimulus (UCS). The nine pairs of animals were given 2000 pairings of the tone with the shock, one of three shock intensities being delivered to each of three different groups of three pairs of animals. Duration of the CS was 1 sec. One-half second after onset of the CS this stimulus was joined by the UCS, both terminating simultaneously after 0.5 second. The onset-to-onset interval between trials with the CS was 10 sec. Each eleventh presentation was the CS alone, this being a test for the development of the conditioned aggressive response. Fighting responses were recorded by an observer who scored any striking or biting movement made by either animal toward the other. Usually such responses were made from a stereotyped fighting posture on the hind legs, which the animals typically maintained through most of each session.

Rates of pain-elicited fighting and the shock intensities bore an inverse relation to one another (Fig. 1). A nonparametric analysis of variance (Kruskal-Wallis; $H=7.19$, $P<0.005$) indicated highly significant differences between total fight responses to the different shock intensities.

The complex nature of pain-elicited fighting is reflected in the change, over 2000 trials, from an initially low response rate of 53 per cent to an average of 88 per cent. This change appears to be largely the result of the extinction of competing responses. As trials progressed, the escape behavior, having no effect upon stimulus events, gradually became less frequent. Although fighting also had no effect upon stimulus events, it increased substantially as trials continued. The fact that unconditioned fighting to 2 mA shock was consistently higher is commensurate with earlier findings which show that 2 mA is the optimum shock intensity for producing pain-aggression in rats (2).

Figure 2 shows that the same relation exists between intensity of shock and rate of conditioned fighting as between intensity of shock and rate of pain-elicited fighting. This is consistent with the common observation (5) that optimum intensities of the UCS for eliciting unconditioned responses are also optimum for the development of conditioned responses. Comparison of Figures 1 and 2 shows that less overlap occurred at each intensity of the UCS among the unconditioned responses; still, nonparametric analysis of variance (Kruskal-Wallis; $H=5.67$, $P<0.05$) of shock intensity and group differences in total conditioned fight responses indicated differences beyond chance expectation.

In addition, animals were observed to fight at a higher rate when they were close together and

* From *Science*, 1966, **152**, 668–669. Copyright 1966 by the American Association for the Advancement of Science.

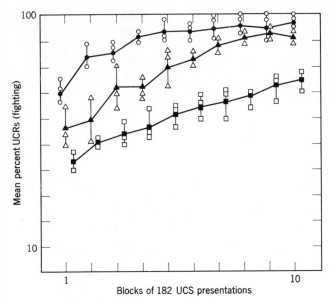

FIGURE 1. Individuals (open symbols) and mean (solid symbols) percentages of UCR's (fighting) are shown for each block of 182 presentations of UCS at each of three different shock intensities. UCS, unconditioned stimulus; UCR, unconditioned response. Circles, 2.0 ma; triangles, 2.5 ma; and squares, 3.0 ma.

facing one another at the time of the onset of shock, in accord with earlier observations (2). It was further observed that, at the onset of shock, if one animal was in a fighting posture, there was considerably greater likelihood that his partner would attack him than if he had all four feet on the grid floor.

The major significance of the study is knowledge of the conditioning phenomenon itself. The finding that organisms respond aggressively to what outwardly may appear to be neutral stimuli, in accordance with their past experiences of pain, suggests a possible explanation for some cases of "apparently unprovoked" aggression.

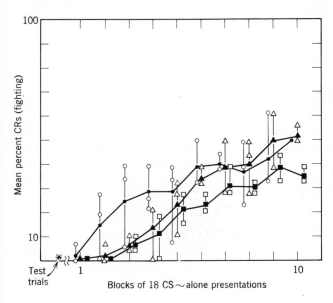

FIGURE 2. After initial test trials demonstrated that no fighting was developed by the tone alone and that no pseudo-conditioning developed (see arrow), tone (CS) plus shock (UCS) trials were instigated. Individual (open symbols) and mean (solid symbols) percentages of conditioned fighting are shown for each block of 18 presentations of CS alone for each of the three shock intensities used. CS. conditioned stimulus; CR, conditioned response. Circles, 2.0 ma; triangles, 2.5 ma; and squares, 3.0 ma.

REFERENCES AND NOTES

1. O'KELLY, L. and STECKLE, L. *J. Psychol.*, **8**, 125 (1939). AZRIN, N., HAKE, D. and HUTCHINSON, R. *J. Exp. Anal. Behav.*, **8**, 55 (1965).

2. ULRICH, R. and AZRIN, N. *J. Exp. Anal. Behav.*, **5**, 511 (1962).

3. AZRIN, N. *Amer. Psychol.*, **19**, 501 (1964). ULRICH,

R., Wolff, P. and Azrin, N. *Animal Behav.*, **12**, 14 (1964).
4. Ulrich, R., Hutchinson, R. and Azrin, N. *Psychol. Rec.*, **15**, 11 (1965).
5. Hilgard, E. and Marquis, D. *Conditioning and Learning* (Appleton-Century, New York, 1940).
6. Supported by grants from the National Institute of

Mental Health and the Illinois Psychiatric Research Authority to a research program developed by Roger Ulrich, then at Illinois Wesleyan University, Bloomington, Illinois, and conducted in that laboratory. The assistance of J. Mabry is gratefully acknowledged.

EFFECTS OF VARIOUS CENTRALLY ACTING DRUGS ON FIGHTING BEHAVIOR OF MICE*

Ralph E. Tedeschi, David H. Tedeschi, Anna Mucha, Leonard Cook, Paul A. Mattis and Edwin J. Fellows[1]

THE rage reaction has been described by Bard and Montcastle (1948) as "a mode of response which in intensity and pattern closely resembles the activity which constitutes the expression of fury in normal members of the same species." Behavior characteristic of anger has been induced in various species of laboratory animals by a variety of experimental methods, including surgical ablation, drugs and electrical stimulation through either implanted electrodes in the central nervous system or foot-shock. The classical studies of Bard (1928) have demonstrated that transection of the brain stem at the level of the hypothalamus in cats evokes a "sham rage" response. Ranson (1934) found that electrical stimulation of the hypothalamus in the unanesthetized or lightly anesthetized cat with brain intact elicits the typical rage behavior pattern. Gaddum and Vogt (1956) have further studied the rage phenomenon in cats after intraventricular injections of morphine and lysergic acid diethylamide. More recently Loewe (1956) and Sturtevant and Drill (1957) have investigated drug antagonism of morphine-induced feline mania. In 1958 Yen and co-workers described the effect of drugs on the isolation-induced aggressive behavior of mice. Numerous investigators have utilized a mild foot-shock to elicit fighting behavior in rats. Miller (1948) studied this phenomenon in experiments involving psychoanalytic displacement.

Rats were placed in a fighting chamber and were trained to strike and fight one another before the foot-shock was turned off. It was found that a single trained animal when placed in the fighting chamber with a celluloid doll would direct its attack at the doll when the foot-shock was turned off. Richter (1950) in efforts to elucidate behavioral differences between domesticated and wild Norway rats subjected the animals to various stress situations. One of the stress situations consisted of subjecting paired rats to foot-shock in a fighting chamber. Covian (1949) studied the role of emotional stress on the survival of adrenalectomized wild rats given replacement therapy. The experimentally induced stress consisted of exposing two adrenalectomized wild rats to footshock in order to elicit fighting behavior. Griffiths and Clifford (1954) studied the self-selection of diet in wild and domesticated Norway rats under conditions of experimentally induced stress which involved observation of rats in fighting pairs during footshock.

The experiments to be described were undertaken to investigate the effects of physiological and pharmacological alterations on a fighting reaction between two mice subjected to foot-shock. This reaction has been termed "fighting behavior."

METHODS. Adult male albino mice (18–28 gm) obtained from Carworth Farms (CF No. 1 strain)

* From the *Journal of Pharmacology and Experimental Therapeutics*, 1959, **125**, 28–34. Copyright 1959 by Williams & Wilkins Co., Baltimore. A preliminary report of this work was presented at the FASEB Meeting in Philadelphia, April 1958.

[1] The authors wish to express their gratitude to Mr. John F. Pauls for assistance in the statistical analysis of the data.

were used as test animals. Mice were not tested more frequently than once every 10 days and were permitted free access to food and water except during the brief test period. Fighting behavior was produced in randomly selected pairs of mice by subjecting them to a foot-shock for a 3-min. period. Unless otherwise specified the parameters for electrical stimulation consisted of an interrupted direct current of 3 mA, 400 V. stimulus intensity of 0.2 sec. duration administered at a frequency of five shocks per second. The frequency of fighting episodes was determined during the 3-min. interval of foot-shock. The apparatus employed consisted of a grid floor composed of parallel stainless steel rods $\frac{3}{16}$ in. in diameter placed approximately $\frac{1}{4}$ in. apart. The pairs of mice were confined during foot-shock by placing them under an inverted circular glass enclosure (2-l. beaker) $5\frac{3}{4}$ in. in diameter and 8 in. high.

Pairs of mice were preselected for drug testing on the basis that they would exhibit at least one fighting episode within a 3-min. period of foot-shock. If the mice failed to respond to the stimulus with a typical fighting reaction they were discarded and a new pair selected. A quantal assay procedure was employed for the analysis of effects of drugs on fighting episodes. Drugs were administered orally to both mice of a fighting pair. Drug-treated pairs of mice exhibiting three fighting episodes or less within 3 min. of foot-shock were designated negative responders, whereas pairs fighting four or more times were considered as positive responders. In this manner a dose-response curve was obtained based on an all or none response. In general, each compound was tested at 3-dose levels using 8 pairs of mice per dose. The dose producing suppression of fighting episodes in 50 per cent of fighting pairs (ED50 F) and 95 per cent fiducial limits were determined by the graphic log-probit method of Litchfield and Wilcoxon (1949). Pairs of mice were tested only once at the time of peak drug action as determined in a preliminary experiment using the same rage test in which different groups of mice were exposed to a single dosage level of drug and the pretreatment time varied. It was important to determine whether certain compounds might exhibit selectivity in suppressing fighting episodes. Therefore, compounds were tested for their effect on spontaneous motor activity (SMA) in mice and also for their ability to protect against maximal electroshock seizures (MES). Motor activity was quantitatively evalu-

ated using the photocell counter technique of Winter and Flataker (1951) as modified by Cook and co-workers (1955). Individual mice were treated with the dose of drug to be tested (six mice per dose at 4-dose levels and six saline-treated controls). Ten minutes before time of peak drug effect[2] each mouse was placed in a photo cell counting chamber. The total count was taken after a 10-min. interval for each individual mouse. The data obtained were then analyzed as follows. A dose-response curve was fitted by computing the regression of mean response on logarithm of dose. From this dose-response curve the dose was estimated which would produce a response equal to 50 per cent of the average control response (DD50SMA). The 95 per cent fiducial limits for this estimated dose were computed using Fieller's theorem (Finney, 1952).

Compounds were tested for their ability to protect against maximal electroshock seizures (MES) using a method essentially the same as that described by Swinyard and co-workers (1952), except that a 25-mA stimulus intensity was employed. Drugs were tested at their time of peak effect using this same test procedure. ED50's and 95 per cent fiducial limits were calculated employing the method of Litchfield and Wilcoxon (1949).

The following drugs were tested for their effects in suppressing fighting episodes, for their effect on SMA and for their ability to prevent MES: chlorpromazine (Thorazine), prochlorperazine (Compazine), trifluoperazine (Stelazine), pheno-

FIGURE 1. A footshock-induced fighting episode in mice. Note that the animals stand face to face on their hind legs and appear to bite one another.

---

[2] The time of peak drug effect was determined in a preliminary experiment in which treated animals

were observed for the time of maximum depression of spontaneous motor activity.

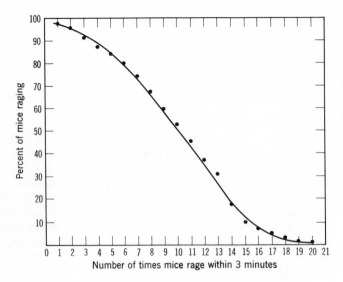

FIGURE 2. Cumulative frequency response curve illustrating the percentage of mice fighting (exhibiting fighting or "rage" episodes) within 3 min. See text for explanation.

barbital sodium, diphenylhydantoin sodium (Dilantin Sodium), mephenesin (Myanesin), diphenhydramine (Benadryl), meprobamate (Miltown), benzilic acid amide (SKF No. 3071) and reserpine.

All drugs were administered orally; the vehicles employed (see Table 2) have been shown not to influence significantly any of the test procedures studied. Doses of drugs are expressed in terms of the free base.

RESULTS. Paired mice reacted to foot-shock by vocalizing and jumping up and down to escape the noxious stimulus. A fighting episode occurred when the mice converged abruptly to close quarters, stood face to face on their hind legs, and sparred and bit savagely at each other. Figure 1 illustrates this reaction. The duration of a rage episode varied from 1 to 4 sec. We considered the demarcation of one fighting episode from the next to occur when the mice returned to the quadrupedal posture and remained apart for 3 or more seconds.

TABLE 1. *The Effects of Repeated Footshock on the Percentage of Mice Exhibiting Fighting Episodes*

Cumulative Time in Minutes	Number of Pairs Fighting/Number of Pairs Tested	Per Cent of Pairs Fighting
1	182/214	85.0
2	200/214	93.9
3	209/214	97.7

The Frequency of Fighting Episodes. Footshock with a 3 mA stimulus intensity elicited fighting episodes in 65 per cent of an untreated

group of 200 randomly selected pairs of mice. The data in Table 1 demonstrate that animals which respond to foot-shock with a fighting episode a first time can be expected to respond in a similar manner after a second exposure to foot-shock. As a result of these observations, pairs of mice were preselected for testing on the basis that they exhibited at least one fighting episode within 3 min.

These data were obtained from 214 pairs of mice preselected on the basis that they exhibited fighting behavior once within a trial 3-min. period of foot-shock. Virtually 98 per cent of these mice exhibited at least one episode of fighting behavior during the second 3-min. period of foot-shock. See text for explanation.

Data for the cumulative frequency response curve in Figure 2 were obtained from a population of 162 untreated, selected, fighting pairs of mice. The figure illustrates the percentage of pairs exhibiting a given number of fighting responses within a 3-min. period. For example, 87 per cent of fighting pairs of mice exhibited four or more fighting episodes. In order to evaluate the effects of drugs on fighting behavior a quantal assay procedure was designed in which the end response was based on these data. Namely, pairs of mice exhibiting three fighting episodes or less after drug administration were designated negative responders, whereas pairs fighting four or more times were considered as positive responders (see METHODS for details).

The Effect of Variations in Stimulus Intensity on the Frequency of Fighting Episodes. The results of this study are illustrated in Figure 3. Paired groups of mice were exposed to the designated stimulus intensities for 3-min. periods. The optimum stimulus intensity to produce fighting episodes was found to be 3 mA. The 2 mA stimulus intensity did not evoke significantly fewer fighting episodes, although there was a trend in this direction. Stimulus intensities greater than 3 mA elicited progressively fewer fighting episodes. Thus, the average number of fighting episodes elicited at a 10 mA stimulus intensity was found to be 0.85 with 95 per cent fiducial limits of 0.33–1.37.

The Effects of Centrally Acting Drugs on Fighting Episodes. The effects of various compounds on the suppression of fighting episodes, the depression of SMA, and protection against MES are summarized in Table 2. A measure of the selectivity of a particular compound in suppressing fighting episodes was obtained by examination of the ratios DD50 SMA/ED50 F and ED50 MES/ED50 F. High ratios were interpreted as an indication of high selectivity and vice versa. It is noteworthy that meprobamate suppressed fighting episodes in doses far below those producing motor depression and in doses less than those protecting against MES. In contrast, in the case of the phenothiazines tested and reserpine, the ratio DD50 SMA/ED50 F was low. Thus, depression of fighting episodes by these compounds occurred only at doses which produced a moderate or marked degree of motor depression. The values of the ratio ED50 MES/ED50 F are shown in Table 2 in column 7. These data serve to controvert any proposed relationship between anticonvulsant activity and suppression of fighting episodes. For example, compounds known to be potent anticonvulsants (phenobarbital and diphenylhydantoin) were relatively weak in suppressing fighting episodes. Conversely, compounds virtually devoid of anticonvulsant activity (chlorpromazine and prochlorperazine) were effective in suppressing fighting episodes.

DISCUSSION. Many of the procedures that have been described in the literature for studying fighting behavior have been hindered by the necessity of employing subjective methods for the measurement of the intensity of the fighting behavior (Bard, 1928; Wikler, 1944; Loewe, 1956). These methods consisted of observing effects such as piloerection, lashing of the tail, biting, spitting, increased spontaneous motor activity, extrusion of the claws, etc., and recording the intensity of each effect using a subjective scoring system. The results obtained with methods involving drug-induced "sham rage" (Sturtevant and Drill, 1957; Loewe, 1956) have been further complicated by the difficulty of accurately delineating the time of

peak fighting behavior. It was difficult to determine whether the fighting reaction was increasing in intensity or was subsiding either spontaneously or as a result of administration of an antagonistic drug. In addition, there are differences of opinion as to whether "sham rage" is elicited at all in the cat by morphine administration. Sturtevant and Drill (1957) and Loewe (1956) are of the opinion that morphine-induced feline mania represents a *bona fide* rage reaction. In contrast Wikler (1944) describes the effects of morphine in the cat as representing "disintegration of adaptive behavior rather than stimulation," and "in no case did the cat exhibit 'rage'."

The procedure described in this manuscript obviates to a great extent many of these problems and permits the objective measurement of the changes in frequency of discrete fighting episodes in fighting pairs of mice. The response for a fighting episode induced in mice exposed to foot-shock is well defined, and this reduces to a great extent the subjective interpretations in measurements. Furthermore, fighting behavior induced by foot-shock eliminates the problems of chemical interaction occurring when fighting is induced by chemical means. This factor takes on an added significance in the study of compounds which suppress fighting episodes, inasmuch as compounds may antagonize the action of morphine itself, for example, rather than the intricate mechanisms involved in the fighting response.

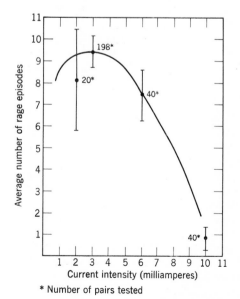

FIGURE 3. The effect of variations in stimulus intensity on the frequency of fighting ("raging") episodes. See text for explanation.

TABLE 2. *Potencies of Various Centrally Active Drugs on the Suppression of Fighting Episodes, Depression of Spontaneous Motor Activity and Protection Against Maximal Electroshock Seizures*

Compound	Vehicle	ED50 F	DD50 Spontaneous Motor Activity	DD50 SMA/ ED50 F	ED50 Maximal Electroshock	ED50 MES/ ED50 F
Chlorpromazine hydrochloride	H$_2$O	mg/kg 6.8 (4.2–10.8)	mg/kg 4.7 (3.5–6.5)	0.69	mg/kg —	—
Prochlorperazine dihydrochloride	H$_2$O	4.6 (2.0–10.7)	5.8 (4.0–9.8)	1.3	—	—
Trifluoperazine dihydrochloride	H$_2$O	0.85 (0.21–3.4)	0.85 (0.3–1.4)	1.0	—	—
Phenobarbital sodium	H$_2$O	37 (27.8–46.3)	—	—	17.4 (15.0–20.3)	0.47
Diphenylhydantoin sodium	tragacanth, aqueous, 1%	44 (30.7–63.5)	—	—	8.6 (7.8–9.6)	0.19
Mephenesin	NaCl, aqueous, 1%	250 mg/kg 20% suppression	204 (109–281)	—	260 (239–283)	—
Diphenhydramine hydrochloride	H$_2$O	70 mg/kg 30% suppression	—	—	26.9 (21.7–33.4)	—
Meprobamate	tragacanth, aqueous, 1%	84 (56–126)	221 (149–300)	2.6	132 (122–143)	1.6
Benzylic acid amide, SKF 3071	tragacanth, aqueous, 1%	96 (45.7–202)	224 (145–448)	2.3	47 (40–50)	0.49
Reserpine phosphate	citric acid, 1.25 g; ascorbic acid, 1.25 g; polyethylene glycol 400, 62.5 ml; water, q.s.ad 500 ml	4.4 (3.3–5.7)	1.2 (0.88–1.5)	0.27	—	—

It should be noted that values were recorded in this table only if they were below the prostrating or lethal dose levels. For example, a value was not recorded in column 6 for the anticonvulsant action of chlorpromazine, inasmuch as significant protection was produced only at prostrating doses. In a similar manner a value was not entered in column 4 for the depression of SMA produced by phenobarbital in view of the fact that it did not produce a significant decrease in SMA in doses below those producing prostration.

Miller's experiments (1948) on psychoanalytic displacement indicated that rats could be trained to strike and fight either one another or an inanimate object. The training procedure consisted of rewarding fighting by cutting off the footshock. It is noteworthy in the present experiments that when either a small rubber mouse or a dead mouse was placed with a live animal and footshock was applied, the live mouse could not be induced to attack the inert or lifeless object. Thus, it was not possible to elicit fighting behavior unless a minimum of two live mice were placed together and subjected to mild foot-shock. It appeared that of a pair of mice which did exhibit fighting behavior each mouse reacted as if its partner was responsible for its suffering.

The optimum parameters of electrical stimulation for producing fighting episodes were 3 mA, 400 V. at five shocks per second. The more intensive footshock incapacitated the mice to such an extent that they could not stand on their hindlimbs and they became prostrate. Foot-shock at current intensities greater than 15 mA induced tonic-clonic convulsions.

Compounds were tested which possessed a wide variety of pharmacological properties; included were anticonvulsants, centrally acting skeletal muscle relaxants, neuroleptics and antihistaminics. In order to determine whether a compound was selectively suppressing fighting episodes, compounds were also tested for their effectiveness in decreasing SMA and protecting against MES.

Of the compounds examined in this battery of screening tests, meprobamate exhibited the highest values for the ratios DD50 SMA/ED50 F and ED50 MES/ED50 F. Meprobamate suppressed fighting episodes in doses far below those which produced depression of SMA, and below those which protected against the tonic hind leg extensor component of the maximal electroshock seizure. It was of interest to find that SKF No. 3071, benzylic acid amide, also suppressed fighting episodes in doses below those which produced a decrease in SMA. However, it differed from meprobamate in that it was more effective as an anticonvulsant than as an inhibitor of fighting behavior. The values for the ratio DD50 SMA/ED50 F for those phenothiazines examined and reserpine were relatively low, indicating that these compounds suppressed fighting episodes at doses which produced moderate to marked motor inactivation. In this connection it is appropriate to mention the experiments of Schneider (1955) who produced "sham rage" in cats according to the method of Bard (1928). He reported that although reserpine abolished "sham rage," the cats were inactive, made no effort to climb or walk and were semiprostrate. Sturtevant and Drill (1957) found that reserpine in doses of 0.1 to 0.5 mg/kg intraperitoneally produced partial antagonism of morphine-induced feline mania if the reserpine was administered after, but not before the morphine. These doses of reserpine alone produced deep tranquillity approaching catatonic stupor. These same investigators also found that chlorpromazine partially antagonized morphine excitement in doses which when administered alone produced mild tranquillity and ataxia. These findings are consistent with the view that reserpine and the phenothiazines (chlorpromazine, prochlorperazine and trifluoperazine) abolish fighting behavior only in doses above or bordering on those which produce overt signs of neurological deficit. It may be concluded, therefore, that these compounds are relatively nonselective in suppressing fighting behavior. The suppression of fighting episodes in mice and of "sham rage" in cats by reserpine contrasts with its enhancing effect on the morphine-induced feline mania reported by Loewe (1956). This apparent discrepancy can possibly be explained by the fact that Loewe administered the reserpine before the morphine. Sturtevant and Drill (1957) found that although reserpine partially antagonized the morphine excitement when given after morphine, it had no such effect when given before morphine; indeed under these latter circumstances all five cats exhibited mania and two of the animals died.

Phenobarbital and diphenylhydantoin both differed from meprobamate in the battery of tests employed in that they did not produce a decrease in SMA in doses below those producing either prostration or lethal effects. Furthermore, the ratios ED50 MES/ED50 F for these compounds were much lower than for meprobamate, i.e., phenobarbital and diphenylhydantoin were more effective as anticonvulsants than in suppressing fighting episodes.

SUMMARY

Fighting episodes were produced in mice by exposing the animals to a mild foot-shock. This method permitted the objective measurement of the frequency of discrete fighting episodes in fighting pairs of mice. The effects of variations in stimulus intensity were studied and the optimum parameters for producing fighting episodes were determined. A quantal type assay procedure was devised in an effort to study the effects of centrally active drugs on the frequency of fighting episodes. Compounds were also studied for their effect in depressing spontaneous motor activity and protecting against maximal electroshock seizures. Reserpine and the phenothiazines that were tested suppressed fighting episodes only in doses which produced a moderate to marked degree of motor inactivation. Diphenylhydantoin and phenobarbital were more effective as anticonvulsants than in suppressing fighting episodes. Of the compounds studied in this battery of tests meprobamate exhibited a particular profile of activity which differentiated it from any of the other compounds studied. This pattern of activity was characterized by a mild degree of anticonvulsant activity, a mild depression of spontaneous motor activity and a more pronounced suppression of fighting episodes.

REFERENCES

BARD, P. *Amer. J. Physiol.*, **84**, 490, 1928.
BARD, P. and MOUNTCASTLE, V. B. *Proc. Ass. Res. nerv. Dis.*, **27**, 362, 1948.

COOK L., WEIDLEY, E. F., MORRIS, R. W. and MATTIS, P. A. *J. Pharmacol. exp. Therapeaut.*, **113**, 11, 1955.

COVIAN, M. R. *J. clin. Endocrinol.*, **9**, 678, 1949.

FINNEY, D. J. *Statistical Method in Biological Assay*, Hafner Publishing Company, New York, 1952.

GADDUM, J. H. and VOGT, M. *Brit. J. Pharmacol.*, **11**, 175, 1956.

GRIFFITHS, W. J. JR. and CLIFFORD, H. T. *Jap. Psychol. Res.*, **1**, 9, 1954.

LITCHFIELD, J. T. JR. and WILCOXON, F. *J. Pharmacol. exp. Therapeut.*, **96**, 99, 1949.

LOEWE, S. *Arch. int. Pharmacodyn.*. **108**, 453, 1956.

MILLER, N. E. *J. abnorm. soc. Psychol.*, **43**, 155, 1948.

RANSON, S. W. Tr. & Stud., Coll. Physicians, Philadelphia., **2**, 222, 1934.

RICHTER, C. P. *Proc. Ass. Res. nerv. Dis.*, **29**, 19, 1950.

SCHNEIDER, J. A. *Amer. J. Physiol.*, **181**, 64, 1955.

STURTEVANT, F. M. and DRILL, V. A. *Nature*, Lond., **179**, 1253, 1957.

SWINYARD, E. A., BROWN, W. C. and GOODMAN, L. S. *J. Pharmacol. exp. Therapeut.*, **106**, 319, 1952.

WIKLER, A. *J. Pharmacol. exp. Therapeut.*, **80**, 176, 1944.

WINTER, C. A. and FLATAKER, L. *J. Pharmacol. exp. Therapeut.* **103**, 93, 1951.

YEN, H. D. Y., STANGER, R. L. and MILLMAN, N. *J. Pharmacol. exp. Therapeut.*, **122**, 85A, 1958.

PART TWO

BEHAVIORAL

INTERDEPENDENCE

CHAPTER SIX

COMMUNICATION

THERE are a few patterns of behavior on which man cannot resist the temptation of comparing himself with lower animals. The previous chapter considered such a behavioral pattern and, with respect to aggression, man apparently does not emerge the superior (Lorenz, 1966). But what man loses in comparison with lower animals in relation to aggression, he rapidly recovers by claiming indisputable superiority in the area of communication. Both comparisons have little, or even negative, heuristic value, and Hockett (1961) aptly observed that the zoologist "learns as much by comparing snails and birds as by comparing either of those with *Homo sapiens*" (p. 392). In 1929, Bierens de Haan, the noted Dutch zoologist, tried to compare human and animal "language." He set down six criteria along which this comparison could fruitfully, he thought, be made. Humans, Bierans de Haan noted, communicate by means of sounds that (a) are *vocal*, (b) are *articulate*, i.e. they are "composed of syllables which are joined into words," (c) have some *conventional meaning*, that is, their meaning is learned and not given at birth, (d) *indicate* something, i.e. they may not only express emotions and feelings, but also refer to situations and objects, (e) all uttered with the *intention* of communicating something to somebody, and (f) can be *joined* into phrases in ever new combinations. Bierens de Haan in examining animal "languages" concludes that they meet the first and to a very small extent the second criterion. In all other respects animal languages are deficient. In a considerably more advanced analysis Hockett (1961) set down thirteen criteria, and, according to his comparison, man emerges not quite so superior.

The question of whether animals have a language has been debated and argued, probably ever since man could argue, and the opinions range from one extreme to the other. Garner (1905) thought that monkeys have their own languages, and thought that he identified their own "words" for "monkey," "bread," "apple," "banana," etc. Also, there apparently was a dog who "knew" 400 human words (Warden & Warner, 1928). The criteria used for determining *if* animals have a language have also ranged from one extreme to the other, depending on how anthropocentric the author was. Robyn Dawes, a colleague, remarked once during the course of our seminar, that one very reliable linguistic difference between humans and lower animals is that the former can and do lie at will.

Even if one cannot have a scintillating conversation with, for instance, bees, there still is a great deal to investigate about their process of communication. Wenner's precise measurement of this communication is reported in the first selection. The two selections that follow also attempt to identify and to describe the physical characteristics of some animal signals and to specify the conditions under which they are emitted and *transmitted*. This task is done for dolphins by Lilly and Alice Miller and for monkeys by Mason and Hollis. Church's study demonstrates that the rat responds dramatically to the affective stress of another. The final experiment in this chapter performed by Marler and Tamura, deals with the *conventional* aspect of communication, and it suggests that it may only be an ecological accident that animal "dialects" are not as varied as human.

REFERENCES

BIERENS DE HAAN, J. A. Animal language in its relation to that of man. *Biological Review*, 1929, **4**, 249–268.

GARNER, R. L. *The speech of monkeys.* London: Heinemann, 1892.

HOCKETT, C. F. Logical considerations in the study of animal communication. In W. E. Lanyon and W. N. Tavolga (Eds.). *Animal sounds and communication*, AIBS Symposium Proceedings, AIBS, Washington, 1961.

LORENZ, K. Z. *On aggression.* New York: Harcourt, Brace and World, 1966.

WARDEN, C. J. and WARNER, L. H. The sensory capacities and intelligence of dogs with a report on the ability of the noted dog "Fellow" to respond to verbal stimuli. *Quarterly Review of Biology*, 1928, **3**, 1–28.

AN ANALYSIS OF THE WAGGLE DANCE AND RECRUITMENT IN HONEY BEES*

Adrian M. Wenner, Patrick H. Wells and F. James Rohlf[1]

INTRODUCTION

THE honey-bee waggle dance (*Apis mellifera* L.) has become one of the most extensively studied behavioral patterns among animals. During this dance within the colony certain signals apparently pass between a successful forager and a potential recruit bee. These signals contain information about direction, distance, odor, and, possibly, richness of a food source remote from the hive. Some of the evidence reported to date includes the following: (1) The direction a dancing bee heads on the vertical comb while waggling its abdomen is well correlated with the direction it has traveled on its way from the hive to the nectar

* From *Physiological Zoölogy*, October, 1967, **40**, No. 4. Copyright 1967 by The University of Chicago. Supported by contract NR 301-800, Office of Naval Research, and by a faculty research grant from the University of California. We thank Miss Danielle Shaw, Miss Janice Gillette, and Mr James Jones for technical assistance; Mr Parlane Reid for use of facilities; Autonetics of Anaheim for converting tape signals into graphs; and Dr Howell Daly for critically reviewing the manuscript. The Western Data Processing Center at the University of California, Los Angeles, provided the computer time and facilities for the extensive data-manipulating required in this study.

[1] Present address: Department of Entomology, University of Kansas, Lawrence, Kansas 66045.

source (von Frisch, 1946). (2) The length of time of a straight run (von Frisch and Jander, 1957) and the length of time a bee produces sound during each waggle phase (Wenner, 1962; Esch, Esch, and Kerr, 1965) correlate well with the distance a bee has traveled from the hive on its outward trip, a correlation further strengthened by evidence that a bee's antennae apparently receive best those sound frequencies produced by dancing bees (Heran, 1959). (3) Odor of a nectar source may cling to the outside of a forager until it dances in the hive (Nixon and Ribbands, 1952) or be transmitted directly by nectar regurgitation (Dirschedl, 1960). (4) Finally, bees reportedly indicate richness of a source by "vigor" in the dance (von Frisch, 1950) or by rate of modulation of the sound signal (Esch, 1963).

Even with the great amount of work already expended on studies of this behavioral pattern we still had no direct evidence at the start of this study as to whether the signals we obtain from the dance maneuver actually guide the recruited bees to a remote nectar source. Furthermore, each of the above-mentioned maneuvers within the dance is in itself a complex action, and we still had no evidence about which component of each of these actions might convey information. Within the dance, for instance, several factors correlate well with distance traveled from the hive (Wenner, 1962), but we still did not know which element, if any, might be used by recruit bees. Finally, we lacked evidence on the relative efficiency of this dance for recruiting completely new bees to a food source.

One of the first questions which arose from our consideration of the above was whether some component of the sound signal also correlates with the direction a bee has traveled from the hive. That is, although one element within the dance maneuver correlates with the direction a forager travels from the hive, this correlation is not necessarily exclusive.

Toward this end, we initiated a pilot study in the summer of 1963 to determine whether some component within the sound signal produced by dancing bees correlates with the direction a foraging bee travels from the hive. Simultaneously with this we collected data on possible modification of the sound signal due to differences in sugar concentration at the food source, being unaware at this time of the study already completed by Esch (1963).

From this initial study, we obtained statistically significant correlations of certain elements within the sound signal both with relative direction traveled from the hive and with sugar concentra-

tion of the food source. Unexpectedly, however, the correlations we obtained (unpublished) could be reconciled neither with what we expected from our earlier studies not with the results published by Esch (1963).

Clearly, because of the presumed importance of the possibility that a sound signal can carry information about direction or sugar concentration of a food source, these conflicting results indicated that we should repeat the experiment on a much larger scale. Accordingly, we designed an experiment to answer the following questions: (1) Do bees modify their sound signal during the dance as a result of imbibing sugar solutions of various concentrations? If so, (*a*) Which component of the sound signal correlates best with a change in sugar concentration? (*b*) Do different types of sugar result in different signals? (*c*) Is sugar concentration or sugar viscosity more important in modifying the signal? (2) Can some environmental factor(s) explain earlier discrepancies?

To answer these questions, we designed experiments to measure all pertinent variables (when feasible) and to compare these variables to each other simultaneously. These experiments later led to an initial study of dance efficiency and, eventually, to other more basic studies of the dance "language."

METHODS

Placement of the two observation hives, structure of the feeding dish, marking of bees, and physical features of the inclosure used in tape-recording sounds during the dance very closely followed a previously described system (Wenner, 1962).

After initially training bees to visit a feeding station (Wenner, 1961), we regularly furnished fresh sugar solution 30 min. prior to each recording session. Use of an aerial photograph of the area in conjunction with a topographic map permitted choosing sites for feeding stations at two equal distances from the hive (400 m) and at the same elevation. The area was at the base of the Santa Ynez Mountains and approximately 7 km due north of the University of California, Santa Barbara. This location and the time of year provided for remarkably uniform weather conditions during the course of the experiment. Skies were generally clear during data-collecting (with one morning of fog), and the wind invariably came from the south. Mean daily temperature gradually increased during the course of the month-long experiment, with only the last day (August 5, 1964) being markedly warmer than other days.

Independent Variables

Independent (environmental) variables under consideration in the study and their method of measurement were as follows:

SUGAR CONCENTRATION. Each sugar concentration (0.8-, 1.5-, and 2.5-M sucrose; 2.5- and 5.0-M fructose; and 2.0-M glucose) was prepared separately with no addition of flavor or odor.

SUGAR VISCOSITY. We obtained calibration curves for relative viscosity at different temperatures under controlled conditions in the laboratory by using samples of each sugar concentration. Values from these graphs compared to outside temperature at the time of sugar uptake by bees furnished the data for later comparison with other variables.

LIGHT INTENSITY. These relative values were read as the amount of reflected light from a uniform gray card by pointing a General Electric DW 68 light meter at this card from a height of approximately 60 cm.

TIME OF DAY. Although we operated at three discrete times of day (morning 0915–1015, noon 1215–1315, and afternoon 1545–1645), except when using fructose, we also tallied the specific time (Pacific Daylight Savings) each bee danced.

For fructose the times were 1000–1030, 1215–1315, and 1430–1530.

ROOM TEMPERATURE AND HUMIDITY. A calibrated recording thermohumidigraph automatically registered these values, but, in addition, temperature within the room was measured each 10 min. by a hand-held thermometer. A later check of the thermohumidigraph showed some drifting during the course of the experiment; for this reason, values for humidity obtained at the feeding station must be considered more reliable.

STATION TEMPERATURE AND HUMIDITY. Readings obtained by using wet-bulb and dry-bulb thermometers each 10 min. later provided the basis for calculations of humidity, corrected according to barometric pressure.

SUN DIRECTION AND ALTITUDE. We obtained these values directly from published tables for each recording time.

WIND SPEED AND DIRECTION. Bees traveling from the hive to the west feeding station had to fly over a low (6-m-high) ridge. During each recording session, one of us stood on this ridge and read wind speed each 10 min. from a hand-held anemometer. We estimated wind direction as to whether it came from any one of four southerly sectors.

Three other independent variables were tallied for later comparison: *sugar type*, *date*, and *absolute direction*, in which most data were gathered from bees visiting a station at about

TABLE 1. *Schedule of Data-Gathering Sessions*[a]

Date	Sugar	Molarity		
		Morning	Noon	Afternoon
7–8–64	Sucrose	2.5	2.5	2.5
7–10	Sucrose	0.8	0.8	0.8
7–13	Sucrose	1.5	1.5	1.5
7–14	Sucrose	1.5	1.5	1.5
7–15	Sucrose	2.5	2.5	2.5
7–21	Sucrose	0.8	0.8	0.8
7–22	Sucrose	2.5	2.5	2.5
7–23	Sucrose	1.5	1.5	1.5
7–24	Sucrose	0.8	0.8	0.8
7–29[b]	Sucrose	1.5	2.5	0.8
7–30	Sucrose	2.5	0.8	1.5
7–31	Sucrose	1.5	0.8[c]	2.5
8–1 to 8–4	Glucose and fructose attempts	—	—	—
8–5	Fructose	5.0	5.0	5.0

[a] Sounds were recorded from at least three different bees at each session.
[b] Bees traveled to stations in two directions on this day.
[c] Data-gathering failed at this time due to no dancing.

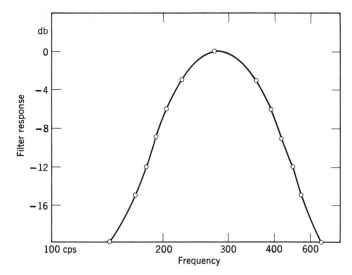

FIGURE 1. Characteristics of the Allison filter system through which recorded honey-bee signals were passed prior to measurement.

295° from the north. On July 29, however, some bees fed at a station about 50° from the north.

We used a hand-held American Microphone Company model D33 microphone to record the sounds produced by each bee as it danced in the brood area of the hive after having been individually marked at the feeding station while imbibing sugar solution. At the beginning of each recording session we also recorded the rate of ticking of an 8-day clock as a test of the dependability of tape speed during recording. Signals so obtained were fed through a custom-built battery-operated preamplifier into a Uher 4000S Report tape recorder, run at 7½ in./sec. Immediately at the conclusion of the entire experiment, all tapes were edited so as to choose five noise-free straight runs from each bee recorded. In almost all cases, these straight runs were consecutive, but occasionally data were compiled from two sections of the same dance sequence. All samples contained the space after the fifth straight run by including the beginning of the sixth straight run.

In an attempt to eliminate environmental factors (other than time of day or sugar concentration), we randomized our data-collecting sessions for sucrose within each replicate according to Table 1. This arrangement provided for essentially four composite replications. In each of the recording sessions on each day, we recorded sound from three bees from which we had never gathered data before, as well as from one or two of the previous bees, when possible. As a result we eventually obtained data from 156 visits to the food source by 100 different bees. We regularly killed bees at the station in a later session if we had already gathered data from them, thereby

achieving a turnover and preventing a few bees from biasing the data. We also attempted to gather data from August 1 to August 4 from bees after they had visited 2.0-M glucose and 2.5-M fructose. Although these attempts failed, a later attempt with 5-M fructose succeeded.

Signals from tapes so gathered and edited were transferred to a Honeywell model LAR 7300 tape recorder run at 60 in./sec., played back at 7½ in./sec. through two Allison variable filters so as to pass dance frequencies but eliminate unwanted noise (see Fig. 1 for characteristics of this filter system), and then fed into a Sanborn Recorder for display and measurement.

Measurements of the trace so produced (Fig. 2) provided some problems. These will be treated separately in the following discussion concerning measurement of the dependent (behavioral) variables.

Dependent Variables

NUMBER OF PULSES IN A STRAIGHT RUN (N). The number of pulses could be directly counted from the trace. Occasionally, whenever a pulse would either be of low amplitude or missing with a space large enough for a complete pulse, we felt it better to count a pulse here and obtain an accurate computation of pulse rate than to ignore the space.

STRAIGHT-RUN TIME (Ts). Failure to include the space after the last pulse can result in a significant error (Wenner, 1962). Consequently, the value of this variable was obtained by the following formula: $Ts = t + [t/(N-1)]$, in which t is the time between the beginning of the first pulse

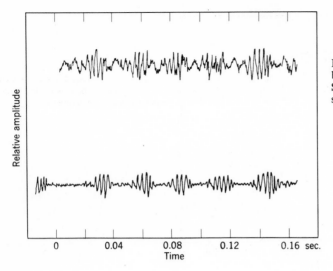

FIGURE 2. Unfiltered and filtered honeybee sound signals as displayed by the Sanborn Recorder. Each display represents part of one straight-run time.

Relative amplitude

0   0.04   0.08   0.12   0.16 sec.
                      Time

and the beginning of the last pulse and *N* is the total number of pulses in the straight run.

CIRCLE-RUN TIME. This, also, was measured indirectly, so as to not include the space after the last pulse of a straight run. We measured the time from the start of one straight run to the start of the next and subtracted the straight-run time (*Ts*) computed above.

RATE OF MODULATION. We computed this by dividing *N* by *Ts*.

NUMBER OF CYCLES PER PULSE. For each straight run we began at the first clear pulse and counted the number of cycles within each third pulse for a total of five pulses. The number used for later comparison was the average number for these five.

FREQUENCY OF SOUND (F). In a similar manner to the computation of modulation rate, computation of sound frequency must include the space after the last positive peak within a pulse. While counting the number of cycles per pulse, we measured the time of the entire pulse. Thus, frequency of sound of a straight run is computed from an average of the same five pulses used for determining the number of cycles per pulse.

After completing all measurements, we transferred the values for dependent and independent variables onto IBM tabulating cards for later analysis.

RESULTS OF DATA ANALYSIS

In studies of the dance elements in the past, it has usually been a practice to compare each dependent variable with each independent variable. With 6 dependent variables and 12–15 independent variables under simultaneous consideration (together with the relationships within the classes of dependent and independent variables), however, such simple comparisons become extremely difficult to present or to reconcile with each other. Consequently, all variables in our study were initially handled as a group with the aid of appropriate statistical analysis as follows: (1) determination of correlation coefficients and graphing of data, (2) principal-components analysis (see Sokal, Daly and Rohlf, 1961, for explanation), (3) complete factor analysis, (4) computation of bits of information within the sound signal, and (5) comparison of data for the two different directions involved.

For ease of presentation, the results of these treatments will be discussed separately.

Correlation Coefficients, Means, and Standard Deviations

Initially we used a standard computer program on file (BMD02D, Correlation with Transgeneration—Version of May 21, 1964) at the Western Data Processing Center at the University of California, Los Angeles. Computer output from use of this program provided us with sums, means, standard deviations, correlations between any two variables, and 540 graphs of relationships between dependent and dependent as well as between dependent and independent variables. The program also permitted computation of logarithms of values for the first four dependent variables and comparison of these logarithmic values with both dependent and independent variables. Since these comparisons of all variables with logarithmic values of the four dependent variables did not increase our understanding of relationships, however, we do not discuss such comparisons further.

The initial determination of correlations between dependent and independent variables for the *total data* resulted in a surprising number of significant correlations (Table 2). Of the 90 possibilities, 23 showed relationships significant at the 1 per cent level and 41 at the 5 per cent level or better.

Clearly we could not attempt a biological interpretation of such a complex set of interactions on the basis of this preliminary computation of correlations. For instance, does the statistically significant correlation between wind direction and frequency of sound produced in the hive during the dance indicate a communication of information about wind direction to potential recruit bees, or might this relationship merely result from the wind direction and frequency of sound being simultaneously correlated with time of day or temperature? This problem was further complicated by the apparent non-linearity of some relationships (notably pulse rate vs. sugar concentration—computation of correlation coefficients by use of this program does not accommodate non-linearity, and the actual degree of association between these two variables is thus greater than implied by the correlations obtained).

To reduce the effect of undue influences resulting from cross-relationships within either dependent or independent variables, we partitioned the data according to time of day (three periods minus the fructose data) as well as to sugar concentration (four concentrations) and repeated the process of comparing all variables with the aid of the same computer programs used before. The computer thus provided us with seven new comparisons of the relationships between variables and provided more basis for interpreting the results than by merely using the initial comparisons. As an example, if a correlation for the data as a whole provides a possibility for communication among bees (e.g., sugar concentration vs. pulse rate), it should persist regardless of the time of day. In fact, real correlations between sugar concentration and elements of the sound signal should emerge stronger after such partitioning of data as a result of elimination of extraneous environmental influences. Likewise, any correlation between time of day and some element of the sound signal should be strengthened by analyzing data for each sugar concentration separately.

TABLE 2. Correlations Between Dependent and Independent Variables as Obtained from an Analysis of Total Data

Dependent Variables	7 Sugar Concentration	8 Sugar Viscosity	9 Light Intensity	10 Time of Day	11 Shed Temperature	12 Station Temperature	13 Shed Humidity	14 Station Humidity	15 Sun Direction	16 Sun Altitude	17 Wind Speed	18 Wind Direction	19 Sugar Type	20 Absolute Direction	21 Date
1. No. of pulses	*	—	*	—	—	—	*	—	—	—	—	*	**	—	**
2. Straight-run time	*	—	—	*	—	—	*	—	**	—	—	*	—	—	—
3. Circle-run time	*	—	*	—	—	*	—	—	—	—	—	—	**	—	**
4. Pulse rate	*	—	—	**	—	**	—	**	**	—	—	—	**	—	**
5. No. of cycles/pulse	—	**	—	**	**	**	**	**	**	—	—	**	*	*	*
6. Sound frequency	—	*	**	—	**	**	**	**	*	*	—	—	—	—	—

* Significant at the 5 per cent level.   ** Significant at the 1 per cent level.

TABLE 3. *An Example of Correlation Coefficients Obtained for All the Relationships Between Each Two Variables*

Independent Variables

Each cell lists, in order: Column 1 (total data, then uniform data); Column 2 (0.8-, 1.5-, 2.5-, and 5.0-M sugar solutions, descending); Column 3 (morning, noon, afternoon). Asterisks (*) indicate missing coefficients.

Dependent Variables	7 Sugar Concentration	8 Sugar Viscosity	9 Light Intensity	10 Time of Day	11 Shed Temperature	12 Station Temperature
1. No. of pulses	−0.11, 0.06; *, *, *, *; 0.10	0.06, 0.05; 0.02, −0.06, 0.17, −0.01; 0.14, 0.09, 0.11	−0.11, −0.07; −0.16, 0.09, −0.13, −0.12; −0.24, 0.03, −0.02	0.02, 0.07; 0.08, −0.05, 0.11, 0.11; *, *, *	−0.02, 0.07; −0.06, 0.12, −0.05, 0.07; 0.05, −0.03, 0.02	−0.03, 0.07; −0.09, 0.02, 0.08, 0.05; −0.10, 0.18, −0.04
2. Straight-run time	−0.09, 0.03; *, *, *, *; 0.02	0.01, −0.01; 0.15, −0.08, 0.23, 0.04; 0.11, 0.03, 0.07	−0.08, −0.07; −0.07, 0.12, 0.14, −0.18; −0.26, −0.13, 0.05	0.11, 0.15; 0.20, 0.04, 0.15, 0.08; *, *, *	0.00, 0.05; 0.01, 0.08, −0.04, 0.01; −0.06, −0.11, 0.04	0.02, 0.08; −0.00, 0.05, 0.07, 0.00; −0.13, 0.08, 0.00
3. Circle-run time	0.11, −0.09; *, *, *, *; 0.06	−0.03, −0.05; 0.16, 0.01, −0.15, −0.33; −0.28, −0.05	0.10, 0.00; 0.12, −0.02, 0.07, 0.23; 0.15, 0.04, 0.09	0.00, −0.12; −0.26, 0.05, 0.04, 0.42; *, *	0.06, −0.10; −0.06, −0.06, 0.12, 0.42; 0.02, 0.00, 0.02	0.09, −0.10; −0.01, −0.02, 0.10, 0.32; 0.08, 0.03, 0.04
4. Pulse rate	−0.11, −0.08; *, *, *, *; 0.22	0.08, 0.10; −0.20, 0.01, −0.15, −0.20; 0.04, 0.14, 0.13	−0.08, −0.01; −0.17, 0.00, 0.01, 0.15; 0.02, 0.32, −0.20	−0.16, −0.14; −0.27, −0.18, −0.06, 0.19; *, *, *	−0.05, 0.07; −0.17, 0.17, −0.04, 0.32; 0.25, 0.18, −0.08	−0.13, 0.01; −0.20, −0.01, 0.02, 0.22; 0.06, 0.24, −0.12
5. No. of cycles/pulse	−0.03, 0.14; *, *, *, *; 0.25	−0.12, −0.21; 0.16, −0.08, 0.04, 0.03; −0.06, −0.09, −0.17	0.07, 0.05; −0.05, 0.02, 0.04, −0.01; −0.06, −0.18, 0.19	0.14, 0.25; 0.22, 0.14, 0.08, −0.28; *, *, *	0.14, 0.19; 0.21, 0.05, 0.07, 0.00; 0.10, −0.10, 0.05	0.17, 0.25; 0.13, 0.14, 0.11, −0.04; 0.12, −0.01, 0.08
6. Sound frequency	−0.03, 0.09; *, *, *, *; −0.06	0.10, 0.12; −0.07, 0.02, 0.01, 0.08; 0.28, 0.14, 0.03	−0.17, −0.19; −0.10, 0.05, −0.28, 0.08; −0.24, −0.17, −0.22	0.08, 0.11; 0.28, 0.16, −0.05, −0.39; *, *, *	−0.17, −0.20; −0.15, −0.03, −0.32, −0.07; −0.20, −0.28, −0.33	−0.16, −0.15; −0.18, −0.11, −0.17, 0.11; −0.35, −0.08, −0.34

Note.—Column 1 in each cell (see 1×8) gives coefficients for *total data* (top) and for *uniform data*. Column 2 in the same cell lists coefficients for 0.8-, 1.5-, 2.5-, and 5.0-M sugar solutions in descending order. Column 3 lists, top to bottom, coefficients for morning, noon, and afternoons. Asterisks indicate missing coefficients brought about by data-partitioning. A complete table is available upon request to the senior author.

184

TABLE 4. *Matrix of Correlation Coefficients for the Total Data* (All Numbers × 10^{-2})

Variables	1	2	3	4	5	6	7	8	9	10	11	12	13	14	15	16	17	18
1. No. of pulses	—																	
2. Straight-run time	91	—																
3. Circle-run time	−22	−20	—															
4. Pulse rate	41	01	−09	—														
5. No. of cycles/pulse	−06	10	−05	−38	—													
6. Sound frequency	16	17	−16	00	06	—												
7. Sugar concentration	−11	−09	11	−11	−03	−03	—											
8. Sugar viscosity	06	01	−03	08	−12	10	50	—										
9. Light intensity	−11	−08	10	−08	07	−17	01	−20	—									
10. Time of day	02	11	00	−16	14	08	−05	−30	03	—								
11. Room temperature	−02	00	06	−05	14	−17	04	−21	63	46	—							
12. Station temperature	−03	02	09	−13	17	−16	11	−23	50	56	84	—						
13. Room humidity	09	09	−04	−01	23	12	−10	−01	−13	−14	−18	−11	—					
14. Station humidity	−03	−04	−04	00	−17	17	17	33	−40	−61	−76	−87	01	—				
15. Sun direction	04	12	00	−15	15	09	−03	−27	−02	99	44	54	−12	−61	—			
16. Sun altitude	−06	−03	07	−08	04	−09	08	−18	88	−02	46	33	−07	−22	−07	—		
17. Wind speed	04	05	01	01	04	08	−08	−18	20	47	35	28	−13	−36	47	18	—	
18. Wind direction	−09	−10	05	01	03	−18	13	04	08	02	27	21	−18	−18	03	−15	−11	—

185

TABLE 5. *Matrix of Correlation Coefficients for the Uniform Data* (All Numbers × 10^{-2})

Variables	1	2	3	4	5	6	7	8	9	10	11	12	13	14	15	16	17	18
1. No. of pulses	—																	
2. Straight-run time	91	—																
3. Circle-run time	−23	−22	—															
4. Pulse rate	35	−05	−03	—														
5. No. of cycles/pulse	−01	12	−08	−32	—													
6. Sound frequency	14	17	−15	−05	04	—												
7. Sugar concentration	06	03	−09	−08	14	09	—											
8. Sugar viscosity	05	−01	−05	10	−21	12	51	—										
9. Light intensity	−07	−07	00	−01	05	−19	−06	−19	—									
10. Time of day	07	15	−12	−14	25	11	−08	−34	05	—								
11. Room temperature	07	05	−10	07	19	−20	−13	−24	64	43	—							
12. Station temperature	07	08	−10	01	25	−15	−14	−27	53	54	82	—						
13. Room humidity	04	00	01	12	12	−00	−24	−03	−24	−12	−31	−26	—					
14. Station humidity	−08	−06	05	−07	−27	24	19	28	−46	−56	−78	−92	10	—				
15. Sun direction	08	15	−12	−13	26	12	−07	−31	00	99	41	53	−08	−56	—			
16. Sun altitude	−02	00	−07	−03	−01	−09	12	−15	84	00	43	31	−12	−21	−05	—		
17. Wind speed	03	06	−02	−02	07	08	06	−18	23	50	46	39	−06	−41	50	20	—	
18. Wind direction	−08	−11	05	05	08	−12	−02	−11	24	04	39	26	−22	−28	04	−05	07	—

Note.—See text for explanation.

TABLE 6. *Numbers of Cases for Various Classes of Data and Sizes of Correlation Coefficients Required for Significance at the 5 per cent and 1 per cent Levels*

	No. of Cases	Significance	
		5 % Level	1 % Level
Total correlation, all data	743	0.088	0.115
Total minus repeats and fructose (uniform data)	439	0.098	0.128
	Sugars Separated		
Sucrose:			
0.8 M	198	0.138	0.181
1.5 M	264	0.138	0.181
2.5 M	236	0.138	0.181
Fructose:			
5.0 M	45	0.288	0.372
	Time of Day Separated		
Morning	236	0.138	0.181
Noon	220	0.138	0.181
Afternoon	242	0.138	0.181

Note.—Values presented are applicable to those numbers listed in Tables 3–5.

As a final maneuver for determining the validity of these correlations, we compiled one set of *uniform data* cards which included only the first values obtained from each bee visiting sucrose at the western station only and used the computer program for comparisons.

We thus eventually had nine statistical comparisons between each pair of variables (Table 3). The first number in the first column within each comparison in Table 3 (see 1 × 8) is the correlation coefficient for all data before partitioning (*total data*); the number below this is the comparable coefficient for the *uniform data* (one direction, one sugar type, and one sample per different bee). The second column lists the correlation coefficients for the four concentrations of sugar (0.8, 1.5, 2.5, and 5.0 M), and the third column lists the coefficients for the three times of day (morning, noon, and afternoon). Tables 4 and 5 are the complete correlation matrixes for the *total data* and the *uniform data*, respectively.

Since the number of cases changes as the treatment of the data varies, numbers within any one comparison cannot be directly compared. Table 6 gives the sample size in each statistical treatment and the size coefficient required for significance at the 5 per cent and 1 per cent levels. Finally, Table 7 presents means and standard deviations for each variable in each circumstance.

Principal-Components Analysis

From an initial study of the correlation coefficients one can discern an interrelationship among certain variables, as would be expected from a consideration of the variables measured (i.e., temperature, time of day, humidity, etc.). Less obvious, however, is the relative strength of a relationship between any two variables within one of these clusters.

We initially used a principal-components analysis (computation of eigen-values and eigenvectors of a matrix of correlation coefficients; Seal, 1964) for determining the relative contribution of each variable to the total variance. We also computed partial correlations in certain cases, but the results of this procedure did not yield a different interpretation from that obtained by the principal-components analysis or by the factor analysis and are not given here.

TABLE 7. *Means and Standard Deviations for All Data*

Variables	Total Data	Uniform Data	Sugar Concentration (M)				Time of Day			Second Direction
			0.8	1.5	2.5	5.0	Morning	Noon	Afternoon	
No. of pulses:										
Mean	28.44	28.87	27.44	30.15	28.39	23.07	28.82	28.32	29.18	31.20
SD	8.28	8.29	6.72	7.98	9.35	7.30	8.41	7.60	8.58	9.64
Straight-run time:										
Mean	0.86	0.86	0.87	0.88	0.85	0.79	0.84	0.86	0.90	0.92
SD	0.22	0.23	0.20	0.21	0.26	0.21	0.22	0.21	0.24	0.19
Circle-run time:										
Mean	1.64	1.56	1.69	1.58	1.61	1.96	1.63	1.64	1.61	1.65
SD	0.43	0.39	0.42	0.38	0.42	0.64	0.38	0.45	0.40	0.56
Pulse rate:										
Mean	33.00	33.51	31.77	34.08	33.49	28.93	34.27	32.87	32.53	33.20
SD	4.10	3.93	3.90	3.83	4.07	2.74	4.07	3.97	3.89	4.82
No. of cycles/pulse:										
Mean	4.31	4.35	4.59	4.19	4.14	4.69	4.07	4.37	4.43	3.94
SD	0.93	0.92	1.02	0.82	0.94	0.70	0.88	0.96	0.94	0.98
Sound frequency:										
Mean	276.00	279.40	271.00	281.60	275.10	268.80	274.60	273.40	280.90	268.50
SD	25.80	26.60	26.10	24.00	27.60	16.70	25.90	27.30	25.00	19.60
Sugar concentration:										
Mean	1.84	1.68	—	—	—	—	1.65	1.65	1.62	1.62
SD	1.04	0.68	—	—	—	—	0.59	0.75	0.68	0.70
Sugar viscosity:										
Mean	499.60	495.90	231.40	532.60	637.40	697.40	589.00	422.50	445.60	468.80
SD	198.00	192.10	93.30	60.20	139.30	53.00	189.00	173.60	185.40	218.00
Light intensity:										
Mean	27.80	26.90	29.20	26.90	27.20	30.20	23.20	36.10	24.30	29.70
SD	6.80	7.30	6.30	4.40	9.00	4.70	5.10	1.50	3.50	5.40
Time of day:										
Mean	1324.00	1321.00	1336.00	1243.00	1330.00	1256.00	—	—	—	1224.00
SD	—	—	—	—	—	—	—	—	—	—
Shed temperature:										
Mean	24.90	24.60	25.30	24.80	24.40	26.60	22.40	26.60	25.60	26.10
SD	2.70	2.80	2.60	2.80	2.60	1.30	2.60	1.60	1.60	1.60
Station temperature:										
Mean	22.20	21.80	22.50	22.00	21.60	24.30	20.20	22.80	23.00	22.80
SD	2.00	2.10	1.90	2.20	1.80	1.30	1.40	1.70	1.60	2.20
Shed humidity:										
Mean	52.00	54.00	57.00	55.20	42.00	63.60	55.00	49.30	49.40	52.50
SD	16.60	14.20	5.60	4.00	25.80	1.80	16.80	18.10	15.20	3.10
Station humidity:										
Mean	62.70	63.00	60.60	61.90	65.20	64.30	70.10	60.50	57.30	62.80
SD	8.40	8.70	7.20	9.10	8.40	4.20	6.80	6.30	6.60	8.60
Sun direction:										
Mean	175.50	175.20	185.90	167.10	176.90	171.10	95.30	164.60	264.30	174.10
SD	70.30	72.40	66.80	75.70	68.20	58.20	3.80	15.70	3.70	65.90
Wind speed:										
Mean	6.70	6.80	6.70	6.50	7.30	4.70	4.80	7.60	8.10	6.60
SD	2.70	3.10	2.20	2.90	2.90	1.00	2.60	1.80	2.40	2.20
Sun altitude:										
Mean	55.50	54.60	57.60	51.60	57.30	60.10	45.60	75.30	46.30	55.00
SD	13.80	13.80	14.80	11.90	14.60	9.50	3.60	1.60	3.30	14.00
Wind direction:										
Mean	2.50	2.40	2.50	2.50	2.50	3.00	2.60	2.30	2.60	2.60
SD	0.80	0.80	0.60	0.90	0.70	—	1.10	0.60	0.50	0.70

Note.—Each column shows such values according to different methods of partitioning data.

Table 8 lists eigenvalues and eigenvectors derived from the matrix of correlation coefficients for the *total data* (from Table 4). The first row of numbers in this table is the computed set of eigenvalues arranged in order of descending value. While the computer program furnished a matrix of normalized eigenvectors to accompany these eigenvalues, a conversion of these eigenvectors so that their sum of squares equaled their eigenvalues was performed by multiplying each eigenvector by the square root of its eigenvalue. Although the computer program furnished 18 such columns, we include only the first 6, since this total set of values accounts for almost 75 per cent of all variance in the data.

The relative contribution of each variable to the total variance of the data, then, can be discerned by inspecting one, two, three, or more of the columns at a time. A sample comparison can perhaps clarify use of this table.

Columns I and II together account for almost 40 per cent of the variance (the sum of eigenvalues 4.51 and 2.46 divided by 18); and the values in these two columns provide an indication of the relative contribution of each variable to this amount of variance. As might be expected, for example, station temperature and station humidity show high values in Column I but have opposite signs.

A better understanding of the total interrelationship among these 18 variables emerges from a consideration of two-dimensional graphs of the values in any two columns. One such comparison appears in Figure 3, where the values in Column II are plotted against those in Column

III. This figure clearly indicates that elements obtained from the sound signal produced by dancing bees (X) generally vary independently from the various environmental parameters (O) measured in this study.

Factor Analysis

Whereas a principal-components analysis permits a two-dimensional (or even a three-dimensional) comparison of variables, it typically expresses the relations in a way which is difficult to interpret. Factor analysis with rotation to simple structure has been found to enable simpler interpretation. Toward this end we submitted the correlation matrixes for the *total data* and for the *uniform data* to a full factor analysis. For this we used another standard computer program (BMD 03M, General Factor Analysis—Version of September 1, 1965) at the Western Data Processing Center of the University of California, Los Angeles. The resulting varimax rotated factor matrixes appear in Tables 9 and 10, respectively.

TABLE 8. *Eigenvalues and Eigenvectors Determined by an Analysis of Principal Components for the Total Data*

	I	II	III	IV	V	VI
Eigenvalues	4.51	2.46	1.89	1.60	1.44	1.26
1. No. of pulses	06	− 74	− 59	23	− 08	− 12
2. Straight-run time	− 02	− 72	− 47	06	− 30	− 13
3. Circle-run time	− 09	36	17	05	06	04
4. Pulse rate	17	− 20	− 33	39	52	− 00
5. No. of cycles/pulse	− 22	− 08	17	− 47	− 54	− 31
6. Sound frequency	15	− 41	04	− 15	− 25	43
7. Sugar concentration	06	30	08	52	− 61	24
8. Sugar viscosity	43	09	− 01	52	− 48	18
9. Light intensity	− 57	49	− 56	− 15	− 07	10
10. Time of day	− 74	− 41	40	08	01	17
11. Room temperature	− 87	16	− 19	16	− 07	− 09
12. Station temperature	− 89	08	− 04	15	− 13	− 15
13. Room humidity	17	− 19	− 06	− 47	− 23	− 28
14. Station humidity	87	09	01	− 05	− 08	23
15. Sun direction	− 72	− 44	43	10	00	15
16. Sun altitude	− 43	44	− 61	− 25	− 12	29
17. Wind speed	− 53	− 22	04	− 03	13	48
18. Wind direction	− 17	22	14	48	− 04	− 57
	(All numbers × 10⁻²					
Cumulative variance (%)	25	39	49	58	66	73)

Note.—For explanation of columns see text.

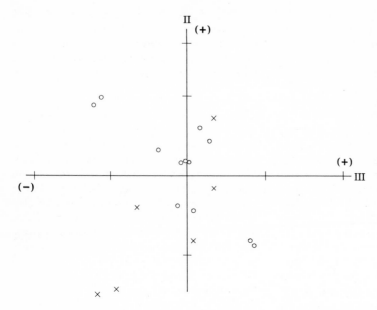

FIGURE 3. A two-dimensional plot showing relative contributions of dependent (X) and independent (O) variables to the total variance shown in columns 2 and 3 of Table 8. This display indicates that the principal axes of dependent (behavioral) and independent (environmental) variables are at right angles to each other. Each graph unit equals 50.

TABLE 9. *Full Factor Analysis of the Correlation Matrix Shown in Table 4 for the Total Data* (All numbers × 10⁻²)

Variables[a]	Factors[b]					
	I	II	III	IV	V	VI
1. No. of pulses	02	**96**	00	− 00	− 18	02
2. Straight-run time	09	**9–**	− 01	05	10	07
3. Circle-run time	02	− 39	− 09	06	− 04	− 10
4. Pulse rate	− 14	33	04	− 15	− 7–	− 12
5. No. of cycles/pulse	12	07	− 06	− 04	**82**	− 05
6. Sound frequency	09	23	15	15	11	*58*
7. Sugar concentration	00	− 13	− 08	**88**	03	− 06
8. Sugar viscosity	− 26	09	15	**78**	− 10	01
9. Light intensity	09	− 10	− **94**	− 05	01	− 10
10. Time of day	**95**	01	08	− 07	07	06
11. Shed temperature	*60*	01	− *60*	01	02	− 37
12. Station temperature	*69*	00	− *47*	03	12	− 39
13. Shed humidity	− 23	23	07	− 22	*50*	08
14. Station humidity	− **72**	− 07	34	24	− 07	37
15. Sun direction	**95**	03	13	− 05	08	05
16. Sun altitude	− 01	− 06	− **94**	− 01	01	15
17. Wind speed	*61*	− 02	− 20	− 08	− 18	35
18. Wind direction	09	− 07	09	17	− 02	− **77**

[a] Behavioral variables (1–6) generally vary independently from the environmental variables (7–18).
[b] Boldface numbers = 70[+]; italic numbers = 40[+].

The results shown in Table 9 (factor analysis of the total data) support the interpretation provided by the principal-components analysis. The heavy loadings (arbitrarily set at 40 or above— see Sokal and Daly, 1961) on values for dependent variables occur in columns other than those which contain heavy loadings on the independent variables.

The factor analysis for the uniform data (Table 10) is remarkably similar to that for the total data, despite the marked differences in the correlation matrixes for the two sets of data (Tables 4 and 5). This is as it should be if this tool provides a true picture of the relationship among variables. As stated in the methods section, the uniform data do not contain the repeat values for individual bees, the data for the second direction of travel of foragers, or the data provided by bees foraging on fructose sugar. Even though the experiments with fructose sugar coincidentally were run on the hottest day of the experimental program, and thus contributed to a large share of the extreme values in the total data, the interpretation provided above remains unchanged: No dependent (behavioral) variable measured can be considered closely related to any of those environmental variables we measured.

An interesting comparison, and one which clarifies the interpretation of factor analysis, is that provided by the general and well-known relationship between cricket chirp rate and temperature. Table 11 provides such a factor analysis furnished by a graduate class in animal behavior at the University of California, Santa Barbara. The students recorded and analyzed the data produced by "calling" crickets (*Gryllus* sp.) from 76 different animals in the field, while simultaneously recording temperature, time of day, wind speed, and date. In this output one can see clearly the relationship between chirp rate and temperature as well as between the various elements in the sound signal. See Walker (1962) for a full discussion of such a comparison.

Information Analysis

Before completion of any other data analysis, a separate analysis of information content in the sound signal from our tape recordings was conducted in another laboratory (Bottlik and Ricker, 1965). By using statistical-decision theory and measure of information (channel information), workers in this laboratory determined that the pulse-rate modulation could be used to differentiate among the three sucrose sugar concentrations with a probability of 0.46 compared to the 0.33 (i.e., one out of three choices) which would be obtained by guessing. Number of pulses,

TABLE 10. *Full Factor Analysis of the Correlation Matrix of Table 5 for the Uniform Data*[a] (All numbers $\times 10^{-2}$)

Variables	Factors[b]					
	I	II	III	IV	V	VI
1. No. of pulses	01	**95**	05	03	−22	02
2. Straight-run time	04	**93**	03	04	09	−05
3. Circle-run time	−11	−*42*	08	14	−*07*	12
4. Pulse rate	−04	18	04	08	−**81**	16
5. No. of cycles/pulse	14	18	02	07	**79**	17
6. Sound frequency	19	19	15	−23	04	−*51*
7. Sugar concentration	−07	10	−03	−**83**	17	−12
8. Sugar viscosity	−28	06	16	−*69*	−23	−11
9. Light intensity	10	−04	−**91**	01	01	28
10. Time of day	**93**	09	08	08	17	01
11. Shed temperature	*53*	09	−*51*	01	−02	*56*
12. Station temperature	*63*	12	−*39*	07	06	*52*
13. Shed humidity	−16	06	16	*56*	00	−32
14. Station humidity	−*63*	−11	29	−18	−03	−*56*
15. Sun direction	**93**	09	13	08	17	01
16. Sun altitude	02	03	−**96**	04	02	−08
17. Wind speed	**71**	−01	−23	−08	−09	−09
18. Wind direction	04	−10	06	−14	02	**76**

[a] Despite an elimination of extreme values from the data, interpretation remains virtually the same as for the factor analysis of the *total* data (Table 9).
[b] Boldface numbers = 70+; italic numbers = 40+.

straight-run time, and number of cycles per pulse were essentially similar to pulse rate in this regard. In addition, however, pulse rate could be used to differentiate among the three times of day about as well as it could be used to differentiate among concentrations of sugar.

TABLE 11. *Full Factor Analysis for Field Data from 76 Different Crickets ('Gryllus' Sp.) Included for Comparison with Tables 9 and 10*[a] *(All Numbers × 10⁻²)*

Variables	Factors[b]		
	I	II	III
1. Chirp rate	88	29	05
2. No. of pulses/chirp	− 04	20	− 83
3. Sound frequency	14	65	− 37
4. Chirp interval	− 90	− 17	17
5. Chirp duration	− 94	− 11	− 14
6. Pulse rate	88	30	− 09
7. Pulse duration	− 90	− 17	16
8. Temperature	62	72	01
9. Time of day	− 81	12	30
10. Wind speed	26	− 31	− 63
11. Date	15	89	20

[a] These data were gathered by N. Barnes, N. Broadston, P. Collard, W. Hand, and A. Wenner as a class project in animal behavior.

[b] Boldface numbers = 70+; italic numbers = 40+.

In light of the satisfactory results obtained with the factor analysis and in line with the discussion which follows in the remainder of this paper, it would appear that determination of the information content of a signal produced by an animal is not a particularly valuable technique for this type of study. In part, this approach fails to give any consideration to the capabilities or state of the receiver.

Comparison of Data for Two Directions

We succeeded in getting only nine different bees to travel to the east feeding station on August 29 for a total of three different times of day with three different concentrations of sugar. Comparison of the values for dependent variables obtained from bees visiting the east station (Table 7) with the values of equivalent variables for bees visiting the west station yielded no significant differences for pulse number, straight-run time, or pulse rate. Although measures of number of cycles per pulse and frequency of sound were significantly different for the two stations, the slight differences which exist do not necessarily indicate that communication of

direction information exists in these elements of the sound signal, particularly since environmental variables such as temperature or wind direction can easily account for such small differences (as discussed in the next section).

DISCUSSION OF DATA ANALYSIS

General Considerations

In a study of this magnitude, one obviously cannot fully discuss all interrelations simultaneously. Yet the treatments of the data in the last section permit us to discuss the results as a whole before turning to particulars.

Failure of the pattern of correlation coefficients to persist after data-partitioning (Table 3) provides strong evidence that none of the described elements in the sound signal can serve in communicating information between a dancing bee and a recruit bee about those environmental parameters measured in this study. This disappearance or marked alteration of correlation coefficients after partitioning of data also suggests that interrelations among environmental variables can account for much of the variance in the data. (See also the discussions of principal-components analysis and factor analysis below.)

The slight differences between means for the measurements of the various dependent variables on such a large sample (Table 7), despite significant correlations, also argue against effective communication of information about environmental parameters (including sugar concentration) among bees. A correlation of 0.12 between any two variables for a comparison in the *total data*, for instance, is significant at the 1 per cent level. When squared, however, this correlation explains only about 1 per cent of the variance in the data—hardly enough for efficient communication.

The principal-components analysis supports this line of reasoning (Table 8). In this type of analysis of the correlation matrix, the eigenvalues decline slowly from column to column; therefore no one factor can be considered responsible for a major portion of the variation in the data, although temperature (or its related factors) is especially prominent in contributing to a large part of the eigenvalues. Similarly, pulse number and straight-run time together account for more variation in the data than any other dependent variables, but they do not emerge from the bulk of the variation as likely vehicles for communication of information about any of the environmental variables measured in this study.

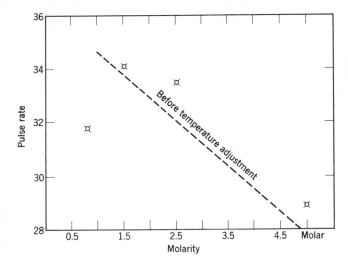

FIGURE 4. Relationship between pulse rate and sugar concentration before temperature correction. The line was fitted without reference to the weight of each point. (See also Figs. 5 and 6.)

A full factor analysis of the data further supports the above interpretation. Some of the dependent variables we measured are closely related to each other, notably pulse number to straight-run time (previously reported by Wenner, 1962) and pulse rate to number of cycles per pulse. Similarly, as might be expected, some of the environmental variables show a strong relationship to one another (Tables 9 and 10). In no case, however, do these results support the notion that bees communicate information as to the richness of the food (variable 7) by altering an element of the sound signal, as reported by Esch (1963).

Specific Considerations

Although no one factor in the sound signal emerges as a possible vehicle of communication about any one of the independent variables measured, this study does contribute to our knowledge of dependent variables and of the effect of environmental influences on the different elements of the sound signal.

Since pulse rate is computed by dividing pulse number by straight-run time, the high correlation between these two determinants of pulse rate (0.91 for the total data) leaves little room for variation of pulse rate with respect to some invironmental variable such as sugar concentration. Likewise, the principal-components analysis and factor analysis indicate that those small variations which are present in the dependent variables may result from temperature variation.

In this connection we can now reconsider the problem of the relationship among temperature, pulse rate, and molarity. It is interesting to note

that Esch (1963) succeeded in raising the pulse rate of the sound signal produced by a dancing bee by exposing this dancing bee to radiation from a heat lamp. From such a result it is reasonable to postulate that different foraging bees, after having flown through air of different temperatures, should produce pulse rates which vary according to how much their temperature has been raised after arriving in the hive and later dancing in the warm, constant temperature of the brood nest (where all bees danced while their sounds were being tape-recorded in this present study). (In this connection, Heran [1956] and Brauninger [1964] have already found an effect of external temperature on the "dance rhythm.") It also follows, then, that a bee flying through a cooler external temperature should have its metabolism modified more after returning to a warm hive than a bee flying through relatively warm air and that a comparison of pulse rate to the *difference* between hive temperature and external temperature might prove more informative than a direct comparison of pulse rate to external temperature or temperature of the dance area.

An inspection of the data from Table 7 regarding a comparison between pulse rate and sugar concentration (Fig. 4) indicates a possible slight relationship between these two variables. On the other hand, a close look at the means for sugar concentration and for temperature in the partitioned data (Table 7) reveals that our attempt to randomize data-gathering failed in this regard. On the day we gathered data from bees visiting 5-M fructose, the temperature was markedly higher than on other days—this explains the 0.11 positive correlation in the *total* data (significant at the 1 per cent level, approximately) between sugar concentration and temperature. A further study of Table 7 reveals that we generally inadvertently furnished higher concentrations of sucrose at cooler external temperatures, on the average, than was the case for low concentrations of sucrose.

Graphs derived from the appropriate means given in Table 7 illustrate the point better. Figure 5 shows the average pulse rate for data partitioned according to sugar concentration compared to the differences in temperature between the brood nest (approximately 35°C; Ribbands, 1953, p. 225) and the feeding station. If a line is fitted to these four points without regard to the weight of each point, such a comparison reveals a 2-pulses/sec. change in pulse rate with a 1°C change in temperature.

It is now possible to compare the pulse-rate–molarity relationship with the temperature-molarity relationship (Fig. 6). The close correspondence between the two sets of points suggests that the little relationship which does exist between pulse rate and molarity can be better explained as a consequence of temperature influence on the dance signal than as a communication of information about sugar concentration. If one corrects each value of the pulse-rate–molarity relationship as shown in Figure 4 with respect to the values obtainable from Figure 5, the pulse-rate–molarity relationship disappears.

The average pulse-rate obtained from the uniform data in this study (33.5 pulses/sec. at an average temperature of 21.8°C) agrees fairly well with that published earlier by Wenner (1962;

31.1 pulses/sec. at an average external temperature of 27.2°C) and with that initially indicated by Esch (1961; about 33 pulses/sec.—"Die Vibrationsstosse dauern ungefahr 15 msec. Ihnen folgen Pausen ungefahr gleicher Dauer"). These values all fail to agree, however, with those published later by Esch (1963), in which the values ranged from 20–30 pulses/sec. Unfortunately in the latter of these two papers Esch failed to publish data on time of day, external temperatures, or dance-floor temperatures extant at the time of recording bee dance sounds. It is not possible, therefore, to compare the results of this study with his results.

The high inverse correlation between straight-run time and circle-run time emerges from this study in a manner similar to that reported earlier (Wenner, 1962). In a study of environmental influences on the number of figure-8 dances per minute, Brauninger (1964) found the temperature coefficient (Q_{10}) to be "surprisingly low" in a consideration of the effect of external temperature on the "dance rhythm." If, however, the basic elements of the dance (straight-run time and circle-run time) vary inversely, one can expect such a coefficient to be lower for the whole than for its parts.

Measurement of the frequency of sound produced during the dance in this study yielded an over-all average of about 280 cycles/sec. for the uniform data (279.4 ± 1.27 cycles/sec., $N = 439$). This is in remarkable agreement with the physiological findings published by Heran (1959), in which he found the antennae most receptive to vibrations of about 274 cycles/sec. in the lateral direction and 285 cycles/sec. in the dorso-ventral direction. This suggests that, if dance sounds are important in communication, they can be received by the antennae of the recruit bee.

DANCE COMMUNICATION—
CURRENT STATUS

Implied Logic of Previous Studies

From our general knowledge of dance elements and behavior of bees during recruitment to a food source, we can list the basic premises and general conclusions involved in earlier studies:

1. Successful foragers dance after returning to the hive.

2. This dance contains information about location, type, and maybe quality of food (see the Introduction). In the case of location and quality of food, this information is quantitative.

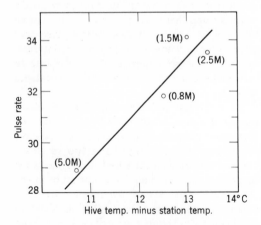

FIGURE 5. Relationship between pulse rate and the difference between hive and station temperatures. Hive temperature is assumed to be 35°C, and the mean station temperatures are from Table 7.

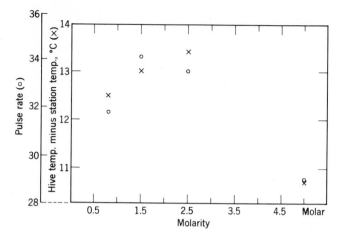

FIGURE 6. Pulse rate (*O*) and hive temperature minus station temperature (*X*) plotted against molarity. Close correspondence between the two sets of points suggests that the relationship displayed in Fig. 4 can best be interpreted as a temperature effect.

3. Bees recruited tend to go only to the food location indicated in the dance (step and fan experiments of von Frisch, 1954).

4. The number of bees visiting a newly replenished source increases exponentially until a plateau is reached with regard to the number of foragers involved.

5. Successful foragers succeed in directing other bees out to a specific food source. A recruit independently finds the source by interpreting and following quantitative location information and qualitative odor information obtained from a dancing bee.

Analysis of Earlier Logic in the Light of New Evidence

The above logic is internally consistent, certainly; but investigators who have worked with honey-bee communication have apparently come to regard the above interpretation as exclusive. The possibility that bees communicate quantitative information via the dance maneuver ("semantic signals" of Lindauer, 1967), however, suffers a most serious defect—there is no direct evidence that a recruit bee uses any of the quantitative information contained in the dance elements which have been measured.

Factor analysis, as applied in the experimental section of this report and as applied by Brauninger (1964), provides a very useful tool for separating spurious correlations from real relationships. Although behaviorists in the past have freely interpreted their findings after measuring only two or three variables, the above analysis again emphasizes the weakness of such a procedure. Perhaps of more importance, the factor analysis led us to evaluate correlations in general.

Even if correlations are real, as many of those

are which have been found between dance elements and various environmental parameters, these correlations are insufficient in themselves to establish that communication actually occurs between bees by use of these elements. Some examples may clarify this point.

We know that a male cricket will change its chirp rate with a change in external temperature (Table 11; Walker, 1962) and that a male cricket is communicating with a female cricket while chirping. From this knowledge, though, we would never conclude that a male cricket is sending temperature information to a female cricket.

Similarly, we know that a bee which flies against a wind on its way to a food source will produce a signal which indicates a greater distance to a food source, on the average, than a bee which flies either with a tail wind or on a still day (von Frisch and Lindauer, 1955; Heran, 1956; Wenner, 1963). The information contained in this signal would then be related to energy consumption or time of flight and would vary within certain limits. If an investigator already knows the distance to the food source and the wind direction, the signal *also* contains information for *him* about wind speed, in that the investigator already knows information which a recruit bee might not get or possess. Clearly, though, we would hesitate to say that a dancing bee communicates information about wind speed on the strength of the fact that such information exists in the signal.

Likewise, any definite correlation we can obtain between an element within the dance maneuver and some environmental factor does not establish, by itself, that communication occurs by this means.

The factor analysis of the data in this study led

us to an even deeper assessment of the evidence for the possibility of communication of quantitative information by dancing among bees. Although correlations do exist between dance information and environmental parameters and between dance information and later disposition of recruits in the field, evidence of a more direct nature would be required now to support the dance-language hypothesis. One particularly satisfactory method would be the experimental direction of bees from a hive to a point source in the field by furnishing the requisite information within the hive in the manner attempted by Steche (1957) and Esch (Anonymous, 1963; Esch, 1967). Only if bees could be predictably directed to a point source never before visited by hive mates would it be possible for one to claim a satisfactory understanding of this recruitment by means of a dance "language."

Our search for more decisive evidence for the hypothesis, via the methods of Steche and Esche, unexpectedly led us to a series of discrete experiments dealing with specific questions which challenged the logical basis of the dance-language hypothesis. As the experimental program progressed over a period of about 3 years, these questions were answered in turn. They are best handled here in the approximate sequence in which they arose.

1. *The role of learning in the dance-language hypothesis*. It would appear that the basic hypothesis rests on the premise that the dance communication process is an "instinctual signal system" (von Frisch, 1962; Skinner, 1966). On the other hand, the results of recent experiments (Johnson and Wenner, 1966) indicate that one must clearly distinguish between those recruited bees which have had previous experience at visiting a specific food source and those bees which have not yet had such experience.

Bees can be conditioned (simple discrimination conditioning) to forage at a familiar site upon presentation of a stimulus previously associated with the presence of food (Ribbands, 1954; Johnson and Wenner, 1966; Wenner and Johnson, 1966). That is, a neutral odor normally provided in (or with) the food, when blown into the hive, results in an immediate increase in the number of experienced bees inspecting a familiar site, even if no food is provided at such a site.

One can also interpret the repeated visits by experienced bees at a specific food source in the field as a manifestation of a "learned response" by the animals (choice discrimination conditioning— Wenner and Johnson, 1966). That is, a bee has

learned by association and reinforcement during its first few trips that it must travel past certain landmarks before it can feed again at a given food source. This food source, in turn, would be recognized by its characteristic odor and color patterns. It would appear, moreover, that such an interpretation is commonplace in the literature without having been stated as such.

An important exception to this neglect of the role of conditioned response in foraging is the important and extensive work by various Russian workers (e.g., see Chesnokova, 1959; Lopatina, 1959; Nikitina, 1959; and Lopatina, Nikitina, and Chesnokova, 1966). Most publications on the subject fail to mention the contributions of these workers. Among other finds the Russian work indicates that bees can be trained to respond sequentially to a chain of stimuli and that this is apparently very important in foraging behavior and recruitment to crops.

The dance-language hypothesis as generally stated (von Frisch, 1950; Lindauer, 1967) does not include a discussion of simple and choice discrimination conditioning in relation to recruitment and exploitation of food sources. Yet, if bees can learn, such learning would undoubtedly play a major role in this exploitation of crops. Not only would such a factor be important for bees which have already visited a specific source, but it is doubtful that any newly recruited forager goes into the field without some prior experience at orientation to landmarks and odor sources.

2. *A successful forager can recruit experienced bees without dancing*. Bees conditioned to feed at a food source inspect the immediate area for other sources once their primary source becomes depleted. If unsuccessful, they apparently return and remain in the hive for varying periods. Each of them routinely leaves the hive, briefly inspects the original and neighbouring sites, and returns to the hive (von Frisch, 1950; Johnson and Wenner, 1966). If food is furnished at this source, the first of the successful monitoring bees fills and begins regular foraging. From that time on the cumulative number of different foraging bees rises exponentially until virtually all of the experienced bees are involved once more (Ribbands, 1955; Johnson and Wenner, 1966). This exponential increase can occur without any of the successful bees dancing upon their return to the hive (Johnson and Wenner, 1966). Apparently such a recruitment of experienced bees occurs as a consequence of conditioned responses (simple conditioning) engendered by the odor(s) brought into the hive on the body of each successful bee.

If two groups of bees are trained to forage in opposite directions from the hive, and if food is provided simultaneously at both sites after a period of depletion, a tally of the arrivals indicates that an exponential increase occurs among the experienced bees at each site. This occurs even if the researcher captures all arrivals at one of the stations but not at the other, thus permitting successful foragers to return from only one of these sites (Johnson, 1967a).

An experienced bee from one group can even attend a dance executed by a successful forager from the other group and go to the site at which it had been successful previously. This occurs despite the fact that the dance provides a different set of information concerning distance and direction (Johnson, 1967a).

3. *Few of the recruited bees which leave the hive succeed in finding a site indicated in the dance.* One can gain an impression from the literature that most of those bees which attend a dancing bee arrive at the feeding site in the field within a minute or so. However, a search of this literature on the dance-language hypothesis has failed to yield any data on the efficiency of recruitment.

Although some of our preliminary studies on this subject (Johnson and Wenner, 1966) approach the problem only indirectly, it would appear that relatively few (as low as 1–2 per cent) of those bees which contact the dancer and leave the hive actually reach the specific location visited by that dancer, even if the site is within 200 m of the hive. (The increasing of odor concentration at such a site increases the efficiency.) Much of the evidence gathered to date indicates a performance level little better than that expected if such bees simply search for the odor of bees, odor of food, odor of location, or odor of all (Johnson and Wenner, 1966). It also follows that any hypothesis of recruitment to a food source must include a discussion of the efficiency of recruitment before it can be very useful for future investigations.

4. *The classic experiments on bee "language" appear to lack essential controls.* One of the strongest sets of evidence for the dance-language hypothesis arises from the use of "step" and "fan" experiments (von Frisch, 1954). In these experiments bees from an experimental hive were allowed to forage at only one of several sites in the same general direction and distance from the hive. Subsequently, these regular foragers could recruit other, "non-experienced" or "experienced" bees from the parent hive. These recruited bees would then be free to arrive at the experimental site or at any of the control sites (sites which contained scent but no food).

A tally of recruits indicated that they had arrived either at or very near the site already visited by regular foragers from their own hive. From the results of this type of experiment von Frisch concluded that the ability of recruited bees to follow the distance and direction information contained in the dance was exceptionally good.

A careful analysis of the design of these fan and step experiments, however, reveals the lack of certain essential controls. Bees recruited by regular foragers apparently preferentially land at *or near* a site at which bees are feeding *because bees are already present.* Several repeats of the step and fan experiments, with a provision of bee visitation at both control and experimental sites (by use of a different color of bees from another hive), resulted in a totally different set of results from that reported by von Frisch and co-workers. In the more carefully controlled situation bees apparently distributed themselves at the various sites according to the geometry of the site placement (Johnson, 1967b; Wenner, 1967). As few as 9 per cent of those recruits captured had landed at the experimental site, for example, once control sites more closely resembled the experimental site.

It would appear, therefore, that the classic experiments of von Frisch lacked essential controls against recruit bees traveling to or near an experimental site in the field by sole use of odor cues obtained from a dancing bee. These would include odor of other bees, odor of location, and odor of food (Johnson, 1967b; Wenner, 1967).

In a less closely related vein, it would also appear that the original experiments on sugar concentration and sugar intake by foraging bees lack necessary controls (P. H. Wells and J. Giachino, in preparation). Contemporary experiments executed with careful controls of pertinent factors showed no differences in load per trip per bee over the concentration range of 0.5-M to 2.5-M sucrose, or scented versus unscented, or the type of sugar presented (fructose vs. sucrose).

5. *A possible role of conditioning in re-location of swarm clusters.* One problem which emerges from the above rationale is that of the relocation of swarm clusters (reportedly by dance communication; Lindauer, 1957). Can simple conditioning function in this relocation?

When a swarm is clustered on a bush or a tree outside a hive before it moves as a unit through the air, certain "scout" bees dance on the outside of this mass of bees. These dances, in turn, contain the necessary information for a human to determine the approximate location for the colony before it moves to its new site (Lindauer, 1957). In light of the above discussion concerning an alternative explanation to communication via "semantic" dance signals (together with other logical considerations), certain objections can be raised to the concept that a swarm relocates by using information contained in the dances of scout bees. Some of these are as follows:

(a) It seems unlikely that all bees in the cluster would contact the relatively few dancers involved.

(b) If all bees in the cluster did get information from dancers concerning the location of the new site, a good percentage would automatically get erroneous information as a result of error both in the production and interpretation of signals.

(c) When a swarm moves through the air, individual bees generally circle through arcs. Thus each bee covers a much greater distance, spends much more time, and utilizes much more energy than is normally required for flying the actual distance which the swarm travels as a unit. All this makes it unlikely that inexperienced swarm bees convert distance and direction information contained in the dance directly into distance and direction traveled. Lindauer (1957, 1961) does not discuss this point.

(d) If (c) is valid, then the moving swarm must either receive a signal to stop or suddenly fail to receive a signal to move.

Conditioning may play an important role in this swarm movement. Conceivably, scouts, while repeatedly traveling between swarm and the future site of the colony, could learn the route well (after each scout has found the site by rapidly searching for the same site as discovered earlier by a scout). Then, as the swarm begins to move through the air, this fairly large group of scouts could effectively lead the queen or the other workers by utilizing a chemical attractant (e.g., see Morse, 1963).

The interpretation provided in this discussion may come as much of a surprise to others as it initially did to us. In retrospect, however, it should come as no surprise that bee behavior is more similar to the behavior of other animals than we previously supposed, even if some bee behavior appears exotic.

SUMMARY

1. We subjected sound signals produced during the honey-bee waggle dance to a statistical analysis to test whether bees modify their signals as a result of imbibing sugar solutions of different concentrations, types, or viscosities and to clarify simultaneous relationships between components of dance sounds and some environmental variables. Dependent variables isolated from the sound signal included pulse number, straight-run time, circle-run time, rate of pulse modulation, cycles per pulse, and frequency of sound. Independent (environmental) variables recorded included sugar concentration, sugar viscosity and type, light intensity, hive and feeding-station temperatures and humidities, time of day, sun direction and altitude, wind speed and direction, date, and absolute direction of flight.

2. All variables were first analyzed as a group. The data then were partitioned into classes for further statistical treatment. Analysis of total and partitioned data yielded nine correlation coefficients for each pair of variables. Of 90 correlation coefficients determined for the total data, 23 were significant at the 1 per cent level and 41 at the 5 per cent level. From our analysis it is clear that environmental factors influenced the signals but that any one significant correlation does not necessarily indicate communication among bees.

3. An analysis of principal components (a first step in factor analysis) and a full factor analysis, techniques suggested earlier by Sokal, Daly, and Rohlf (1961) and applied to data by Sokal and Daly (1961), sorted out which variables of those measured were largely responsible for variation in the data. This method of analyzing data proved to be especially valuable as an aid in separating spurious correlation from real relationships.

4. Temperature or its related variables account for much of the variance in the data. The slight relationship obtained in a comparison of pulse-rate modulation and sugar concentration can be explained as a relationship between pulse-rate and temperature.

5. The factor analysis of the data led to a critical over-all look at the evidence for communication by means of the waggle dance, an evaluation which revealed basic discrepancies in the dance-language hypothesis. The results of discrete experiments which arose as a consequence of questions raised by the factor analysis are summarized. This summary includes the following points:

(a) Bees can be conditioned to visit a site in response to a reinforced stimulus presented in the hive.

(b) One must separate a discussion of the performance of those recruits which have had experience at foraging at a particular food source from those recruits which have not had such experience.

(c) At least some of the complex behavior previously ascribed to communication by dancing may be explained in terms of simple discrimination conditioning.

(d) The efficiency of finding a food source by naïve bees recruited by dancing is apparently very low.

(e) A repeat of the classic "step" and "fan" experiments of von Frisch and co-workers with more stringent controls yields results not consistent with the dance-language hypothesis. These experiments indicate that the original experiments lacked necessary controls.

(f) The fact that simple conditioned responses are readily instilled in bees also provides a possible explanation for relocation of swarm clusters.

REFERENCES

ANONYMOUS, 1963. Bee beep. *Time* (Magazine). 31 May: p. 54.

BOTTLIK, I. and RICKER, G., 1965. Information analysis of sonic signals in honey bee communications. Final technical report, May, 1964–February, 1965. No. C5-494/3111, Contract Af 33 615 1336. (Autonetics, Anaheim, Calif.)

BRAUNINGER, H., 1964. Über den Einfluss meteorologisher Faktoren auf die Entfernungsweisung im Tanz der Bienen. Z. vergl. Physiol. **48**: 1–130.

CHESNOKOVA, E. G., 1959. Conditioned reflexes established in bees through a chain of visual stimuli [in Russian]. Tr. Inst. Fiziol. Pavlova **8**: 214–220.

DIRSCHEDL, H., 1960. Die Vermittlung des Blütenduftes bei der Verständigung im Bienenstock. Inaugural dissertation. Ludwig - Maximilians-Universität München. 56 pp.

ESCH, H., 1961. Über die Schallerzeugung beim Werbetanz der Honigbiene. Z. vergl. Physiol. **45**: 1–11.

———, 1963. Über die Auswirkung der Futterplatzqualität auf die Schallerzeugung im Werbetanz der Honigbiene (*Apis mellifica*). Zool. Anzeiger **26** (Suppl.): 302–309.

———, 1967. The evolution of bee language. Sci. Amer. **216**: 96–104.

ESCH, H., ESCH, I. and KERR, W., 1965. Sound: an element common to communication of stingless bees and to dances of the honey bee. Science **149**: 320–321.

FRISCH, K. VON, 1946. Die Tanze der Bienen. Österreichische zool. Z. **1**: 1–48.

———, 1950. Bees, their vision, chemical senses, and language. Cornell Univ. Press, Ithaca, N.Y.; Methuen, London.

———, 1954. The dancing bees. Methuen, London.

———, 1962. Dialects in the language of bees. Sci. Amer. **207**: 78–87.

FRISCH, K. VON and JANDER, R., 1957. Über den Schwanzeltanz der Bienen. Z. vergl. Physiol. **4**: 1–21.

FRISCH, K. VON and LINDAUER, M., 1955. Über die Fluggeschwindigkeit der Bienen und über ihre Richtungsweisung bei Seitenwind. Naturwissenschaften **42**: 377–385.

HERAN, H., 1956. Ein Beitrag zur Frage nach der Wahrnehmungsgrundlage der Entfernungsweisung der Bienen (*Apis mellifica* L.). Z. vergl. Physiol. **38**: 168–218.

———, 1959. Wahrnehmung und Regelung der Flugeigengeschwindigkeit bei *Apis mellifica* L. *Ibid.* **42**: 103–163.

JOHNSON, D. L., 1967a. Communication among honey bees with field experience. Anim. Behav. (in press).

———, 1967b. Honey bees: do they use the direction information contained in their dance maneuver? Science **155**: 844–847.

JOHNSON, D. L. and WENNER, A. M., 1966. A relationship between conditioning and communication in honey bees. Anim. Behav. **14**: 261–265.

LINDAUER, M., 1957. Communication in swarm-bees searching for a new home. Nature **179**: 63–66.

———, 1961. Communication among social bees. Harvard Univ. Press, Cambridge, Mass.

———, 1967. Recent advances in bee communication and orientation. Annu. Rev. Entomol. **1**: 45–58.

LOPATINA, N. G., 1959. Conditioned reflexes of bees to complex stimuli [in Russian]. Pchelovodstvo **36**: 35–38.

LOPATINA, N. G., NIKITINA, I. A. and CHESNOKOVA, E. G., 1966. Significance of conditioned reflexes in the development of the signal activity of honey bees [in Russian, English summary]. Zh. obshche. Biol. **27**: 605–614.

MORSE, R., 1963. Swarm orientation in honeybees. Science **141**: 357–358.

NIKITINA, I. A., 1959. Conditioned signal reflexes in honeybees [in Russian]. Nauchnoe Soobshchestvo Inst. Fiziol. Pavlova **1**: 55–57.

NIXON, H. and RIBBANDS, C. R., 1952. Food transmission in the honeybee colony. Roy. Soc. (London), Proc., B. **140**: 43–50.

RIBBANDS, C. R., 1953. The behaviour and social life of honeybees. Bee Research Association, London.

———, 1954. Communication between honeybees. I: The response of crop-attached bees to the scent of their crop. Roy. Entomol. Soc., London, Proc., A. **29**: 141–144.

———, 1955. Communication between honeybees: II. The recruitment of trained bees, and their response to improvement of the crop. *Ibid.* **30**: 26–32.

SEAL, H. L., 1964. Multivariate statistical analysis for biologists. Wiley, New York. 207 pp.

SKINNER, B. F., 1966. The phylogeny and ontogeny of behavior. Science 153: 1205–1213.

SOKOL, R. R. and DALY, H. V., 1961. An application of factor analysis to insect behavior. Univ. Kansas Sci. Bull. 42: 1067–1097.

SOKAL, R. R., DALY, H. V. and ROHLF, F. J., 1961. Factor analytical procedures in a biological model. Univ. Kansas Sci. Bull. 42: 1099–1121.

STECHE, W., 1957. Beiträge zur Analyse der Bienentanze (Teil I). Insectes soc. 4: 167–168.

WALKER, T. J., 1962. Factors responsible for intra-specific variation in the calling songs of crickets. Evolution 16: 407–428.

WENNER, A. M., 1961. A method of training bees to visit a feeding station. Bee World 42: 8–11.

——, 1962. Sound production during the waggle dance of the honey bee. Anim. Behav. 10: 79–95.

——, 1963. The flight speed of honey bees: a quantitative approach. J. Apicult. Res. 2: 25–32.

——, 1967. Honey bees: do they use the distance information contained in their dance maneuver? Science 155: 847–849.

WENNER, A. M. and JOHNSON, D. L., 1966. Simple conditioning in honey bees. Anim. Behav. 14: 149–155.

VOCAL EXCHANGES BETWEEN DOLPHINS*

John C. Lilly and Alice M. Miller[1]

IN a previous article (1) it was shown that the individual bottlenose dolphin (*Tursiops truncatus*) emits several classes of complex and varied sounds. At least one of these classes (click-creakings) is used in finding food, ranging, and navigating; other classes of sounds may be used for communication between individuals. These latter classes are (i) click trains (not creakings), (ii) whistles, (iii) quacks, blats, and squawks, and (iv) combinations, such as click trains or quacks plus whistles. In this report the first experiments on the possibilities of communication between two dolphins are presented. The techniques and apparatus were those described previously (1). Emphasis is placed here on the elicitation of vocal exchanges and on the formal description of these exchanges.

Special experimental conditions are set up (Table 1).

CONDITION 1. The animals are in solitude, confined and isolated to the extent given in Table 1. This is the "near-zero exchange" state (2, 3). The "dullness" and "evenness" of the situation is purposefully maximized. (These conditions are analogous to solitary confinement of a human being in a small box.)

CONDITION 2. One set of physical constraints is attenuated, and more three-dimensional move-ment is allowed. (These conditions are analogous to solitary confinement of a human being in a large room.)

CONDITION 3. Each animal is allowed to hear and reply (in water or air, or in both) to one other unseen, untouched, untasted dolphin (Fig. 1); the dolphin-to-hydrophone distance is controlled, and vocal emissions from the two animals are separated in the recordings. A hydrophone is placed near the rostrum of each dolphin. The animal is so held that it cannot move its head more than a few inches from the hydrophone. The water space is so shallow (10 to 15 in.), so narrow (15 in.), and so short (slightly over one body length) that the animals cannot swim. Each one rests on the bottom most of the time unless it is disturbed by the presence or intervention of a human being.

CONDITION 4. The animals are no longer confined to the extent of "enforced resting," and each animal has the option of swimming. The distance between the animal and the hydrophone is controlled to a lesser degree than in condition 3 but is still limited to a maximum distance of a few feet. Play with floating "toys" is allowed.

CONDITION 5. Free bodily contact, biting, mutual play with "toys," racing, courting, mating, stealing and giving food, and so on are all allowed. The animal-to-hydrophone distance is

[1] The authors are affiliated with the Communication Research Institute of St Thomas, U.S. Virgin Islands, and Miami, Florida. Dr Lilly is director of the institute.

TABLE 1. *The Most Frequent Forms of Vocalization and Vocal Exchanges (One and Two Dolphins). This Classification Applies Best to Newly Captured Animals. After Several Weeks Unpredictable, Complex Vocalizations Appear That Are Inconsistent With the Classification Given Here*

Condition	Vocal Emission (Initial or Response)	
	Dolphin A	Dolphin B
1. One in solitude, confined*, isolated†, (near-zero exchange)	Whistles and clicks	No response
2. One in solitude, free-swimming‡ (objects-and-background-exchange only)	Whistles, clicks, creakings§	No response
3. Pair, confined, isolated, limited to acoustic exchange path, intraspecies only	Whistles; clicks and/or whistles; clicks	Whistles; clicks and/or whistles; clicks
4. Free-swimming pair, isolated, limited to intraspecies acoustic exchange	Whistles; clicks and/or whistles; clicks; creakings§	Whistles; clicks and/or whistles; clicks (see 3)
5. Free-swimming pair, free body contact, free acoustic and mating exchanges of all kinds, interspecies isolation	Whistles; clicks and/or whistles; clicks; creakings§; quacks, squawks, blats, etc., with or without whistles	Whistles; clicks and/or whistles; clicks (see 3); whistles; clicks or quacks, etc., with or without whistles
6. Condition 1, 2, 3, 4, or 5 plus presence or intervention of human being	Same exchanges as in 5 with increase in quacks, etc.; airborne sounds suddenly increase in frequency of occurrence and in amplitude	Same exchanges as in 5 with increase in quacks, etc.; airborne sounds suddenly increase in frequency of occurrence and in amplitude
7. Confined in oceanaria in colony	Not yet examined experimentally	Not yet examined experimentally
8. Fre-swimming, unisolated, unconfined in the sea in colony	Not yet examined experimentally	Net yet examined experimentally

* Confined in a box 7½ ft. by 15 in. in water 10 to 15 in. deep. † Separated by a barrier to prevent bodily contact with another dolphin or with human beings. ‡ Swimming in one of three pools with, respectively, (i) rough rock sides, 70 by 20 ft., with water up to 12 ft. deep; (ii) vinyl sides and bottom 10 by 8 ft., water 22 in. deep; (iii) Fiberglas, smooth sides and bottom 9 by 7½ ft., water up to 30 in. deep. § Creakings do not occur as responses to another dolphin's vocalizations but do occur during feeding, navigation in murky water for novel objects, and so on—that is, as "sonar" for recognition and ranging. ‖ No other species of animals entered or were near the pool, especially no human beings.

not controlled, and there is some confusion between emissions from one animal and those from the other. After a period of study and observation, the individual emitting whistles, blats, or quacks, can be identified, but identification is less easy in the case of clicking.

CONDITION 6. Human beings are present and "intervene" through feeding, "rewarding," "punishing," operant-conditioning (especially in production of specific kinds of emissions), play, vocal interspecies "exchanges," transporting, direct brain stimulation, and other measures.

CONDITION 7. In the oceanaria, captive colonies can be observed. To date, no experiments have been undertaken to study possible exchanges between individuals in such colonies. (Such experiments have been proposed as possible controls for the experiments described here; there are many technical difficulties, such as that of identifying the vocal emissions of specific individuals.)

CONDITION 8. At sea, the difficulties of experimenting with wild animals and studying their exchanges are increased by the difficulties of finding and staying near the animals for a significantly long period of time. To determine the effects of capture and of captivity on the vocalizations, control experiments should eventually be carried out at sea, with the dolphins in their most free state.

RESULTS

From the standpoint of measurement of physical acoustical quantities, conditions 1 and 3 give the best results; condition 3 gives the best physical recordings of exchanges. From the standpoint of the health and vigor of the animals in captivity, conditions 5 and 6 are best and give the greatest variety of vocalizations. We have a few data for condition 7 but none for condition 8. Most of the results given here are for conditions 3 and 4.

In conditions of solitude the animals' vocal behavior is different from their vocal behavior when they are in pairs. When they are maintained under condition 1 for a few hours a day (the

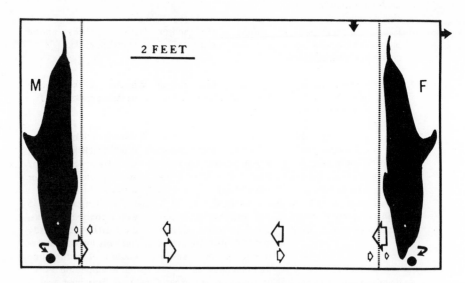

FIGURE 1. Configuration for recording vocal exchanges between a pair of isolated and confined dolphins. The animals are resting on the bottom and are shown in lateral elevation, the male (*M*) from his left side, the female (*F*) from her right. (Dotted line) foam-covered barrier preventing the dolphin from entering the rest of the pool; (black rectangle) inner walls of the salt-water pool; (open arrows, decreasing in size) acoustic-energy transfer from the source-dolphin to the sink-dolphin; (curved solid arrows) energy received by the hydrophone from the dolphin nearest it (attenuated signals are received from the far dolphin); (straight solid arrows) water inflow and outflow. Spoor may possibly be transferred from *M* to *F*; however, the flow of water is greatly impeded by the barriers, so that any exchanges of spoor that may occur probably have little value as signals. Other configurations eliminate spoor-trace clues completely, since no water can move from one dolphin to the other.

condition of near-zero exchange), there are usually no creakings. The slow click trains gradually decrease and finally cease. Whistles become less and less frequent over a period of days to weeks when there is a continued lack of "response" from the environment and from other animals, including man. To keep the dolphins healthy, exposure to condition 1 must be limited to short periods.

In condition 2 creakings occur as needed for food finding, exploring new objects, and navigating in muddy water or at night. Whistles and slow click trains gradually cease in a few days or weeks; they are elicited immediately by (i) presentation of the spoor of other dolphins in the water; (ii) visual or acoustic stimuli from another dolphin; (iii) human stimuli of various sorts; or (iv) presentation of toys or novel objects. In other words, a change to conditions 3, 4, 5, 6, or 7 increases the rate of occurrence of vocalizations.

Under conditions 3, 4, or 5, where the dolphins are in pairs, spontaneous vocalizations (Table 1) occur fairly frequently, in bursts lasting from a few seconds to many minutes. In a typical 24-hour day there is a total of at least 4 hours of vocalization, and on many days there is more than this. Under condition 5 (with freedom to swim, to make body contacts, and to mate), a male and a female emit various sounds steadily for periods up to 20 or more minutes, concurrently with play, courtship, and mating behavior.

Under conditions 3 and 4, where members of a pair are in acoustic and vocal contact, definite vocal "exchanges" are demonstrable (Fig. 2). These exchanges are briefer and rarer than the vocalizations that occur under condition 5 (body contact). A "monologue" or a "solo" by one or the other animal may precede, follow, or be unrelated to, an exchange, in the same few minutes; most monologues, however, occur in a close time relation to an exchange. These monologues differ from those of the same animal in solitude (see Table 1, conditions 1 and 2): they are more frequent and more varied in amplitude and frequency.

As shown in Table 1, when the dolphins are in pairs (conditions 3, 4, and 5), the two animals produce alternating emissions (Fig. 2). Rare interruptions of one by the other, or "overlaps," do occur. A special phenomenon, called a "duet"

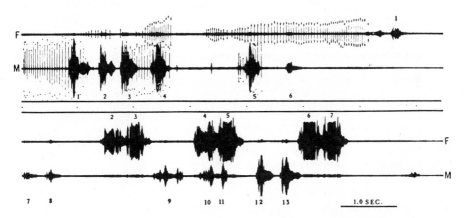

FIGURE 2. A graphic record of a vocal exchange between two dolphins. (Top trace in each pair) emissions of the female (F); (bottom trace in each pair) emissions of the male (M). The upper pair of traces shows a click-and-whistle exchange; the lower pair, a continuation of the same record without the clicks. (Dots between pairs of traces) seconds of elapsed time; elapsed time for the whole record, 15 sec. For reproduction, the peaks of the clicks of the female were marked with black dashes; the tips of those of the male, with black dots. Whistles are numbered in sequence for each animal. Other disturbances in the base line are, in most cases, water noises (see text).

(Fig. 2: F4 and M10; F5 and M11), also occurs: the two animals whistle simultaneously, sometimes matching frequencies and time-patterns so exactly that the relatively low-frequency difference between their simultaneous whistles can be heard.

Alternations without interruptions, overlaps, or duets are the most frequent exchanges. Such alternations consist of whistles or slow click trains, or both (Table 1 and Figs. 2 and 3).

If two dolphins are transferred from condition 3 (confined and isolated with only an acoustic path between them) to condition 4 (free to swim but with no body contact), they make creaking sounds. The creaking can be related to detection, ranging, and recognition of novel objects, to finding food, to pursuit games with a toy, or to navigation in the dark or murkiness to avoid rough walls or other obstacles (1, 3).

If a dolphin is allowed to touch the body of another dolphin (Table 1, condition 5), it makes another set of sounds—squawks, quacks, blats, barks, and so on—both under water and in the air. [Graphic results of analyses of some such individual sounds are shown in the previous paper (1).] If a human intervenes with one of a pair of dolphins limited to acoustic exchange (thus changing condition 3 to condition 6), the dolphin barks, squawks, and quacks, apparently at the human, and it may whistle simultaneously

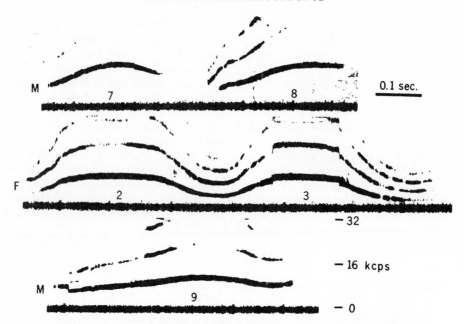

FIGURE 3. Sonograms of a portion of an exchange between two dolphins. (M) Emissions of the male; (F), of the female. Emissions are numbered to correspond to the numbering of the amplitude trace in Figure 2. The fundamental (f) and the first two overtones (2f and 3f) may be seen on these sonograms. Additional sonograms with twice this frequency scale show that energy in the third, fourth and fifth overtones decreases rapidly as compared to that of the fundamental and the first two overtones. There was some enhancement of the higher frequencies in recording the sonograms (6).

every so often, apparently in exchange with the other dolphin. When two dolphins are swimming freely together (condition 5), they exchange such complex mixtures of sounds, but deciding which dolphin emits which sounds is extremely difficult.

After several weeks in captivity in shallow water (18 to 30 in.), a dolphin begins to emit each and every class of sounds in air, including clicks and whistles in addition to quacks.

In Figure 2 are shown some results obtained in studies made under condition 3—the best condition for controlling and distinguishing the underwater vocal emissions of the dolphins. The upper trace is that of the female dolphin F of Figure 1; the lower trace, that of the male M of Figure 1. This particular exchange opens with the male's slow click train, followed by four whistles (Fig. 2, M1, 2, 3, and 4). During whistle M1, the male stops clicking shortly after the female begins to click, and the female maintains the train during whistles M1, 2, 3, and 4. During M3, the male starts clicking and continues to click and to whistle (M4) throughout the rest of the female's click train. He keeps his click train carefully out of step with hers, and stops his when she stops hers. The female starts clicking again before his

whistle M5; the male joins in with a few clicks, whistles once (M5), and stops clicking. The female continues her clicks for a total interval of 3 sec. The male whistles once again, faintly (M6). The female moves her head (water noises appear at the end of the train) and whistles faintly (F1). He answers with two faint whistles (M7 and 8). She suddenly whistles loudly twice (F2 and 3); he replies with a long single whistle of medium intensity (M9). She responds with whistles of high intensity (F4 and 5), and he joins her (M10 and 11) in a precise "duet," matching her time pattern, and goes on to "reply" more loudly with two whistles (M12 and 13). She answers loudly (F6 and 7), and both become silent.

INDIVIDUAL DIFFERENCES

Inspection of these amplitude-time graphs shows some differences between the whistles of these two animals. The male tends to become silent briefly between his pairs of whistles (M1 through 8 and 12 and 13). The female tends to "fill in" between what correspond to his pairs of whistles with sound. He builds his average intensity to a fast peak and drops it to a lower level in the last half of the record. A male may have one or more fast notches and fast peaks in his record. A female also does this (other records); she may often start at a lower level, rise suddenly to a higher one, and drop back to the lower one (Fig. 2, F6 and 7). She also may have one or more deep fast notches and peaks in the amplitude record.

In analyzing these sounds for their frequency, these amplitude variations can be correlated with the frequencies emitted. Figure 3 shows sonograms of frequency versus time for emissions M7 and 8 (top traces), F2 and 3 (middle traces), and M9 (bottom traces). In general all of the time pattern of the fundamental (frequency f) is shown; the second harmonic ($2f$) and third harmonic ($3f$) sometimes do not show because of their low intensity. At other times all the harmonics up to the sixth ($6f$) have sufficient intensity to be recorded. (Almost all of the frequencies are integral multiples of the fundamental.)

Inspection of the traces of the male show that, in general, the recorded harmonics are enhanced during his amplitude peaks—that is, when the amplitude is high the harmonic content is high (M8, M9). The traces of the female show in general that her high amplitudes correspond to her highest frequencies of the fundamental. The harmonic content of the female's emissions tends to be more constant than that of the male during a given emission. The frequencies covered by his fundamental are from 6 to 15 kc/sec.; her fundamental varies from 3.5 to 11 kc/sec. Some of her low-frequency and her low-amplitude emissions correspond to his short silences between pairs of whistles (Figs. 2 and 3, F2 to F3).

The separation of the emissions of whistles into countable integral units is based on measurements of occurrences and durations and on graphical records of isolated single whistles, on repetitions of similar (but not necessarily identical) patterns of frequency variation in pairs and triplets, and on the qualities of their total effect as heard (slowed down) by experienced observers. When the female whistles, she emphasizes the relatively flat, high-frequency portion and de-emphasizes the low-frequency "slump," somewhat the way some people fill in between words with a low-intensity "aaaah." (Listening to records obviously does not give us the same acoustical experience that the dolphins have; the dolphins have a much wider hearing range than we do and they may have special resonators in the hearing side as well as on the transmitting side which may make the

hearing experience of sounds an entirely different kind of experience for them than for us.)

Usually, but not always, the duration of the whistles is of the order of 0.2 to 0.4 sec. (see Fig. 2). Under special conditions yet to be thoroughly determined, extremely short (0.1 sec.) or extremely long (2 to 3 sec.) whistles, or both, occur. Most whistle transmissions are in the middle range.

Trains of clicks cover the full range of duration of the whistles and sometimes continue as long as 15 sec. without pause. The most frequently observed durations for click trains are close to those of the longer whistles (0.5 to 1 sec.).

The amplitudes of the middle-frequency components (less than 40 kc/sec.) of each click are varied by the dolphin in systematic ways. Sonograms and detailed high-speed oscillograph recordings show each click to be a complex train of sine waves whose components vary in frequency and in amplitude with time within the train. From click to click there is more controlled variation in the middle and low range of frequencies (1 to 20 kc/sec.) than in the high range (20 kc/sec. to apparatus limits at 64 kc/sec.). The clicks are not "white noise" in the range below 20 kc/sec. The lower-frequency portion of the train lasts up to 5 msec. and can have mean frequencies as low and as high as the whistle frequencies, with variations ranging from $\frac{1}{2}$ to 2 times the mean value. The high-frequency portion of the train (above 20 kc/sec.) is very brief (0.1 msec.) and may, with certain kinds of frequency analyzers, appear to be white noise (4).

If one listens to slowed tape recordings (slowed to $\frac{1}{16}$ of normal speed) the complex tonal variations can be perceived within each click, from one click to the next, and from animal to animal. The clicks of "creakings" (Table 1) are higher pitched, shorter, and "harder-sounding" than those of exchanges (4). The "sonar" click is usually one of high frequency; the exchange click, one of lower frequency.

CONCLUSION

In this report (5) we have presented something of what dolphins transmit in their exchanges—signals plus noise. A few tentative, simple "meanings" have been found ("distress," "attention," "irritation," and so on); however, most of the exchanges are not yet understood.

Note added in proof.—Since this report was submitted we found that dolphins can emit ultrasonic clicks independently of sonic clicks and vice versa.

REFERENCES AND NOTES

1. LILLY, J. C. and MILLER, A. M. *Science*, **133**, 1689 (1961).
2. LILLY, J. C. *Psychiat. Research Repts.*, **5**, 1 (1956).
3. —— and SHURLEY, J. T. *Psychophysiological Aspects of Space Flight* (Columbia Univ. Press, New York, 1960).
4. SCHEVILL, W. and LAWRENCE, B. *J. Exptl. Zool.*, **124**, 147 (1953); *Breviora*, **53**, 1 (1956): KELLOGG, W. N., KOHLER, R. and MORRIS, H. N., *Science*, **117**, 239 (1953); KELLOGG, W. N., *ibid.*, **128**, 982 (1958); ——, *J. Acoust. Soc. Am.*, **31**, 1 (1959).
5. This research has received support from the Air Force Office of Scientific Research, the Coyle Foundation, the U.S. Department of Defense, the National Institute of Mental Health, the National Institute of Neurological Diseases and Blindness, and the Office of Naval Research. We thank K. N. Stevens of the Massachusetts Institute of Technology for the use of a Kay Sonograph, J. C. Steinberg of the University of Miami for the use of a hydrophone set, and Herbert Gentry of Orlando, Fla., for the use of a Precision Company wide pass-band tape recorder.
6. The sonograms were made with a Kay Electric Company Sonograph.

COMMUNICATION BETWEEN YOUNG RHESUS MONKEYS*

William A. Mason and John H. Hollis

THE relationship between the social behavior of the rhesus monkey (*Macaca mulatta*) and its demonstrated ability to solve complex discrimination problems is obscure. It is probable, however, that the factors involved in performance on complex discrimination tasks are relevant and important in the daily social interactions of macaque monkeys in their native habitat.

A prominent feature of the social relations of rhesus monkeys is the element of conflict, resulting on the one hand from the presence of a dominance hierarchy in which aggression—either overt or threatened—plays an essential role; and on the other, from the existence of strong cohesive factors—reflected in such activities as play, grooming and sexual behavior—which attract the individual to the other members of the group. Possibilities for conflict are amplified by the fact that these activities frequently occur between animals that differ widely in size, strength and dominance. (Chance, 1956.) It is apparent, therefore, that the nature of social life among rhesus monkeys places a premium on the ability to identify individual status characteristics and to respond appropriately to transient changes in companions of mood, motivation or intent.

Examples of other forms of complex social discrimination occurring with reference to spatial locus of companions, are given by Chance and Mead (1952).

Studies of rhesus monkeys raised in a restricted environment (Mason, 1960; 1961a; 1961b) indicate that the ability to make such discriminations is at least partly dependent upon previous social experience and suggest that many of the deficiencies and aberrations observed in the social relations of laboratory-reared monkeys result from a failure to make effective discriminative responses to social stimuli.

If primate social adjustment requires a high level of learning capability, as the foregoing considerations suggest, this would help to explain the discrepancy noted by some students of primate behavior (e.g., Harlow, 1958) between intellectual achievement as revealed by laboratory investigation and the apparent requirements of the natural environment. Moreover, since this view implies that a major selective pressure during the evolution of primate discrimination learning ability was social in nature, one might expect that such ability will approach its fullest and most effective expression in a social setting.

* From *Animal Behavior*, 1962, **10**, 211–221. Copyright 1962 by Baillière, Tindall & Cassell, London. This research was completed at the University of Wisconsin. Support was provided through funds received from the Graduate School of the University of Wisconsin, Grant G-6194 from the National Science Foundation, and Grant M-722 from the National Institutes of Health.

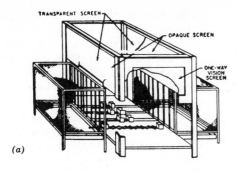

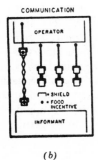

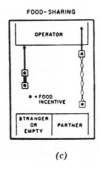

FIGURE 1. *A*. Schematic drawing of the communication apparatus. The relevant dimensions are: restraining cages 18 in. deep, 28 in. wide, 24 in. high; table 36 in. by 42 in.; outer runways separated from inner runways by 5 in., center-to-center; separation between inner runways, 8 in., center-to-center. *B*. Arrangement of apparatus for communication testing. *C*. Arrangement of apparatus for food-sharing tests.

The present experiments were designed within the above context. They approach the broad problems of communication, social co-ordination and social control from the background provided by the constructs and procedures employed in studies of discrimination learning by individual animals. Accordingly, their scope is restricted to communication, defined herein as any observable behavior by which information is transmitted from a social source to a social recipient (Newcomb, 1953, p. 141). The primary objectives of these researches were to describe the acquisition and transfer of communication behavior in young laboratory-reared monkeys.

GENERAL METHOD

Subjects

The *S*s were 12 rhesus monkeys ranging in age from 15 to 18 months at the beginning of this research. They were born in the laboratory, separated from their mothers within 18 hours after birth and subsequently housed in individual cages. Approximately 11 months before the start of the first experiment six of these animals were housed in pairs and these living arrangements were maintained throughout the present research

were maintained throughout the present researches.

Apparatus

The basic apparatus, shown schematically in Figure 1, consisted of two barred restraining cages 28 in. wide, 24 in. high, 18 in. deep, separated 3 ft. by a table. A super-structure mounted on the table supported opaque and transparent screens which could be lowered in front of each restraining cage. To prevent the *S*s from seeing the *E*, opaque panels were placed at the ends of the restraining cages and a one-way vision screen was used.

Four pairs of food carts were mounted on fixed runways on the table and each pair of carts was connected by an expandable rack so that movement of one cart simultaneously extended the other in the opposite direction. Handles were attached to one side of each pair of carts. In communication training the carts were shielded to prevent the monkey having access to the handles (operator monkey) from seeing the food, but permitting its partner (informant monkey) to see it (Fig. 1*b*); in food-sharing tests only the two outermost pairs of carts were used and the shields were removed making the food plainly visible to both animals (Fig. 1*c*).

General Procedure

The Ss were trained individually to operate the unshielded food-carts before starting Experiment 1, and further individual training was given with the shielded carts before the first communication problem.

Unless otherwise indicated Ss were tested in fixed pairs; the members of three pairs were cagemates and the members of the three remaining pairs lived alone. The experiments are presented in chronological order.

The following procedures were observed on individual trials. Before a trial both opaque screens and the transparent screen in front of the operator were lowered, (the second transparent screen was never used), the appropriate carts were baited, and the one-way vision screen was lowered. In Experiments 1, 3 and 4 the trial was initiated by raising both opaque screens simultaneously and by raising the transparent screen in front of the operator 5 sec. later. In Experiment 2, the opaque screen in front of the informant remained down unless raised by the operator. The non-correction technique was employed throughout.

EXPERIMENT 1:

COMMUNICATION AND FOOD-SHARING BEHAVIOR

Method

APPARATUS AND PROCEDURE. The experiment was conducted in six phases, as shown in Table 1.

TABLE 1. *Design Experiment 1*

Food-sharing I. Twenty sessions (20 trials/session), partner *vs.* empty cage; 20 sessions, partner *vs.* stranger.

Communication I. Original training. One member of each pair designated operator, the other informant. Twenty sessions (24 trials/session).

Communication II. Reverse roles. Sixty sessions (24 trials/session). Control tests (48 trials) with informant absent.

Communication III. Return to original role condition (Communication I). Sixty sessions (24 trials/session). Control tests (48 trials) with informant absent.

Communication IV. Alternation. Each S given 10 sessions (24 trials/session) in each role.

Food-sharing II. Repeat Food-sharing I.

The first phase tested food-sharing behavior. The arrangement of the apparatus for the food-sharing tests is shown in Figure 1c. It should be noted that there were two pairs of carts, all carts containing food which was visible to the operator, and that the restraining cage opposite the operator was divided into two compartments, 14 in. wide. Each S served as operator in 20 food-sharing test sessions (20 trials/session) in which one compartment of the partitioned restraining cage contained the partner and the other was empty; and in an equal number of sessions in which the partner was presented with a stranger, selected from among the other animals in the group. Every animal served as stranger twice with each operator. At 5-trial intervals during the food-sharing tests the partner's location was reversed by rotating the compartmented cage. Thus, the operator was rewarded for every response, and by selecting the appropriate cart, could also deliver food to its partner on every trial.

Communication training began upon completion of the food-sharing phase. For communication training there were four pairs of carts and the carts were shielded to prevent the operator monkey from seeing the food. Only one pair of carts was baited, and the location of the food varied randomly, with the restriction that each pair of carts was rewarded with equal frequency in every 24-trial session. Communication training occurred in four phases. During the first phase, one member of each pair was arbitrarily designated the operator monkey and the other, the informant. Each pair was tested for 20 sessions, a total of 480 trials. At the conclusion of this phase, operator-informant roles were reversed and each pair received 60 sessions (1,440 trials) under the reversed-role condition. To check on possible utilization of nonsocial cues, every operator was given 48 control trials between sessions 46 and 47 of this phase in which the procedures were identical in all respects to those observed during the regular communication training except that the informant was not present. Upon completion of this phase of communication training, the Ss were returned to their original roles and were given 60 additional training sessions (1,440 trials). A second 48-trial control series was completed during this phase. The final phase of communication training consisted of an *alternation series* of 20 sessions (480 trials), in which each member of a pair served as operator on half the sessions and as informant on the remainder. At the conclusion of communication training, the food-sharing

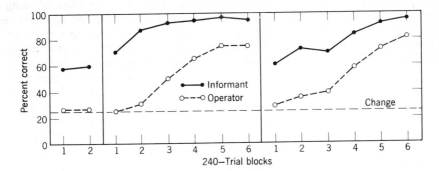

FIGURE 2. Performance during the first three phases of communication training. Results of the original training (communication 1) are shown in the panel on the left. The centre panel shows performance following reversal of operator–informant roles (communication 2), and the panel on the right shows performance following return to the original role condition (communication 3).

Results

COMMUNICATION. Performance curves for the operator monkey during the first three phases of communication training are given in Figure 2. For purposes of comparison, data on position of informants are also included. The primary purpose of the first phase was to provide a baseline against which subsequent transfer effects could be assessed. Although the informants were appropriately positioned on approximately 60 per cent of the trials during the first phase of communication training, the operators showed no evidence of learning.

Initial performance during the second phase of communication training (in which the operator-informant roles were reversed) averaged 25 per cent during the first block of 240 trials and was only slightly better during the subsequent trial block. Performance improved progressively with further training and during the last 480 trials all pairs were substantially above chance. Positioning responses by the informants started at 70 per cent and improved with training; the changes in their performance and in the performance of the operators were significant at the 0.01 level as determined by Friedman tests. During the control series, performance dropped to chance, indicating that the presence of the informant was essential for success.

In the third phase of communication training, also graphed in Figure 2, the monkeys were returned to the original role condition. The principal purpose of this training phase was to investigate the possibility of transfer across roles. During the intensive training given in the preceding phase each pair achieved performance levels consistently above chance. It might be expected that in the course of this training the S had acquired incidentally some proficiency in the complementary role, and if so, that this would be reflected in a high initial level of correct response on return to the original role. As can be seen from Figure 2, however, a comparison of data for the first and third phases of communication training provides little evidence to support this expectation. With further training, performance improved for both operators and informants and these effects were significant at the 0.001 level (Friedman).

The alteration phase of communication training was introduced primarily to give all Ss equivalent experience in each role immediately before repeating the food-sharing tests. Performance during this phase averaged 79 per cent correct.

FOOD-SHARING. The first food-sharing tests gave no evidence of preferential responses to the partner when presented with either the stranger or the empty cage; under the former condition there was a slight and statistically nonsignificant preference for the stranger (52 per cent), which could be dismissed entirely if it were not for the results of the second food-sharing series, and for the fact that 9 of the 12 Ss displayed this tendency although in no case was the differential greater than 18 per cent.

The tendency to respond preferentially to the

stranger was somewhat stronger following communication training. For the nine animals showing a preference for the stranger, the differential ranged from less than 1 to 32 per cent. For the group as a whole the stranger was preferred on 54 per cent of the trials and the difference between partner and stranger was significant at the 0.05 level (Wilcoxon). The difference between the first and second phases of food-sharing was not statistically significant.

In contrast with the slight effect of communication training on food-sharing behavior when two social stimuli were present, a substantial increase in food-sharing responses was obtained for the condition in which the partner was presented with an empty cage (see Fig. 3). During the first 80 trials of the second food-sharing test series the partner received 72 per cent of total responses, as compared with 51 per cent in the first series; although some reduction occurred with further testing, at no point in the second series did responses to the partner fall below 60 per cent. The preference for the partner over the empty cage was significant at the 0.01 level, as was the increase in responses to the partner as compared with the first series (Wilcoxon).

LIVING ARRANGEMENTS. During the first phase of communication training the monkeys living together were slightly more efficient than animals housed individually. Inasmuch as neither group departed significantly from chance, it is not surprising that they did not differ significantly from each other. The effect of living arrangements on performance during subsequent phases of communication training is shown in Figure 4, and it is evident that learning was more rapid for pairs living together. In the final phase of communication training (alternation) the superiority of the pairs caged together was pronounced: as operator each member of this group achieved a total score of at least 90 per cent correct, whereas no animal housed individually surpassed 83 per cent (range 36 to 83). The overall level of correct responses for the last three phases of communication training was 61 per cent for pairs housed together, as compared with 46 per cent of animals living apart, and this difference was significant at the 0.05 level as determined by t tests performed on the data of individual operators ($N = 12$) or on the pooled operator data for each pair ($N = 6$). There was no evidence that food-sharing behavior was consistently affected by living arrangements.

<h1 style="text-align:center">EXPERIMENT 2:</h1>

<h2 style="text-align:center">STIMULUS-PRODUCING RESPONSES</h2>

Method

APPARATUS. The communication apparatus (see Fig. 1a) was modified by installing in the table at front of the operator's cage a 7-in. vertical level with a 2-in. brass ring at the top which was accessible to the operator when his transparent and opaque screens were raised. By pulling this lever toward him the operator automatically raised the counterweighted opaque screen in front of the informant's cage.

PROCEDURE. On the basis of performance during Experiment 1, the most efficient operator-informant combination was determined for each of the original pairs and these roles were maintained throughout the experiment. Preliminary adaptation consisted of a series of 1-min. trials in which the operator's screens were raised, but the informant's screen remained down unless the vertical lever was pulled. During adaptation the informant was present but the food carts were empty and their handles were removed. Adaptation continued until each operator made 20 lever responses, and all operators met this criterion within 30 trials. Formal training was conducted in three phases: During the first phase each pair received ten sessions of 24 trials each, in accordance with the standard communication training procedures. The vertical lever was present, but pulling it was without consequence inasmuch as

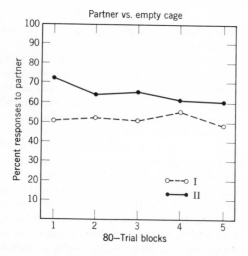

FIGURE 3. Food-sharing responses to the partner when presented with an empty cage. Performance preceding and following communication training is indicated by 1 and 2 respectively.

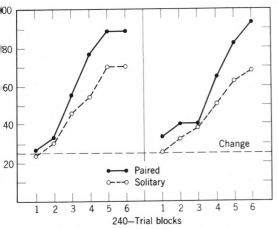

FIGURE 4. Performance during phases 2 and 3 of communication training of animals caged together (paired) and animals housed individually (solitary).

the opaque screen in front of the informant was always raised 5 sec. before the lever was made accessible to the operator by raising his transparent screen. The purpose of this phase of the experiment was to establish baseline measures for lever-pulling, and for communication performance. During phase II the operator's opaque and transparent screens were raised according to the standard procedure, but the informant's screen remained down. The operator was allowed 2 min. to respond to one of the food-carts and this response terminated the trial. The response could occur as soon as the carts were accessible, or it could be deferred until the lever had been pulled thereby revealing the informant. The development of this stimulus-producing response was the principal concern of the present experiment. Testing continued for 20 sessions (480 trials). Upon completion of this phase each pair was given 240 trials as in the first phase of the experiment to determine the persistence of lever-pulling when this response was no longer relevant to communication performance.

Results

Communication performance and lever responses are presented for each phase of the experiment in Figure 5. Responses to the lever were infrequent during the first phase of testing. During the next phase, in which the informant's screen remained down unless the lever was pulled, a progressive increase in the frequency of lever responses occurred over the first nine sessions, and with the occasional minor reversals this trend continued to the end of the phase; as would be expected, there was close correspondence between the curves describing lever responses and selection of the correct cart. As compared with the first phase of testing, the increase in

the percentage of lever responses was significant at the 0.05 level (Wilcoxon). Differences between phases I and III were not statistically significant.

The superior communication performance of monkeys caged together was reflected in the present experiment in a substantially higher incidence of lever responses during phase II (87 per cent *vs.* 13 per cent) which was significant at the 0.01 level as determined by *t* test. Interpretation of this outcome is complicated somewhat by the fact that the operators living with their partners also made more lever responses during the first (12 per cent *vs.* 5 per cent) and third (14 per cent *vs.* 5 per cent) phases of the experiment in which the lever was nonfunctional, although in these instances the differences did not approach statistical significance.

EXPERIMENT 3:

TRANSFER TESTS

Method

APPARATUS. The standard communication situation was used. For presentation of inanimate stimuli on transfer tests the informant's cage was replaced by a shelf at table level on which the stimuli were presented.

PROCEDURE. Base-line level of communication performance was established by giving every *S* five sessions of 24 trials with its customary informant. The transfer tests were administered upon completion of this measurement period. All 12 *S*s served as operators only and were tested under the following stimulus conditions: (1) *Monkey.* The discriminative stimulus was another monkey as in the regular communication tests. The two monkeys used as informants,

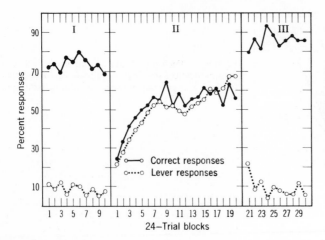

FIGURE 5. Performance in Experiment 2. Panel 1 shows the percentage of responses to the rewarded pair of carts (closed circles) and to the lever (open circles) during the baseline measurement period (lever nonfunctional). Panel 2 gives comparable data for the period in which lever responses revealed the informant. Panel 3 shows performance upon return to the baseline condition.

however, were not members of the experimental group and had not been tested previously with these animals. Prior to serving as stimulus animals they were given separate training sessions until they showed efficient and reliable positioning responses. In the test series each of these animals appeared as informant three times with every operator. (2) *Puppet*. A monkey hand puppet, 14 in. high and 5 in. wide, motor driven to provide continuous, slight, eccentric movements about a fixed base. (3) *Plaque*. A stationary plaque, covered with terry-cloth and constructed to approximate the size, form, and colour of the puppet. Before a trial in which the puppet or plaque appeared E placed them on the shelf behind the rewarded food-cart while the operator's opaque screen was down. With the monkey stimuli, the standard procedures were observed. The three transfer stimulus conditions were presented in counterbalanced order. Each S was given six sessions of 24 trials within each stimulus condition.

Results

Performance during the base-line period and on the subsequent transfer tests is shown in Figure 6. It is evident that the differences among test conditions were substantial and persistent ($p = 0.001$, Friedman). There was almost perfect positive transfer from the original partner to the new monkey informants and the difference between performance during the base-line period and the equivalent transfer period did not approach statistical significance. Although some improvement in performance occurred with further training, this effect fell short of statistical significance. Performance with the puppet started at 50 per cent and improved progressively

($p = 0.01$, Friedman), With one exception, total individual scores for this condition were intermediate between *monkey* and *plaque*. In spite of the gross physical similarity of puppet and plaque, each S made its lowest score with the latter, and there was little evidence that performance improved with training. At no point in testing did group performance with the plaque exceed 45 per cent, which contrasts sharply with the high scores of 69 and 82 per cent with puppet and monkey, respectively. As is suggested by the curves, performance levels on the first 72 trials with the puppet and the plaque were significantly lower than the base-line level ($p = 0.02$, Wilcoxon). Under all conditions the Ss housed with a companion were slightly superior but the effect of living arrangements was not statistically significant.

EXPERIMENT 4:

DISCRIMINATION BETWEEN INFORMANTS

Method

APPARATUS AND PROCEDURE. The standard communication situation and test procedures were used except that the informant's cage was partitioned as in the food-sharing tests (Experiment 1), and two informants were present. Thus, on each trial food was available in one of the four pairs of carts, and successful performance required the operator to select the appropriate informant and the appropriate cart. To maintain adequate food motivation in the informants, two pairs of monkeys were used: the two animals serving as test stimuli in the preceding transfer experiment, and a second pair of young monkeys given preliminary training as informants before formal training began. The 12 experimental *S*s served as operators only; because of persistent refusal to work one animal was dropped from the experiment and its data were not included in the analysis.

To establish base-line performance, each *S* was given three sessions of 24 trials with its original partner serving as informant. On the discrimination series each operator was given a total of 16 24-trial sessions, eight sessions with each pair of informants.

Results

Percentage of correct responses for base-line and discrimination phases are shown in Figure 7. All *S*s experienced initial difficulty with the discrimination task and performance on the first 72 trials averaged 47 per cent correct, as compared with 78 per cent for the baseline period ($p = 0.01$, Wilcoxon). It should be noted, however, that even though this was a large drop, performance remained substantially above chance expectancy (25 per cent). Moreover, examination of individual records for the first 72 discrimination trials indicated that 65 per cent of total responses were to the correct informant. Figure 7 indicates continuous improving performance by the operators ($p = 0.001$, Friedman) and on the final 72 discrimination trials performance was not significantly inferior to base-line levels. The improvement, however, cannot be attributed entirely to increasing efficiency of the operators inasmuch as the positioning responses of the informants showed a corresponding change, rising from 79 per cent correct positioning responses in the first 48 trials to 98 per cent in the last 48 ($p = 0.001$, Friedman).

DISCUSSION

Communication Learning and Food-sharing

The findings presented in this paper clearly establish that one monkey can secure from the behavior of another specific information regarding the location of food; further, they show that this form of communication does not appear spontaneously in the laboratory-reared macaque but requires considerable training. These results contrast with the findings of Wolfe and Wolfe (1939) that monkeys did not discriminate between the behavior of their partners in the presence and in the absence of food. A factor in their experiment which probably militated

against a successful outcome was that the operator's rewards were not directly dependent upon correct interpretation of its partner's responses. In the present situation, of course, successful performance required the joint participation of both members of the pair and the results demonstrate that under such conditions the behavior of both operator and informant became progressively more efficient.

As an arrangement for testing discrimination learning the communication apparatus is similar in many respects to the Wisconsin General Test Apparatus (Harlow, 1949), and to the situation used with success in studies of primate observational learning (Crawford and Spence, 1939; Darby and Riopelle, 1959). A number of experiments have shown that learning is most efficient when there is minimal separation of discriminative stimulus and reward (Gardner and Nissen, 1948; Jarvik, 1953; Jenkins, 1943; McClearn and Harlow, 1954; Murphy and Miller, 1955). The importance of spatial contiguity is clearly illustrated by the experiment of Murphy and Miller (1958) which showed that monkeys were unable to form a simple black-white discrimination in 400 trials if reward and stimulus were separated vertically by as much as 7 in. In the communication apparatus the separation between the reward and the discriminative stimulus, as measured by the distance between operator and informant, was approximately 3 ft. Moreover, the difficulty created by spatial separation of cue from the site of reward and response was probably compounded by the presence of four response possibilities, in contrast with two in conventional discrimination learning experiments. Although

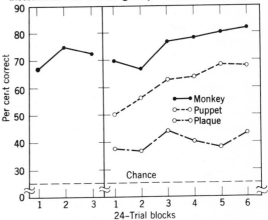

FIGURE 6. Performance in Experiment 3. Percentage of correct responses with the customary informant (left panel) and during transfer tests in which a monkey, a mechanical puppet, and a stationary plaque served as discriminative stimuli (right panel).

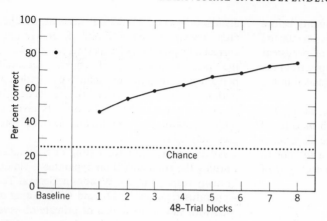

FIGURE 7. Performance in Experiment 4. Percentage of correct responses with customary informant under standard communication training conditions (left panel), and when *S*s were required to discriminate between two informant monkeys (right panel).

communication learning was not rapid, neither was it unduly prolonged, and it was definitely superior to the outcome one might expect by extrapolation from Murphy and Miller's (1958) results.

A primary objective of the first experimental series was to obtain information on transfer of communication training. Two types of transfer effects were measured, inter-role and inter-situational. The failure to obtain evidence of transfer when the animals were switched to the complementary roles indicates that communication learning was role-specific. Although each animal became more proficient in the role in which it was trained, it apparently learned little or nothing that would facilitate performance when roles were reversed. The results of the food-sharing tests, while indicating marked transfer across situations, are consistent with the thesis of specificity of training suggested by communication performance. The tendency to share food with the partner following communication training was evident only in the partner-empty test condition, and probably represents the continuation of control by the partner in a situation in which the social stimulus was no longer a relevant cue. When two monkeys were present, however, the familiar partner was no longer a controlling factor; on the contrary the majority of animals showed a slight preference for the unfamiliar monkey.

The objective results of communication training in the first series of experiments can be explained most parsimoniously in simple stimulus-response terms. Without further comment, however, such an explanation would not do justice to the complex forms of social interaction which were observed. For example, in most pairs attempts by informants to secure food by direct reaching extinguished early in training and their behavior increasingly became oriented toward the oper-

ators. Operators showed a similar growth in attentiveness to informants. The apex of communication skill in the present *S*s was achieved by an operator whose partner would on occasion cease to make effective positioning responses, but would at these times perform a distinctive "back-flip" whenever the rewarded cart was selected. Initially, this reaction occurred after the operator had made an unequivocal response and therefore it could not serve as a useful cue. Eventually, however, the operator evolved a mode of coping with this problem which was both illuminating and successful: it would touch the handle of a cart but not pull it (thus not meeting the criterion of a response) unless the informant performed the back-flip thereby indicating that the rewarded cart had been selected. This observation as much as more theoretical considerations prompted the decision to investigate stimulus-producing responses.

Stimulus Producing Responses

The theoretical role of stimulus-producing responses in discrimination learning has been discussed by Wyckoff (1952), and the acquisition of such responses in chimpanzees has recently been described by Kelleher (1958). In the present context stimulus-producing responses were used to provide information concerning the relationship between operator and informant. The results indicate that operator monkeys will defer a well-practised instrumental act until they have performed another response which produces the informant. It is clear, however, that although the operators had received hundreds of communication training trials in both operator and informant roles, and had been given preliminary training in the performance of stimulus-producing responses, the appropriate pattern of delaying the cart-response until the informant had been revealed did not appear abruptly but in gradual increments over the course of testing.

Transfer Tests

The transfer data clearly establish that experienced operator monkeys show a high degree positive transfer to new trained informants. Unequivocal transfer effects were also obtained with the puppet and the plaque, although performance was consistently below levels achieved with social stimuli.

To the extent that position was a useful cue, the puppet and the plaque, which were always correctly placed and remained in the correct position throughout the trial, should have been at least as effective as the monkey. The fact that they were not clearly indicates that efficiency of communication performance is influenced by variables other than position of the discriminative stimulus. The superiority of the moving puppet over the plaque suggests that attentional factors (cue-distinctiveness) may have been important and this interpretation receives support from data (Mason and Green, unpublished) which establish the differential effectiveness of puppet and plaque in eliciting affective-social responses from rhesus monkeys. However, the consistent inferiority of the puppet to the monkey may indicate more than a difference in cue-distinctiveness. Our observations suggest the possibility that the informant, in contrast with plaque and puppet, was not limited to indicating the location of food but had the additional capability of anticipating the operator's errors (such as reaching for the handle of an empty food-cart) and behaving in a manner which could disrupt an incipient incorrect response before its completion. Thus the monkey discriminative stimulus had the possibility of sending "no" as well as "yes" signals. The extent to which this possibility was actually utilized is, of course, problematic, but we received the strong impression that it was a contributing factor.

Discrimination Between Informants

By requiring the operator to discriminate between two informants the communication problem was made more difficult in at least three respects: first, the area available to the informant was reduced from the normal cage width of 28 in. to 14 in., thereby restricting the magnitude of his gross responses; secondly, on half of the trials the informant could not receive food although he was free to assume any position within his compartment—and the most charac-teristic response on the "no-reward" trials was to remain at the front of the cage; finally, since each informant was equally often "correct" and "incorrect," success required the operator to select the appropriate informant and cart solely on the basis of behavioral differences between informants. In spite of a substantial initial drop in communication efficiency, performance by the end of 384 discrimination trials was not significantly inferior to levels achieved after extensive training on the standard communication task.

Comparison may be made with the study of Riopelle and Copelan (1954) of discrimination reversal to a sign. In their experiment rhesus monkeys capable of solving conventional discrimination and discrimination-reversal problems in one trial were trained to perform discrimination reversals contingent upon a change in the color of the tray bearing the test objects. In the present experiment, of course, the "reversal" sign consisted of changes in the behavior of the informants. Even though they were being trained for the first time on "discrimination reversal" problems in this experiment our Ss achieved in less than 400 trials a level of proficiency which required 700 reversals for Riopelle and Copelan's highly sophisticated monkeys.

Effects of Living Arrangements

Because restrictions on social experience seem to impair discrimination of social cues (Mason, 1960, 1961b), one might expect that monkeys caged together will show more efficient communication performance than animals living alone from birth and the data presented here are in keeping with this expectation. We have unpublished data demonstrating strong mutual attraction between young monkeys living together and it is probable that whatever communication skills were established between cagemates prior to formal training, were augmented by the heightened attention-getting properties of the familiar companion. This is probably especially true of the investigation of stimulus-producing responses (Experiment 2) in which the differential in lever-pulling responses favored the operators housed with their partners by 74 per cent.

SUMMARY

1. The communication performance of 12 rhesus monkeys was investigated in a situation in which the rewards of both members of a pair

of monkeys could not exceed chance levels unless the operator monkey responded to cues provided by the informant monkey which indicated the location of food. Each member of the pair was trained in both operator and informant roles in different phases of the experiment. Communication performance improved progressively to levels consistently above chance. However, communication learning appeared to be specific to the role in which the individual was trained, and when roles were reversed no evidence of transfer was obtained. Tests of food-sharing behavior showed a substantial increase in the tendency to share food with the partner following communication training. This occurred however, only when the partner was the only social stimulus present; if another monkey was also present there was no evidence of preferential responses to the partner. In all phases of communication training, monkeys which were housed together performed more efficiently than did monkeys housed individually.

2. The acquisition of stimulus-producing responses was investigated by causing an opaque screen to remain in front of the informant unless the operator monkey pulled a vertical lever at the front of its restraining cage. Initially, operators responded immediately to the food-carts, but with further testing there was a steady increase in the tendency to defer the response to the food-carts until the lever had been pulled, revealing the informant monkey.

3. Transfer of communication training was tested with new monkey informants, and with two inanimate stimuli, a mechanical puppet, and a stationary plaque. The latter two objects were placed behind the rewarded food-carts before each trial. There was clear evidence of positive transfer to each of these conditions, but marked differences among conditions were obtained. Performance with the monkeys averaged 76 per cent correct, as compared with 62 and 40 per cent, with the puppet and the plaque, respectively.

4. To test the ability of trained operator monkeys to select the appropriate informant on the basis of behavioral cues, the communication situation was arranged so that two informant monkeys were present on all trials. However, on any trial only one of these informants could be rewarded, and the operator's rewards were contingent upon delivering food to this informant. Efficiency of discrimination began at approximately 45 per cent (chance = 25 per cent) and improved progressively to levels above 75 per cent.

REFERENCES

CHANCE, M. R. A. (1956). Social structure of a colony of *Macaca mulatta*. *Brit. J. anim. Behav.*, **4**, 1–13.

CHANCE, M. R. A. and MEAD, A. P. (1952). Social behaviour and primate evolution. *Soc. exp. Biol. Sympos.*, **7**, 395–439.

CRAWFORD, M. P. and SPENCE, K. W. (1939). Observational learning of discrimination problems by chimpanzees. *J. comp. Psychol.*, **27**, 133–147.

DARBY, C. L. and RIOPELLE, A. J. (1959). Observational learning in the rhesus monkey. *J. comp. physiol. Psychol.*, **52**, 94–98.

GARDNER, L. P. and NISSEN, H. W. (1948). Simple discrimination behavior of young chimpanzees: Comparisons with human aments and domestic animals. *J. genet. Psychol.*, **72**, 145–164.

HARLOW, H. F. (1949). The formation of learning sets. *Psychol.*, *Rev.*, **56**, 51–65.

HARLOW, H. F. (1958). The evolution of learning. In G. G. Simpson and A. Roe (Eds.) *Behavior and evolution*. New Haven: Yale. Pp. 269–290.

JARVIK, M. E. (1953). Discrimination of colored food and food signs by primates. *J. comp. physiol. Psychol.*, **46**, 390–392.

JENKINS, W. O. (1943). A spatial factor in chimpanzee learning. *J. comp. Psychol.*, **35**, 81–84.

KELLEHER, R. T. (1958). Stimulus-producing responses in chimpanzees. *J. exp. anal. Behav.*, **1**, 87–102.

MASON, W. A. (1960). The effects of social restriction on the behavior of rhesus monkeys. I. Free social behavior. *J. comp. physiol. Psychol.*, **53**, 582–589.

MASON, W. A. (1961). The effects of social restriction on the behavior of rhesus monkeys: II. Tests of gregariousness. *J. comp. physiol. Psychol.*, **54**, 287–290 (a).

MASON, W. A. (1961). The effects of social restriction on the behavior of rhesus monkeys: III. Dominance tests. *J. comp. physiol. Psychol.*, **54**, 694–699 (b).

McCLEARN, G. E. and HARLOW, H. F. (1954). The effect of spatial contiguity on discrimination learning by rhesus monkeys. *J. comp. physiol. Psychol.*, **47**, 391–394.

MURPHY, J. V. and MILLER, R. E. (1955). The effect of spatial contiguity of cue and reward in the object-quality learning of rhesus monkeys. *J. comp. physiol. Psychol.*, **48**, 221–224.

MURPHY, J. V. and MILLER, R. E. (1958). Effect of the spatial relationship between cue, reward, and response in simple discrimination learning. *J. exp. Psychol.*, **56**, 26–31.

NEWCOMB, T. M. (1953). Motivation in social behavior. In *Current theory and research in motiva-tion.* Lincoln: Univ. Nebraska. Pp. 139–161.

RIOPELLE, A. J. and COPELAN, E. L. (1954). Discrimination reversal to a sign. *J. exp. Psychol.,* **48**, 143–145.

WOLFLE, D. L. and WOLFLE, H. M. (1939). The development of cooperative behavior in monkeys and young children. *J. gent. Psychol.,* **55**, 137–175.

WYCKOFF, L. B., JR. (1952). The role of observing responses in discrimination learning. *Psychol. Rev.,* **59**, 431–442.

EMOTIONAL REACTIONS OF RATS TO THE PAIN OF OTHERS*

Russell M. Church

IT has been observed that animals and people are often responsive or "sympathetic" to the emotional states of others. Of course, emotional reactions of one individual do not always result in sympathetic responses by another, so the general problem is to determine some of the conditions affecting the development of sympathetic responses.

Conditioning theory provides a straightforward explanation for the development of sympathetic responses (Allport, 1924). According to the theory a particular unconditioned stimulus elicits the emotional response; for example, electric shock elicits a pain or fear response. Now if *S* experiences the pain reactions of another *S* to a shock and is himself shocked, on subsequent trials he should show fear at the pain reactions of the other *S*. The present experiment is a test of this hypothesis, using the depression in the rate of bar pressing for food reinforcement as the measure of degree of fear (Estes and Skinner, 1941).

METHOD

Subjects

The *S*s were 32 albino rats of Sprague-Dawley strain, about 100 days old at the beginning of the experiment. Half the *S*s were male and half were female.

Apparatus

The apparatus in which *S*s could press a lever for food consisted of two identical lever boxes. The adjacent sides of the two boxes were ¼-in. transparent Lucite, and the boxes were set 1 in. apart. Each compartment was 7½ in. by 8 in., 9 in. high. The front and back were stainless steel; the sides and top were ¼-in. Lucite. The floor was composed of 16 stainless-steel bars. The lever was made of stainless steel, rounded and smoothly finished, ½ in. thick and 2 in. wide.

The apparatus in which *S*s were given fear conditioning was a grill box with a hardware-cloth partition dividing the box into two identical compartments. Each compartment was 12 in. by 12 in., 10 in. high. The front, back, and sides were wood; the top was wire-mesh screen. The floor of each compartment was composed of 20 bus bars, ⅛ in. in diameter.

The shock was 2,250 V. a.c., 60 cy., with 15-meg. resistance in series with the rat. This shock reliably elicited considerable motor activity and loud, high-pitched squeaks.

Procedure

TRAINING SUBJECTS TO PRESS THE LEVER. On the first day *S*s were permitted to explore the lever box for 10 min., and the number of lever presses was recorded as the operant level. The *S*s were then given magazine training and shaping procedures and allowed to press the bar 50 times for 100 per cent reinforcement. On the second day *S*s were given 50 presses for 2:1 fixed ratio and 50 presses for 4:1 fixed ratio. On days 3 to 13 *S*s were given 10 min. of training for a 4:1 fixed ratio of reinforcement.

* From the *Journal of Comparative and Physiological Psychology*, 1959, **52**, 132–134. Copyright 1959 by the American Psychological Association. This investigation was supported by a research grant (M-1812) from the National Institutes of Health, Public Health Service.

ADAPTATION OF THE INITIAL REACTIONS TO A SHOCKED SUBJECT. On days 14 and 15 Ss were given 10 min. of training for a 4:1 fixed ratio of reinforcement, and the two panel lights were turned on during minutes 3 and 7 of day 15. On day 16 and 17 Ss were given the same procedure as on day 15, except that another rat in the adjacent box was shocked during minutes 4 and 8 (i.e., the minute after the panel lights were turned on), and the panel lights remained on while the leader rat was shocked.

EMOTIONAL CONDITIONING. The Ss were divided into three groups, matched on a number of aspects of their previous performance. The experimental group ($N = 16$) was given three 1-sec. shocks a day for two days in the grill box. These occurred on minutes 2, 5, and 8 of a 10-min. session on days 18 and 19. Preceding each of these shocks another rat was shocked for 30 sec., and both shocks terminated simultaneously. The shock-control group ($N = 8$) received the same six 1-sec. shocks, but these were not associated with shock to another rat. The no-shock control group ($N = 8$) did not have any experience in the grill box.

TEST FOR EMOTIONAL REACTION AT PAIN RESPONSE OF OTHER RAT. During the next ten days (days 20–29) Ss were given 10 min. per day in the lever box. On minutes 3 and 4 the two panel lights were on, and during minute 4 the rat in the duplicate compartment was shocked.

The Ss were 22-hr. deprived at the time of running. They were housed in community cages,

and water was accessible at all times in the home cages. Lab rat food pellets (4 mm, 45 mg) obtained from P. J. Noyes Co., Lancaster, N.H., were used as reinforcement for lever pressing.

Records were made of the number of responses made by each S during each minute of the experiment. The measure of anxiety to a stimulus (the dependent variable) was the decrease in the rate of response during the presentation of that stimulus.

RESULTS

Original Response to the Pain of Others

Figure 1 indicates that Ss showed a radical depression in the rate of bar pressing (from over 30 responses a minute to 0 responses) on the first time another rat was shocked. This depression in rate adapts very quickly, i.e., in two or three further exposures to a shocked rat there is no further depression in the rate of bar pressing. The three groups were matched on the basis of their performance, so there are no significant differences between them.

Conditioned Responses to the Pain of Others

The experimental group had been exposed to six trials in which Ss had been shocked for 1 sec. following 30 sec. of shock to another rat. These Ss showed a radical depression of rate in the lever box when exposed to another rat being shocked for 1 min. (Fig. 2). The no-shock control group, which had not been exposed to the fear-conditioning procedure, did not show this depression in rate, i.e., the emotional responses to the pain of another remained adapted out. Although the number of responses made by the experimental group increased on successive days, even on the tenth day there was a significant difference between the experimental group and the no-shock control group (C_1) in the number of responses made while another S was shocked (Mann-Whitney test, $p = 0.01$).

The shock-control group, that had been given six shocks in the grill box not associated with shock to another rat, showed an effect intermediate between the experimental and the no-shock control group. Both on the first day and over the ten-day period they had a significantly greater number of responses than the experimental group and significantly fewer responses than the control group (Wilcoxon's extension of the U Statistic, $p < 0.01$) (Mosteller and Bush, 1954).

None of the groups showed any depression of

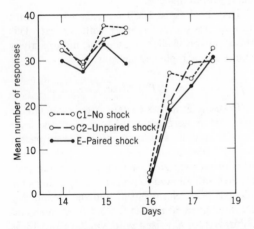

FIGURE 1. Extinction of the original depression in the rate of lever pressing during the pain responses of another rat. Data are from minutes 3 and 7 of day 14 (with no additional stimulus), minutes 3 and 7 of day 15 (with the panel lights on), and minutes 4 and 8 of days 16 and 17 (with another rat shocked in the adjacent box).

rate to the two panel lights that preceded the shocking of the leader rat.

DISCUSSION

The results of this experiment give empirical support for the conditioned-response interpretation of some cases of "sympathy." Thus, if the painful responses of others have been followed by pain to S, then S will show fear. If, on the other hand, this contingency has not been established, S will not show fear.

Perhaps the most interesting result is that a group of Ss that had been exposed to the painful stimulus (shock-control group) showed greater fear to the pain of others than a control group that had not been exposed to this painful stimulus. There are two possible explanations of this result. The first is that this was a result of "sensitization." That is to say, a group of Ss that have been shocked may be more responsive to all stimuli, including the pain responses of others. As a matter of fact, the group was run just to guard against this contingency. However, there are a number of objections to this interpretation. There is no report of such a result on sensitization in any of the dozen or more published studies of the "conditioned emotional reaction," and in an unpublished study the author has failed to obtain this effect using a loud intermittent buzzer as the conditioned stimulus. That is to say, the shock-control group and the no-shock control groups were identical and neither showed any effect of the stimulus, but the experimental group showed the characteristic depression of rate to the stimulus. Furthermore, the Ss in the shock-control group were not generally sensitized to all external stimuli, i.e., they did not show a greater depression of rate to the panel lights on the first test day than did the no-shock control group. Thus, if they were sensitized, they must have been selectively sensitized to a class of stimuli.

An alternative explanation for the fact that the shock-control group showed greater fear to the pain of others than a no-shock control group is that the shock-control Ss *conditioned themselves* to be responsive to the emotional reactions of others. That is to say, these Ss were shocked, and simultaneously they exhibited pain responses (jumping and squeaking). Thus, when they were later exposed to the pain responses of others, they might show conditioned fear. Of course, the interval between the conditioned stimulus (fear responses) and the unconditioned stimulus (shock) is not ideal, i.e., they occur simultaneously. How-

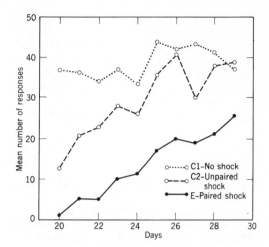

FIGURE 2. Extinction of the conditioned emotional reaction to the pain responses of another rat. Data are from the fourth minute of the ten test sessions (when another rat was shocked continuously in the adjacent box).

ever, in the case of the learning of fear there is reason to believe that this simultaneous conditioning can occur and can have a large effect (Mowrer and Aiken, 1954).

The original response of the Ss to the shock of others may have been previously conditioned in the fighting behavior occasionally observed in the community cages, or it may have been a response to a dramatic stimulus (external inhibition). In any case, it was an extremely marked effect that lasted for only a very brief time before it became adapted or extinguished.

Contrary to expectation, the experimental group did not develop any anticipatory fear to the two panel lights that preceded the shock by 1 min. A procedure is still be be found to produce the "higher-order conditioning" of the conditioned emotional response. The problem is to find a set of conditions under which Ss will learn to fear the objects that frighten a leader.

SUMMARY

Prior to any emotional conditioning rats showed a depression in the rate of bar pressing during the pain responses of another animal, but this depression rapidly adapted. Then 16 experimental Ss were given six 1-sec. shocks, each preceded by 30 sec. of shock to another rat. Following such emotional conditioning Ss showed a dramatic depression in the rate of bar pressing to the pain response of another rat. This depression, considered a measure of anxiety, gradually extinguished, but it was still significantly present after ten days. Eight unshocked control Ss did not show this depression in rate to the pain of another rat. The difference between the experimental and unshocked control group was considered support for a conditioned-response interpretation of some cases of "sympathy."

A second control group consisted of eight Ss that were shocked six times for 1 sec. unassociated with the pain response of another rat. Like the experimental group, they did show a depression in rate of response, although not as great as that of the experimental group. This latter finding was interpreted to be the result of self-conditioning rather than sensitization.

REFERENCES

ALLPORT, F. H. *Social psychology*. Boston: Houghton Mifflin, 1924.

ESTES, W. and SKINNER, B. Some quantitative properties of anxiety. *J. exp. Psychol.*, 1941, **29**, 390–400.

MOSTELLER, F. and BUSH, R. Selected quantitative techniques. In G. Lindzey (Ed.), *Handbook of* *social psychology*. Vol. I. Boston: Addison-Wesley, 1954.

MOWRER, O. H. and AIKEN, E. Contiguity vs. drive-reduction in conditioned fear: Variations in conditioned and unconditioned stimuli. *Amer. J. Psychol.*, 1954, **67**, 26–38.

CULTURALLY TRANSMITTED PATTERNS OF VOCAL BEHAVIOR IN SPARROWS*

Peter Marler and Miwako Tamura

THE white-crowned sparrow, *Zonotrichia leucophrys*, is a small song bird with an extensive breeding distribution in all but the southern and eastern parts of North America (1). Ornithologists have long remarked upon the geographical variability of its song. Physical analysis of field recordings of the several vocalizations of the Pacific Coast subspecies *Z. l. nuttalli* reveals that while most of the seven or so sounds which make up the adult repertoire vary little from one population to another, the song patterns of the male show striking variation (see 2).

Each adult male has a single basic song pattern which, with minor variations of omission or repetition, is repeated throughout the season. Within a population small differences separate the songs of individual males but they all share certain salient characteristics of the song. In each discrete population there is one predominant pattern which differs in certain consistent respects from the patterns found in neighboring populations (Fig. 1). The term "dialect" seems appropriate for the properties of the song patterns that characterize each separate population of breeding birds. The detailed structure of syllables in the second part of the song is the most reliable indicator. Such dialects are known in other song birds (3).

The white-crowned sparrow is remarkable for the homogeneity of song patterns in one area. As a result the differences in song patterns between populations are ideal subjects for study of the developmental basis of behavior. If young male birds are taken from a given area an accurate prediction can be made about several properties of the songs that would have developed if they had been left in their natural environment. Thus there is a firm frame of reference with which to compare vocal patterns developing under experimental conditions. Since 1959 we have raised some 88 white-crowned sparrows in various types of acoustical environments and observed the effects upon their vocal behavior. Here we report on the adult song patterns of 35 such experimental male birds. The several types of acoustical chamber in which they were raised will be described elsewhere.

In nature a young male white-crown hears

* From *Science*, 1964, **164**, 1483–1486. Copyright 1964 by the American Association for the Advancement of Science.

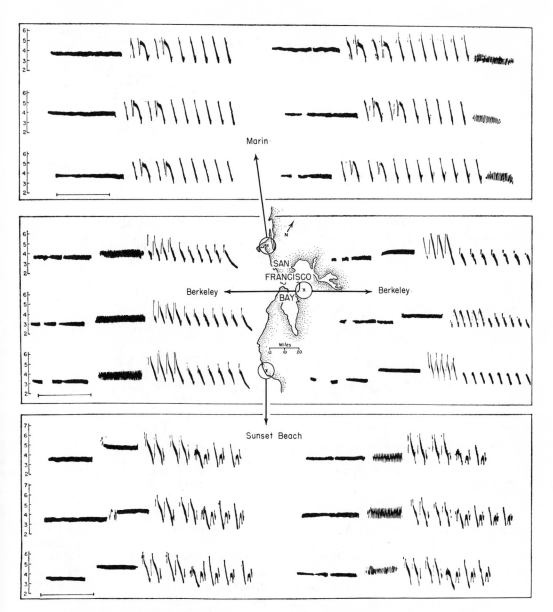

FIGURE 1. Sound spectrograms of songs of 18 male white-crowned sparrows from three localities in the San Francisco Bay area. The detailed syllabic structure of the second part of the song varies little within an area but is consistently different between populations. The introductory or terminal whistles and vibrati show more individual variability. The time marker indicates 0.5 sec. and the vertical scale is marked in kilocycles per second.

abundant singing from its father and neighbors from 20 to about 100 days after fledging. Then the adults stop singing during the summer molt and during the fall. Singing is resumed again in late winter and early spring, when the young males of the previous year begin to participate. Young males captured between the ages of 30 and 100 days, and raised in pairs in divided acoustical chambers, developed song patterns in the following spring which matched the dialect of their home area closely. If males were taken as nestlings or fledglings when 3 to 14 days of age and kept as a group in a large soundproof room, the process of song development was very different. Figure 2 shows sound spectrograms of the songs of nine males taken from three different areas and raised

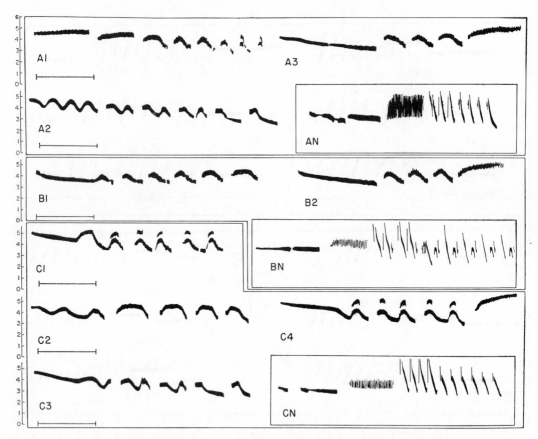

FIGURE 2. Songs of nine males from three areas raised together in group isolation. *A1* to *A3*, songs of individuals born at Inspiration Point, 3 km. northeast of Berkeley. *B1* and *B2*, songs of individuals born at Sunset Beach. *C1* to *C4*, songs of individuals born in Berkeley. The inserts (*AN, BN* and *CN*) show the home dialect of each group.

as a group. The patterns lack the characteristics of the home dialect. Moreover, some birds from different areas have strikingly similar patterns (*A3, B2,* and *C4* in Fig. 2).

Males taken at the same age and individually isolated also developed songs which lacked the dialect characteristics (Fig. 3). Although the dialect properties are absent in such birds isolated in groups or individually, the songs do have some of the species-specific characteristics. The sustained tone in the introduction is generally, though not always, followed by a repetitive series of shorter sounds, with or without a sustained tone at the end. An ornithologist would identify such songs as utterances of a *Zonotrichia* species.

Males of different ages were exposed to recorded sounds played into the acoustical chambers through loudspeakers. One male given an alien dialect (8 min. of singing per day) from the third to the eighth day after hatching, and individually isolated, showed no effects of the training. Thus

the early experience as a nestling probably has little specific effect. One of the group-raised isolates was removed at about 1 year of age and given 10 weeks of daily training with an alien dialect in an open cage in the laboratory. His song pattern was unaffected. In general, acoustical experience seems to have no effect on the song pattern after males reach adulthood. Birds taken as fledglings aged from 30 to 100 days were given an alien dialect for a 3-week period, some at about 100 days of age, some at 200, and some at 300 days of age. Only the training at the age of 100 days had a slight effect upon the adult song. The other groups developed accurate versions of the home dialect. Attention is thus focused on the effects of training between the ages of about 10 to 100 days. Two males were placed in individual isolation at 5 and 10 days of age, respectively, and were exposed alternately to the songs of a normal white-crowned sparrow and a bird of a different species. One male was exposed at 6 to 28 days, the other at 35 to 56 days. Both developed fair copies of the training song which was the home dialect for one and an alien dialect for the other. Although the rendering of the training song is not perfect, it establishes that the dialect patterns of the male song develop through learning from older birds in the first month or two of life. Experiments are in progress to determine whether longer training periods are necessary for perfect copying of the training pattern.

The training song of the white-crowned sparrow was alternated in one case with the song of a song sparrow, *Melospiza melodia*, a common bird in the areas where the white-crowns were taken, and in the other case, with a song of a Harris's sparrow, *Zonotrichia querula*. Neither song seemed to have any effect on the adult

patterns of the experimental birds. To pursue this issue further, three males were individually isolated at 5 days of age and trained with song-sparrow song alone from about the 9th to 30th days. The adult songs of these birds bore no resemblance to the training patterns and resembled those of naive birds (Fig. 3). There is thus a predisposition to learn white-crowned sparrow songs in preference to those of other species.

The songs of white-crowned sparrows raised in isolation have some normal characteristics. Recent work by Konishi (4) has shown that a young male must be able to hear his own voice if these properties are to appear. Deafening in youth by removal of the cochlea causes development of quite different songs, with a variable broken pattern and a sibilant tone, lacking the

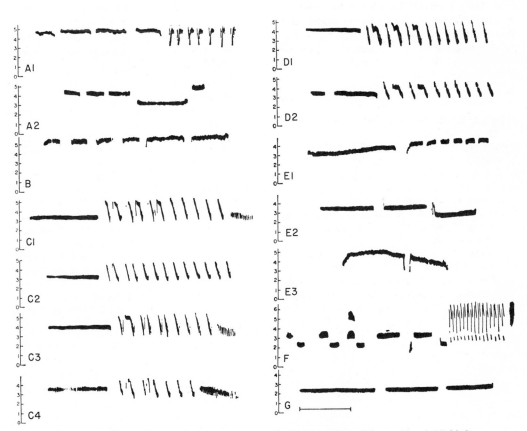

FIGURE 3. Songs of 12 males raised under various experimental conditions. *A1* and *A2*, birds raised in individual isolation. *B*, male from Sunset Beach trained with Marin song (see Fig. 1) from the third to the eighth day of age. *C1* to *C4*, Marin birds brought into the laboratory at the age of 30 to 100 days. *C1*, untrained. *C2* to *C4* trained with Sunset Beach songs; *C2* at about 100 days of age, *C3* at 200 days, *C4* at 300 days. *D1*, bird from Sunset Beach trained with Marin white-crowned sparrow song and a Harris's sparrow song (see *G*) from the age of 35 to 56 days. *D2*, Marin bird trained with Marin white-crowned sparrow song and a song-sparrow song (see *F*) from the age of 6 to 28 days. *E1* to *E3*, two birds from Sunset Beach and one from Berkeley trained with song-sparrow song from the age of 7 to 28 days. *F*, a song-sparrow training song for *D2* and *E1* to *E3*. *G*, a Harris's sparrow training song for *D1*.

pure whistles of the intact, isolated birds. Furthermore, there is a resemblance between the songs of male white-crowned sparrows deafened in youth and those of another species, *Junco oreganus*, subjected to similar treatment. The songs of intact juncos and white-crowns are quite different. Konishi also finds that males which have been exposed to the dialect of their birthplace during the sensitive period need to hear themselves before the memory trace can be translated into motor activity. Males deafened after exposure to their home dialects during the sensitive period, but before they start to sing themselves, develop songs like those of a deafened naive bird. However, once the adult pattern of singing has become established then deafening has little or no effect upon it. Konishi infers that in the course of crystallization of the motor pattern some control mechanism other than auditory feedback takes over and becomes adequate to maintain its organization. There are thus several pathways impinging upon the development of song patterns in the white-crowned sparrow, including acoustical influences from the external environment, acoustical feedback from the bird's own vocalizations, and perhaps nonauditory feedback as well.

Cultural transmission is known to play a role in the development of several types of animal behavior (5). However, most examples consist of the reorientation through experience of motor patterns, the basic organization of which remains little changed. In the development of vocal behavior in the white-crowned sparrow and certain other species of song birds, we find a rare case of drastic reorganization of whole patterns of motor activity through cultural influence (6). The process of acquisition in the white-crowned sparrow is interesting in that, unlike that of some birds (7), it requires no social bond between the young bird and the emitter of the copied sound, such as is postulated as a prerequisite for speech learning in human children (8). The reinforcement process underlying the acquisition of sound patterns transmitted through a loudspeaker is obscure.

REFERENCES AND NOTES

1. Banks, R. C. *Univ. Calif. Berkeley Publ. Zool.*, **70**, 1 (1964).
2. Marler, P. and Tamura, M. *Condor*, **64**, 368 (1962).
3. Armstrong, E. A. *A Study of Bird Song* (Oxford Univ. Press, London, 1963).
4. Konishi, M. (In preparation.)
5. Etkin, W. *Social Behavior and Organization Among Vertebrates* (Univ. of Chicago Press, Chicago, 1964).
6. Lanyon, W. In *Animal Sounds and Communication, AIBS Publ. No. 7*, W. Lanyon and W. Tavolga, Eds. (American Institute of Biological Sciences, Washington, D.C., 1960), p. 321; Thorpe, W. H., *Bird Song. The Biology of Vocal Communication and Expression in Birds* (Cambridge Univ. Press, London, 1961); Thielcke, G., *J. Ornithol.*, **102**, 285 (1961); Marler, P., in *Acoustic Behaviour of Animals*, R. G. Busnel, Ed. (Elsevier, Amsterdam, 1964), p. 228.
7. Nicolai, J. *Z. Tierpsychol.*, **100**, 93 (1959).
8. Mowrer, O. H. *J. Speech Hearing Disorders*, **23**, 143 (1958).
9. Konishi, M., Kreith, M. and Mulligan, J. co-operated generously in locating and raising the birds and conducting the experiments. W. Fish and J. Hartshorne gave invaluable aid in design and construction of soundproof boxes. We thank Dr. M. Konishi and Dr. Alden H. Miller for reading and criticizing this manuscript. The work was supported by a grant from the National Science Foundation.

CHAPTER SEVEN

COOPERATION

ALTHOUGH cooperation can occur among a great number of individuals, for purposes of simplicity, experiments dealing with this form of behavioral inter-dependence restrict observation to just two individuals at a time. In particular, these experiments observe two individuals whose responses are characterized by a mutual reinforcement contingency; that is, the reinforcement of both individuals is the same or nearly the same and it is contingent upon the response of just one. While both individuals are engaged in such responding, each one is able through his own responses to bring about the reinforcement for the other or for both. For social psychologists working with human subjects, the most popular research question about cooperation has always been about the conditions which enhance or inhibit cooperation. For social psychologists working with lower animals, the question has been *whether* animals are capable of cooperation, and whether they can *learn* to cooperate. It has sometimes been argued that cooperation is such a complex phenomenon that one simply would not expect animals lower than man to be capable of engaging in it in a deliberate way. Cooperative relation, it has been argued, requires taking the point of view of the other. It requires the maintenance of a response under a delayed reward. It requires an ability to communicate intentions and to coordinate efforts. It requires, in short, the beginnings of a culture.

It is not necessary for heuristic purposes that cooperation have all these require-ments. If we regard cooperation simply as a form of behavioral interdependence that is characterized by the above mutual reinforcement contingency, we would eventually be able to determine by means of experimental research what role in cooperation is played by a sensitivity to the needs or states of the coactor, by the delay of reward, by coordination, by the communication of intention, etc.

In his classical study of cooperation among chimpanzees, Crawford links co-operation to communication and, in particular, to the animals' ability of com-municating their intentions to one another. It is this ability which helps them to influence each other's behavior in the anticipation of some mutual benefit. As we note from his experimental design, the situation confronting the animals is an extremely complex one, the task requiring individuals to undergo nearly five hundred trials to achieve criterion. Only a suggestion of what may be deliberate cooperative behavior was therefore observed by Crawford. When the experimental situation is simplified (still, however, maintaining the mutual reinforcement contingency that is the defining property of cooperation), cooperative behavior is readily observed, and the experiment by Daniel shows the development of such behavior quite clearly. A promising method for the study of cooperation is found in the experiments of Miller, Banks and Ogawa. Their methodology allows the systematic manipulation of communication among the cooperating monkeys. By systematically reducing the animals' access to each other one can eventually discover the signalling cues that are critical in enhancing cooperation. Further research by these workers has in fact gone in this direction, in one case employing the television screen in the place of a live coactor. The chapter on cooperation ends with what is a special case of cooperation—altruism. The question is raised by Rice and Gainer if animals, in this instance albino rats, would aid another just by being exposed to his signs of distress. Because the matter is somewhat controversial, we are also including an experiment by Lavery and Foley whose results identify the rat as somewhat less virtuous a creature.

THE COOPERATIVE SOLVING BY CHIMPANZEES OF PROBLEMS REQUIRING SERIAL RESPONSES TO COLOR CUES*

Meredith P. Crawford

A. INTRODUCTION

ONE of the most important problems concerning the phylogenetic development of human language is the task of making explicit the differences between human language and animal communication. A number of writers have listed various differences (1, 4, 6); most of them agree that human language is unique in being conventional, symbolic, and intentional. The majority of the data on animals have been obtained through naturalistic observation, in which the function of the cry or gesture in the social situation is more or less a matter of inference. When Bloomfield (2) pointed out that the "fundamental linguistic situation" is that of coöperation between two individuals, A and B, in which A stimulates B to do something impossible for A to accomplish, but which is adaptive with respect to A, he suggested what has proven to be a fruitful approach to the experimental study of communication, even in animals, in situations where its linguistic function may be made clear (see also 5). Gestural communication between chimpanzees in problem situations requiring coöperative behavior for solution has already been described (3). The present report offers a further analysis of such communication in a type of problem situation more complex than those originally used.

In her discussion of the origins of human language, DeLaguna (4) contends that an important step in linguistic development was the separation of the functions of predication and command, both of which inhere in the typical animal cry, which serves merely as a signal to set off patterned or type responses. In describing the results of the experiment in coöperative problem solving with the box-pulling technique (3), the

* From the *Journal of Social Psychology*, 1941, **13**, 259–280. Copyright 1941 by The Journal Press, Provincetown, Mass.

present writer argued that the solicitation with manual gestures displayed by some of the subjects only served to stimulate the partners to do something, but did not tell them what to do. Evidence for this view came from instances in which a subject might respond to solicitation by offering to groom, to walk in tandem with the solicitor, or by handing him some bit of material, and only after repeated solicitation did he hit upon the one correct pulling response. By contrast, the present experiment required the solicited animal to do two things, in a particular order. We were interested to see in how far a solicitor would not only urge the partner to do something, but would indicate which one of the alternative responses should be made. In other words, the experiment was designed as a test of the animal's ability to make, in effect, the predicative statement, "*This is the correct one.*"

The general method was to place around the outside of a cage four electrically interlocking devices which had to be manipulated in a particular sequence in order to release food from a vendor when the sequence was completed. After certain subjects were individually trained in the serial habit, the cage was divided by a grille partition, and a trained animal placed on one side, with an untrained one on the other, each having access to two devices and a food vendor. Should solicitation take place, would the soliciting animal show the untrained partner which of the two devices available to him should be chosen first?

A secondary reason for the choice of this type of problem situation was to provide an account of the development of coöperative behavior between two chimpanzees without human tuition. In the box-pulling experiment (3), after it appeared in preliminary observation that pairs of animals would be very unlikely to hit *upon simultaneous pulling* by chance, as a mode of problem solution, active steps were taken by the experimenter to train them to pull together on their ropes. Early in the present experiment subjects could and did obtain food by entirely independent activity, since no limit was set upon the time which should intervene between the operation of successive stimuli. Coöperative behavior, characterized by watching and responding appropriately to the partner's behavior, and by solicitation, was offered during the course of testing by the animals as a new and more efficient mode of problem solution, entirely without the intervention of the experimenter.

B. PRELIMINARY EXPERIMENT[1, 2]

In working out the methodology for this study a preliminary experiment was conducted in the Northern Division of these laboratories in New Haven. Four interlocking devices, each requiring the subject to make a slightly different type of manipulation, were arranged to be operated in a fixed spatial sequence. Two subjects, Bimba and Bula, who were later used in the main experiment, were given individual training, during which each mastered the correct sequence. They were then placed together in a divided reaction cage for 135 trials. Indications of the development of coöperative behavior included watching of one animal by the other and occasional acts of solicitation. When Bimba and Bula were each paired with untrained subjects (Alpha and Beta), neither made any attempt to direct her partner's activity, during the limited number of trials which were given. When this study was later continued in the Southern Division at Orange Park, the methodology of the main experiment was adopted.

C. METHOD AND PROCEDURE FOR THE MAIN EXPERIMENT

1. Apparatus

A diagrammatic floor plan of the apparatus is given in Figure 1. On the platform around the outside of the restraint cage were placed four identical *stimulus presentation boxes*, A, B, C, and D. Detailed drawings of one of these are presented in Figure 2. In each box was a movable stimulus holder, into which a square, colored stimulus plate could conveniently be fitted. As presented to the subject within the cage this colored stimulus was flush with the white face of the presentation box, and could easily be pushed by the animal for a distance of about 2 in. into the box, where it remained fastened. Each stimulus holder was equipped with a noiseless electric lock and an electric circuit breaker. These were wired through a control switchboard so that when the stimulus in one box was pushed in, an electric circuit was broken and the stimulus holder in

[1] The writer wishes to express his thanks to Drs V. Nowlis, A. H. Riesen and F. M. Fletcher who assisted in observing pairs of subjects.

[2] Reported before the *A.A.A.S.*, Section I, December 27, 1937.

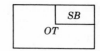

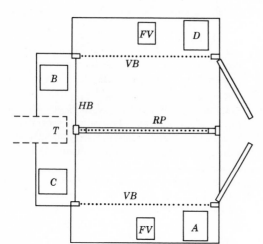

FIGURE 1. Floor plan of apparatus showing restraint cage with sides of vertical grille bars, *VB* and *VB*, and end of horizontal bars, *HB*, divided along the middle by a removable partition of grille bars, *RP*. The stimulus presentation boxes, *A*, *B*, *C* and *D*, and the food venders, *FV* and *FV*, were placed on the platform outside the cage. The positions of the track used for box-pulling, *T*, and the observation table, *OT*, with the switchboard, *SB*, thereon are indicated.

another one of the boxes was unlocked, without making any sound which might afford the subject a secondary cue. Any one of the four colored stimulus plates could be placed in any of the four boxes, but the boxes were always controlled at the switchboard in such a way that the stimuli could be pushed only in the order (1) *yellow*, (2) *green*, (3) *red*, and (4) *blue*,[3] regardless of their location. When the terminal stimulus, blue, was operated, food was released from one or both of the food vendors (*FV*). A sliding door was mounted in front of each stimulus presentation box and food vendor, and all doors were raised and lowered simultaneously from the experimenter's observation table.

2. Subjects

Seven chimpanzees served as subjects: five adolescent animals and two chimpanzee children. The adolescent group included the females Bimba, Bula, Alpha, and Kambi and the male Frank, who were between 7 and 8 years old at the time of experimentation. Each had been used in laboratory investigation almost its whole life, and was well adapted to experimental routine. The four females were subjects in previous problems in coöperation (3), and Bimba, Bula, and Alpha served in the preliminary experiment reported above. Tom and Helene were twins, about 4 years old when used in this study. They had much less

experience in experimental work, and had never worked before in a problem requiring coöperation, yet both adapted readily to this experimental situation. They had lived together most of their lives, and were good friends.

3. Individual Training

Before pairs of subjects were tested for coöperative behavior, certain animals were given an intensive period of individual training to acquaint them with the correct sequence of the color stimuli. Bimba, Tom, and Frank were selected, and about 3 months were spent in allowing them to learn the problem by trial and error. For this purpose the central grille partition was removed, giving the single animal access to all four boxes, and only one food vendor was in place. Table 1 summarizes the results, showing the number of trials given, the largest number of perfect runs during the last 40 trials, and the per cent of

TABLE 1. *Results of Individual Training*

Name	Trials	Correct Runs Out of 40	% Correct Responses
Bimba	480	33	93.0
Tom	488	23	85.1
Frank	440	13	76.9

---

[3] For the reader's convenience, the stimulus plates will be referred to hereafter by number, representing their serial positions, rather than by colors.

correct responses, counting each response to each box as a unit (160 units in 40 trials), during the last 40 trials.

Three significant points about the individual training deserve brief mention. (*a*) During the individual training for the preliminary experiment, in which subjects learned to go to four devices in a fixed spatial order, Bimba required 60, and Bula 80 trials for mastery. In comparison with the data of Table 1, this illustrates how much easier for the chimpanzee are problems requiring response to spatial than to non-spatial cues. (*b*) Bimba's record, and, to a lesser extent, those of Tom and Frank indicated a backward elimination of errors; early in training all subjects tended to make first choices of No. 4 most often, which was gradually eliminated in favor of No. 3, No. 2, and No. 1. The first stimulus plate (yellow) was also often chosen early in training, perhaps because it was almost twice as bright as the other three, which were of approximately equal brightness. (*c*) On almost every trial after about the 320th, as soon as the doors were raised, Bimba sat quietly and looked from one stimulus plate to another, spending sometimes as much as 10 sec. in this preliminary survey before proceeding to push the stimuli in order without hesitation. Tom and Frank made their first choices quickly, then hesitated and looked about between succeeding responses. Only Bimba's behavior suggested what might be called a type of "foresight."

4. Procedure for Coöperative Tests

Because of his slow learning, Frank was discarded as a subject before coöperation tests began. Two pairs of subjects served for the major part of the coöperative work: Bimba was paired with her living-cage companion, Bula, who received no individual training other than tuition in pushing in the stimuli; and Tom was paired with his twin, Helene, who received no preliminary individual training. Bimba and Bula were paired for short test series with two other adolescent females, Alpha and Kambi.

The coöperative testing, like the individual training, was carried on in daily experimental sessions of eight trials, on each of which the four stimuli were placed in different boxes. In varying the location of each colored stimulus only those permutations of four boxes were used in which colored stimuli adjacent in the serial order would not be placed on the same side of the partition; i.e., if Nos. 1 and 3 were on one side, Nos. 2 and 4 would be on the other; never did Nos. 1 and 2 fall on the same side. This was done to give the

animal on either side of the grille as many occasions as possible to operate a stimulus in immediate response to its partner's manipulation of a preceding stimulus.

Since a major interest of the investigator was focused on solicitation, all efforts were made to stimulate such behavior. On most regular trials with Bimba and Bula, and with Tom and Helene, the subjects went rapidly to work at boxes on their own sides of the partition, and the sequence was completed and food delivered in a few seconds, giving little occasion for solicitation. To delay the responses of one of the partners, various techniques were used individually or in combination

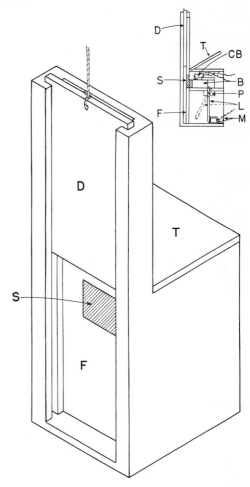

FIGURE 2. View of a stimulus presentation box with insert showing same in cross-section. *T*, hinged top; *D*, vertical sliding door operated by rope; *S*, stimulus plate as it appears to the subject before being pushed in, flush with the face of box *F*. The stimulus is fitted to a block, *B*, which moves along a track. The upper end of the lever, *L*, pivoted at *P* prevents movement of the block when held in place by the electro-magnet, *M*; when the stimulus is pushed in current to a succeeding box is interrupted by circuit breaker, *CB*.

during regular test sessions. These *distraction techniques* were: (*a*) Previous to an experimental session one animal was fed a quantity of the kind of fruit which was to be used as incentive object during the session. (*b*) A "toy" was presented to one of the two subjects in order to distract its attention from the problem. A variety of discarded metal objects, pieces of rope, etc., particularly things which had some moving part, served this purpose. They were often fastened by cord or chain to one side of the cage, so that the animal across the partition could not take it away from the one for whom it was intended. "Toys" had to be changed frequently, since the subjects lost interest in them on repeated presentation. (*c*) No food was given one of the subjects on completion of each trial of a session. On some sessions the food vendor itself was removed from one side of the cage. It was thought that after successive unrewarded trials the subject's response to the stimulus plates would become temporarily extinguished.

When none of the subjects had solicited after a number of the long delays induced by these special techniques, resort was made to the following *special training procedures* which either supplemented individual training or tended to evoke solicitation in situations other than the color sequence problem.

1. *Individual training* was repeated under two conditions. (*a*) The same conditions were used which prevailed for original individual training. (*b*) A single animal was placed on one side of the divided cage and supplied with a stick with which it might manipulate the stimuli on the other side of the cage by reaching with it through the partition. In this training the subject was acquainted with the interrelationship of the boxes on both sides of the partition, and with the use of an instrument, *inanimate* in this case, for operation of the distant boxes. It was thought possible that this training might transfer to the use of an *animate* instrument, the partner, for manipulation of the distant boxes.

2. *Food-sharing tests.* Utilizing a technique developed by Nissen and Crawford (8), two animals were placed in the cage with center grille partition in place, and a large quantity of food, usually two regular meals, was piled on one platform available to only one animal. The food (including potato, tomato, onion, carrot, cabbage, and certain fruits) was divided into a number of small pieces of roughly equivalent size, so that the number of pieces transferred from one half of the partition to the other could be scored. Often the animal in the compartment on the other side of the cage begged, with manual gestures or otherwise, its partner for some of the food. Such gestural behavior has been shown to be essentially similar to the solicitation displayed in problems requiring coöperation (3).

3. *Box-pulling tests.* The box-pulling set-up used previously (3) with Bimba and Bula was duplicated in its essentials in this experiment cage. A platform or track was extended from the platform of the cage between boxes *B* and *C* (see

TABLE 2. *Summary of Experimentation with Bimba and Bula*

Group No.	Inclusive Sessions	Dates	Problem	Total Trials	Delays Caused by Bimba	Bula	Remarks
1	1–32	11/30/37–2/17/38	Color sequence	215	9	7	Watching developed; no solicitation
2		1/24–2/9/38	Bimba and Bula used in color sequence with Kambi and Alpha				Watching; no solicitation
3		2/18–3/8/38	Bimba and Bula trained alone in use of stick				
4	33–38	2/9–2/17/38	Color sequence	26	3	5	Watching; no solicitation
5		3/12–3/14/38	Bimba and Kambi color sequence				
6		3/18–3/26/38	Food sharing				
7	39–44	3/28–4/2/38	Color sequence	30	8	4	Watching; solicitation by Bula
8		4/4–4/25/38	Box-pulling				Solicitation by Bimba
9	45–61	4/26–5/16/38	Color sequence	97	0	5	Solicitation by Bimba
10	62–64	5/17–5/19/38	Color sequence		7	0	Solicitation by Bula
11	65–69	5/25–5/31/38	Color sequence (mesh screen partition)		0	11	No solicitation when mesh in place
12		6/13–6/21/38	Bimba and Kambi color sequence				Solicitation by Bimba

TABLE 3. *Summary of Work With Tom and Helene*

1	2	3	4	5	6
					Helene
No.	Sessions	Dates	Problems	Trials	Delays
1	1–25	11/30/37–1/22/38	Color sequence	160	12*
2			Individual training for Tom		
3	26	2/14/38	Color sequence	8	1
4		3/10/38	Stick training for Tom		
5	27–31	3/11–3/16/38	Color sequence	34	12
6		3/17–3/26/38	Food sharing		
7	32–73	3/28–5/16/38	Color sequence	298	60
A		5/17/38	Food sharing		
8	74	5/18/38	Color sequence	8	4
B		5/18/38	Food sharing		
9	75–85	5/19–5/31/38	Color sequence	60	19
C		6/1–6/3/38	Food sharing		
10	86–93	6/4–6/13/38	Color sequence	40	3

* Both subjects seemed equally responsible for many of the delays during the early trials.

Fig. 1) on which a box filled with weights might slide. Two ropes were attached to the box, one leading to each side of the partition, with which two animals, one on each side of the partition, might pull in the box with coöperative effort. The weight of the box varied between 450 and 550 lb., requiring from 180 to 220 lb. pull on the ropes to move it.

D. RESULTS

In Tables 2 and 3 are presented summary accounts of the color-sequence and the special training procedures used with each pair of subjects, Bimba and Bula, and Tom and Helene. Experimental sessions are grouped according to the type of testing done, and the number of sessions with inclusive dates for each group are indicated. The total number of trials or completions of the sequence and delivery of food are shown, together with the number of delays of more than 30 sec. caused by each partner. These delays indicated roughly the effectiveness of the distraction techniques which were used on a number of sessions, beginning about the middle of the first group with each pair of subjects. In the "Remarks" column of Table 2 entries are made which characterize Bimba and Bula's progress toward coöperative behavior. Tom and Helene's performance changed so little that similar remarks in Table 3 would be of no value.

1. Bimba and Bula

During the first 16 sessions the animals were

allowed to proceed in trial and error fashion, and no special techniques were employed to delay one animal. The development of a rapid and rather efficient type of joint performance took place during these trials, in which coördination seemed to depend both on visual and auditory cues obtained from the partner's manipulation of the stimuli. On no occasion of delay did either animal solicit its partner to continue work.

Excluding those trials for which distraction techniques were used, the time required by the two subjects to complete the sequence decreased from an average of 11.5 sec. on the first two sessions to a minimum of 5 or 6 sec. on trials near the end of this group. Toward the last sessions the animals began to perform sequences without making any errors in the serial order of stimuli. Beginning on session 17, and continuing to the end of the group, from one to six perfect performances were made by the animals during the eight trials of each session. Because of individual differences in tempo of work and use of distraction techniques, detailed analysis of error scores would be of little significance.

Since only Bimba received extensive individual training, it was interesting to see how her choices of stimuli in the social situation compared in correctness with those of Bula. During the first 11 sessions Bimba made only an occasional error, i.e., on her first choice she almost always chose that stimulus on her side of the partition which came earliest in the sequence. Thereafter the correctness of her choices began to decline until session 17, when gradual improvement started. Bula's choices rose from an initial chance score to a fair level of accuracy toward the end of the group. Thus, there is some evidence that Bimba's initial individual training transferred to the social situation.

The most interesting aspect of the behavior was the watching. Even on the first session each animal watched the other at various times, although during most of a typical trial each proceeded to her own stimuli without regard for the other subject. Of most importance were those instances in which one animal observed her partner push in a stimulus and then turned to push in the next succeeding one herself. Bimba did not seem as well motivated with respect to food as did Bula, since she was often slow in beginning her work, so that Bula sometimes pushed at both of her stimuli two or three times before Bimba attempted to push one of hers.

Since Bula and Bimba showed no solicitation when paired with each other or when used with Kambi and Alpha, as indicated in Group 2 of the table, the first special training procedure (training in use of a stick) was introduced. Although each subject learned to manipulate stimuli with the stick on the far side of the partition in correct sequence, neither solicited either partner to operate these same boxes. Kambi was again employed with Bimba for two sessions during which Bimba watched but did not solicit.

Food-sharing tests constituted the second special training procedure used for eight sessions with Bula and Bimba. On four the food was given to Bula and on four to Bimba. Food passed from one cage to the other by various methods including (a) picking up by the unfed animal of food lying within reach and unguarded on the floor of the fed animal's side, (b) direct reaching by the unfed animal into the partner's hands or mouth and forcibly taking food, (c) unsolicited passing, in which the animal possessing food handed some of it to the other "voluntarily," or at least, so far as could be observed, not in response to any command or begging by the unfed animal, and (d) begging, in which one subject reached through the bars with palm up, arm flexed at wrist and elbow, hand moving in beckoning gesture, in a manner suggesting human begging. All of the begging was done by Bula and all of the direct reaching or commandeering by Bimba, who was decidedly the dominant animal. It is interesting to note that on one session in the midst of Bula's begging an interval of play-fighting occurred during which the animals were obviously on friendly terms and mutually enjoyed the sport.

In Group 7 return was made to the color-sequence problem, and during the second delay by Bimba, Bula solicited for the first time. Her characteristic behavior will be described below. She did not solicit any more during this group of sessions, nor did Bimba when Bula delayed. Resort was finally made to the third special training procedure, a repetition of the box-pulling test in which both subjects had solicited a year before the present work. Distraction techniques were used with Bula during these tests (Group 8) which resulted in many delays by Bula; Bimba often watched Bula and coördinated her pulling with the latter's. Also, on certain sessions between the fifth and the twelfth, when the box was baited, Bimba took her rope and braced herself ready to pull, then bent over, and reaching outside the front grille, pushed Bula's rope toward her. This sometimes served to get Bula started pulling.

Suddenly, on session 18, when the experimenter least expected it, Bimba solicited Bula after a long delay. She reached through the grille partition and touched Bula four times on the neck and shoulder. Bula did not respond by pulling, whereupon Bimba threw one of the rope ends toward the box. It was immediately replaced by the experimenter, and Bimba began to solicit again whereupon Bula picked up her rope and the two animals pulled in the box. Of the 16 subsequent trials given during the remaining three sessions, Bimba solicited Bula during 11, after quietly watching Bula for not more than 15 sec. on any one. Her gestures were identical with those displayed two years before: she reached directly for Bula's neck or shoulder, pushed her lightly toward the front of the cage and sometimes patted her on the back or shoulder when she did not immediately begin to pull.

Although Bula was fed at the beginning of each session on the color-sequence problem in Group 9, it was not until the sixth (No. 50) that she caused any delays. On this session Bimba seemed highly motivated and carefully watched Bula during the two delays of less than a minute. The first two trials of the next session were completed without delay, but on the third Bula delayed and Bimba solicited for the first time in the color-sequence problem. Part of the original record, dictated immediately after the trial, is presented below.

Session 51, Trial 3, CBAD (Bimba in AC). [Bimba had opened No. 1, reached through the grille and pushed in No. 2 on Bula's side, forced open No. 3 on her own side, and had been watching Bula at intervals until . . .]

At 7 min. I added a quarter of an orange to Bimba's food vendor. Bimba watched carefully and then went to the east end and reached in behind

Bula toward No. 2, already open. She withdrew her arm, moved a pace toward the west end, returned, reached in, and touched Bula on the neck as she had done in the box-pulling problem. She seemed to push Bula toward No. 4 in Box *D*. Bula began to move in that direction and Bimba followed her along, reaching in again through the bars when Bula was about half way to it, touching her on the neck. Bula arrived at *D* and pushed No. 4 but it did not stay open. Bula started to withdraw her hand when Bimba reached in again, touched Bula firmly on the neck, and seemed to push her again toward No. 4. Bula pushed No. 4 and it opened, releasing food from both vendors (8 min. 10 sec.).

On many of the 12 subsequent occasions of solicitation, Bimba seemed to direct Bula toward a particular one of the two boxes on her side. This she did by grasping Bula at the arm or shoulder, turning her in a certain direction, and gently pushing her forward. Such behavior is illustrated in the following protocol. (*Note.*—For study of directional behavior special settings of stimuli were used in which colors adjacent in the series were placed on the same side of the partition, as in the following case.)

Session 53, Trial 3, CADB (Bimba in AC). Bimba pushed Nos. 2, 1, and 2, then reached through the partition to No. 4 in Box *B*. She persisted in pushing at this stimulus until 25 sec., when the experimenter moved the vertical door of Box *D* (by pulling a rope at the control table), calling Bimba's attention to No. 3. Bimba then grasped Bula's shoulder, who had been watching Bimba work at No. 4, and turned Bula completely around. She took hold of Bula's neck and shoved her toward No. 3. After Bula had pushed No. 3 she turned back and pushed No. 4, as Bimba watched her.

It is extremely interesting that, although Bimba had the ability to solicit, she reached for one of the stimuli on Bula's side and tried to push it in herself.

The three sessions of Group 10 were undertaken to study Bula's solicitational behavior, and for this purpose the three distraction techniques were employed with Bimba, which resulted in several delays of more than 30 sec. Bula only watched her partner during the first two, but on the third she solicited in clear and unmistakable fashion. She reached for Bimba with bent wrist, palm up, arm moving in a beckoning gesture, bobbing up and down on flexed legs and occasionally whimpering. There was no clear element of directiveness in Bula's gestures, although sometimes her hand was placed between her partner's face and the stimulus which was to be operated.

The sessions in Group 11 were used to see whether solicitational behavior would be affected by placing a wire screen ($\frac{1}{2}$-in. mesh) over the grille bars of the partition, which prevented an animal from reaching through. The screen was in place during sessions 65 and 66, but Bimba only watched Bula during the five delays, which lasted from 2 to 10 min. Bimba solicited in the usual fashion when the screen was removed during session 67, but made no attempt to do so when it was replaced on sessions 68 and 69. A similar series of tests was given these same subjects in the box-pulling situation, and again Bimba made no attempt to stimulate her partner to work when the screen was in place. Perhaps some type of solicitation, specially adapted to the situation in which the barrier prevented reaching, would have been developed had the tests been continued longer. From the present limited data, Bimba's solicitation appears to have been a stereotyped response useful only in situations where direct reaching was possible.

Shortly after the last work with Bimba and Bula, Bimba was tested with the previously untrained female Kambi for four sessions. While there were many delays caused by Kambi, Bimba solicited her on only two occasions. In one of these it seemed clear that Bimba was trying to get Kambi to push a particular one of her two stimuli, as may be seen in the original record.

Session 3, Trial 3, DCBA (Bimba in BD). Bimba pushed No. 1 and tried No. 3, then reached through the grille and pushed No. 2, but not far enough for it to remain in. Bimba went to Kambi and placed her hand on Kambi's back, and pushed her gently in the direction of No. 2. Kambi did not move. Bimba withdrew her arm and went to No. 3, on her own side, and forced it open by bracing herself against the grille partition and pushing hard. Bimba then returned to Kambi and placed her hand on Kambi's neck, so that Kambi's head was between her hand and No. 4. Bimba pushed toward No. 4. (This was in clear contrast to the position of Bimba's hand when No. 2 was the next stimulus in sequence.) Instead of moving toward No. 4, Kambi turned to play-fight with Bimba, and it was not until after the play that Kambi pushed No. 4 as Bimba watched her (3 min.).

2. Tom and Helene

During the first few sessions each animal's attack on the boxes was entirely independent of that of its partner. Not until session 8 did any watching occur, and even as late as session 16 it was not at all clear that the watching was in any way related to an attempt to establish coördination. Thereafter, each began to watch the other and to respond by pushing one of its own stimuli, although not always the one next in sequence. It looked as though stimulus operation by one animal merely facilitated like behavior in the partner.

During the early sessions Tom often reached for the stimuli on Helene's side, completely ignoring Helene, yet he often initiated play-fights with her. It appeared that Tom was aware of Helene as a play-partner, but her possible function as a co-worker, or an "animate instrument" which he might induce to help him, seemed never to occur to him. As late as session 10, Tom seemed not to have learned that when Helene pushed No. 4, when it happened to be on her side, food would be delivered from the vendor on his side. With respect to their serial position, Tom's choices of stimuli during this group started, and remained at a chance level. Each subject tried to push the stimulus to which it happened to be nearest when the doors were raised. Nothing resembling solicitation occurred during this group of sessions; the subjects worked rapidly and independently, trying the stimuli available to each in quick succession until they both were pushed in. Often the sequence was completed within 4 to 6 sec.

During the remainder of the experimentation with these subjects little change took place in their behavior, and nothing appeared which could clearly be called solicitation. While Tom mastered the manipulation of the stick, and was able to operate the stimuli on the far side of the partition with it in perfect sequence, when he returned to work on the color-sequence problem with Helene he made no clear attempts to get Helene to perform a similar instrumental function.

Certain interesting similarities appeared between Tom's behavior in the food-sharing tests of Group 6 and subsequent tests with the color-sequence problem. Twenty-one passes of food by Helene were scored as unsolicited because, so far as the observer could tell, they were not made in direct response to any sort of threat, command, or appeal by Tom. The 15 passes scored as direct reaches were all made in response to some overt behavior from Tom. Most commonly, Tom reached through the grille toward the food in Helene's hand or mouth. His palm was upturned, but his forearm did not move in the fashion characterizing begging in other animals. He also bounced up and down on flexed legs, shook the grille bars, and followed Helene back and forth in the cage. During one session some of Tom's direct reaches were accompanied by a whimper, which quickened Helene's response in striking fashion. During another Tom did not reach, but stood at the grille holding the bars with both hands and swayed. It is quite possible that this posture represented a threat toward Helene, to which she was responding in what were classified as unsolicited passes. During the four sessions on which Tom was given the food he guarded it zealously, threatening Helene whenever she reached for a piece. She made no attempt to reach during the last two sessions, and seemed to ignore both Tom and the food.

When the color-sequence tests were resumed in Group 7, Tom sometimes behaved as he did in the food-sharing situation: he followed Helene back and forth along the partition, held the bars and looked at her, and sometimes shook them or stamped his feet. Helene did not show any clear or immediate response to these acts, so that they cannot be classified as solicitation, although they may have served to stimulate Helene in some indirect or delayed fashion, as was suggested in the food-sharing tests. Tom watched Helene and responded to her stimulus manipulation with increasing frequency. While Helene often looked at Tom, there was little evidence that she attempted to coördinate her behavior with his. During session 69, Tom reached in toward Helene many times and often went to the stimulus next in sequence on his side immediately after these reaches. On one trial Tom pushed in No. 1 on his side, proceeded toward No. 3, stopped, turned to look at No. 2 across the partition, and reached toward Helene who was starting toward No. 2. Tom watched her push it and then went to No. 3. Behavior was not significantly different on the three sessions of food-sharing, A, B, and C, nor in Groups 8, 9, and 10 of the color-sequence testing. Occasionally Tom paced the floor, held the bars of the grille partition while looking at Helene, and sometimes knocked on the floor of her cage. Of course it is possible that Tom intended to solicit by some of this behavior, but was unsuccessful. The observer could not with certainty so classify the behavior until Helene had consistently responded to it a few times.

E. DISCUSSION

The individual training given Bimba and Tom before they were paired with their untrained partners, Bula and Helene, did not immediately fit them to direct a coöperative problem solution. Beyond the tendency shown by Bimba to choose that stimulus on her side which occurred earliest in the color sequence, no clear transfer of training from the individual problem situation to the social could be detected. This was somewhat surprising after Bimba's mastery of the sequence of colors and display of what looked like "foresight" during individual training.

Watching, and response to the partner's operation of a stimulus by operation of the succeeding stimulus appeared only after a certain amount of experience in working together. Its earlier appearance with Bimba and Bula than with Tom and Helene may have been due to one or both of two circumstances: (a) the former were older by 5 years than the latter, and were presumably more sophisticated socially, (b) Bimba and Bula served in the preliminary experiment and in other experiments on coöperation, in which they learned to watch each other work. In this problem situation, as in the other situations already studied, watching seemed to be learned, or at least subject to much practice, in the specific situation where it was observed. In watching, the subjects made anticipatory responses to certain sequences of stimuli which always preceded the successful manipulation of the stimulus plates, after the first, on the watcher's side of the cage. The watcher simply attended to the relevant cues of the problem situation, and the behavior would presumably have been the same whether this stimulus sequence involved animate or inanimate objects.

The only behavior of a peculiarly social nature relevant to problem solution was the solicitation. As was suggested before (3, pp. 68–69), the use of solicitation can be thought of as the selection of one of many tools, instruments, or roundabout pathways which might be open to an animal for problem solution. Its appearance was therefore similar to the sudden occurrence of a new mode of attack by an individual animal in a problem situation. We are not able to explain why Bimba and Bula, both of whom had solicited so often in the box-pulling experiment in previous years, were so slow to hit on it in the color-sequence problem, or even when the box-pulling tests were repeated. Apparently the solicitational response requires for its elicitation a delicate balance of both internal and external stimuli, and changes in the external situation of which the observer was unaware may have been significant to the apes. Once solicitation appeared it was rapidly reinforced and became a facile mode of response.

The younger animals, Tom and Helene, never responded in a fashion that might clearly be called solicitational. Even in the relatively naturalistic situation of the food-sharing tests, where Tom clearly dominated Helene and apparently commandeered food by various indirect methods, his gestural behavior was of minor importance. He reached directly toward the food in Helene's hand, but he never showed the rich patterning of gestures displayed by Bimba and Bula. In Tom we find a striking example of an animal who knew how the sequence could be completed, since he did so when given a stick, but who did not show the ability to stimulate Helene to finish it. Perhaps neither Tom nor Helene were socially mature enough to have developed solicitation as a means of controlling the behavior of other animals. The responses which Tom did direct toward Helene seem to have been learned or become mature within the previous year, for when he and other young animals were tested in the food-sharing situation in the spring of 1937, none of them obtained food, by any method, from the partner who was fed.

At times Bimba's solicitation was clearly directive. For example, by pulling her partner away from one device, turning her in another direction, and pushing her toward the other stimulus presentation box on that side she made it quite clear (at least to the observer), which of the two stimuli she wanted her partner to push in. She behaved in this manner both with a partner with whom she had worked successfully (Bula), and with one with whom she had not (Kambi). While it did not by any means completely tell the other animal what to do, it represented the first step, by sending her in the right direction of attack. Bula's begging type of gestures were not so clearly directive.

The significance of these observations in relation to DeLaguna's theory of the phylogenetic origin of language has been anticipated in the Introduction. She argues that in animal communication the functions of commanding and predicting are not separated, but may appear as two aspects of the same gesture or cry. This was also the case in the behavior reported here. The interesting point, however, is that the directive elements of the solicitation, that is, the bodily orientation of the partner, were so clear and so human-like. Perhaps it is significant that a soliciting animal never pointed, but always used what seems to be the more primitive method of pushing or pulling the partner. While one chimpanzee has often been seen to beckon another to come toward it, the writer is aware of no report of the use of pointing. His own casual attempts to direct an ape's attention by pointing alone have met with little success. Of course, primates are quick to follow the direction of another's gaze.

It may be that an important transitional step in the development of language behavior lies between the direct orientation of one animal by another through bodily manipulations and indirect orientation through pointing toward a distant object.

In the above paragraphs we are thinking primarily in terms of the "evocative" (speaker-hearer relation) function of language (7). Another type of animal behavior which may represent an evolutionary forerunner of human language is related to the "representative" function of language, and is being studied by experiments on learning and problem-solving behavior of individual animals. The important question about the use of "symbolic" processes in complex problem solutions, such as instrumentation, multiple choice, delayed response, has recently been critically reviewed by Yerkes and Nissen (9). They summarize the evidence which shows that, under ideal conditions and after a great deal of training, chimpanzees make responses to non-spatial cues which seem to be mediated by symbolic processes. Perhaps it is not too much to expect that a combination may be effected of these two approaches to a study of a type of infra-human primate behavior which may be representative of a stage in the linguistic development of early man. For example, chimpanzees might learn to direct partners in a coöperative solution of a problem with gestures which, because of previous training, had acquired generalized significance to animals within an experimental group. This would be a study in the acquisition by an animal of a simple, intentional, predicative, and conventionalized language.

F. SUMMARY

Four identical stimulus presentation boxes were placed around the outside of a restraint cage which could be reached by the animals within the cage. In the movable stimulus holder on the face of each box could be placed any of four colored stimuli. The stimulus holders could be interlocked in any order, so that the yellow, green, red, and blue stimuli had to be operated in sequence, regardless of location. When the terminal stimulus, blue, was pushed, food was released from a vendor. The six chimpanzee subjects included two children and four adolescents. As preliminary training, individual subjects learned by trial and error to operate the four stimuli in correct sequence. For coöperative work the cage was divided by a grille, so that two stimulus boxes and a food vendor were available to each subject on either side. Obviously animals could obtain food by working persistently but independently at the two stimuli available to each. Coöperative behavior was said to occur when a subject, (a) watched the partner and responded to its manipulation of a stimulus by pushing the succeeding one on its own side, and (b) solicited the partner with manual gestures to push one of its stimuli. The two adolescents with whom most testing was done watched and solicited. Both of the children watched but neither solicited. One adolescent seemed to direct her partner toward a particular one of the two devices in the partner's cage by turning her in the proper direction and pushing her toward it. The significance of these directive aspects of solicitation for a consideration of the phylogeny of language behavior is discussed.

REFERENCES

1. BIERENS DE HAAN, J. A. Animal language in relation to that of man. *Biol. Rev.*, 1929, **4**, 249–268.
2. BLOOMFIELD, L. Language. New York: Henry Holt, 1933. Pp. ix + 564.
3. CRAWFORD, M. P. The cooperative solving of problems by young chimpanzees. *Comp. Psychol. Monog.*, 1937, **14**, No. 68. Pp. 88.
4. DELAGUNA, G. A. Speech: Its Function and Development. New Haven: Yale Univ. Press, 1927. Pp. xii + 363.
5. ESPER, E. A. Language. In *A Handbook of Social Psychology*. C. Murchison (Ed.). Worcester:

Clark Univ. Press, 1935. Pp. 417–460.
6. GROOS, K. Zum Problem der Tiersprache. *Z. Psychol.*, 1935, **134**, 225–235.
7. McGRANAHAN, D. V. The psychology of language. *Psychol. Bull.*, 1936, **33**, 178–216.
8. NISSEN, H. W. and CRAWFORD, M. P. A preliminary study of food-sharing behavior in young chimpanzees. *J. Comp. Psychol.*, 1936, **22**, 383–419.
9. YERKES, R. M. and NISSEN, H. W. Pre-linguistic sign behavior in chimpanzee. *Science*, 1939, **89**, 585–587.

COOPERATIVE PROBLEM SOLVING IN RATS*

William J. Daniel

INTRODUCTION

SEVERAL experiments (2–5) have been presented which have more or less successfully demonstrated cooperative behavior in the higher apes. A few experiments (1, 6, 7), observational in character, have indicated this behavior in children. Only one experiment, that of Wolfle and Wolfle (9) has attempted to study cooperative behavior genetically by comparing the behavior of apes and children in nearly identical experimental situations.

It has generally been believed that a study of cooperative behavior in animals as far down the evolutionary scale as the rat is rather fruitless. Only one such experiment (8) has come to the writer's attention. One of the three experiments which constitutes that monograph was designed to test for cooperative behavior in the rat. This experiment was negative; and aside from the films of Mowrer, no other attempt to obtain cooperative behavior in rats has been reported. The experiment reported here represents an apparently successful attempt at obtaining cooperation and one which relies primarily on quantitative data.

PROBLEM

In the experiment described below we wanted to know if it is possible to arrange an experimental situation in such a manner that two animals can assist one another in obtaining food and at the same time escape electric shock.

The experimental situation consisted of a grid box with an electrically insulated platform at one end which, when a rat stepped on it, would remove the charge from the grid. There was also a food crock flush with the grid and beyond the reach of a rat on the platform. This situation is represented schematically in Figure 1.

Our problem is concerned with the behavior of *two* rats in this situation. Will one rat go to the platform and remain on it, thus enabling the other rat to feed? Will the feeding rat leave the food crock and go to the platform, enabling the rat to leave the platform and feed? Finally, will they exchange positions in such a manner that both are adequately fed and both escape or minimize shock? In short, will cooperative behavior be obtained when two rats are put into a double motive situation if the satisfaction of both of these motives is contingent upon the behavior of both animals?

ANIMALS

Heterozygous albino rats were used, ten males and two females ranging from 90 to 107 days of age at the start of the experiment.

APPARATUS

The Experimental Situation (Fig. 1)

The experimental situation consisted of a paraffined wood cage $22\frac{1}{2}$ in. long by 12 in. wide by $4\frac{1}{2}$ in. high with a grid floor and a glass top. In the center of the cage a food crock, flush with the grid, was placed 8 in. from the edge of the platform thus making it impossible for a rat to feed from the platform.

This grid cage was mounted on a set of stilts thus facilitating the replenishing and replacing of the food-crock. A small wooden wall stop was mounted over the platform at the end of the cage forcing all rats to remain beyond its center of gravity and making it impossible for a rat to administer shock without leaving the platform. The tilting of the platform also completed a light

---

* From the *Journal of Comparative and Physiological Psychology*, 1942, **34**, 361–368. Copyright 1942 by the American Psychological Association. A report of part of this work was presented at the Chicago meetings of the American Psychological Association in Evanston, Ill., September 1941.

This study represents the substance of a thesis submitted to the faculty of the University of North Carolina in partial fulfilment of the requirements for the degree of Doctor of Philosophy in the department of Psychology and was done under the direction of Dr A. G. Bayroff of that department.

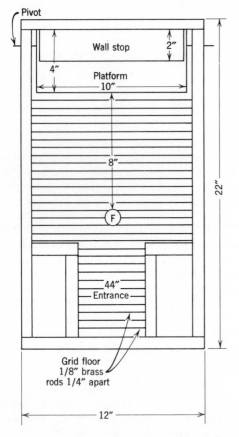

FIGURE 1. Diagram of apparatus. F=food crock.

circuit so that a 40-watt bulb flashed whenever a rat received a shock. This facilitated an objective counting of the number of shocks administered by each rat.

The rats were dropped onto the grid through a small glass door on the top of the apparatus. Directly beneath this door a small entrance alley 6 in. long and 4½ in. wide served to orient the animals in the proper direction, that is, facing directly towards the food-crock and platform.

The Shocking Circuit

The grid was wired in series with a high resistance shocking circuit and the platform automatically shorted out the grid when a rat stepped on it.

The essential problem here was to apply an electrical stimulus to the rats, the physical constancy of which we could be reasonably assured. Our circuit was of such a high external resistance that the added resistance of one or two rats gave the same meter reading as when a

copper wire was placed across the grid. The transformer of this shocking circuit applied 3,750 volts to the rectifier tube and the current at the shock grid terminals could be varied from 100 microamps to 5 milliamps. The average shock intensity of 250 microamps required a circuit resistance of 3,400,000 ohms.

PROCEDURE

The Preliminary Training

The aims of the preliminary training were three-fold:

1. To train the rats to feed in the experimental situation.
2. To train the rats to go to the platform when the grid was electrified.
3. To develop this discrimination, basic to the solution of the problem to be presented in the social situation, to the point at which the rats immediately made the response appropriate to the situation when the situations were varied in an irregular order.

The following schedule was maintained:

1. The rats were unfed for 24 hours.
2. One rat was placed in the grid cage alone with the shock off and the food-crock in place. It remained there for two 450-sec. trials, and was weighed before and after each day's trials. This procedure was continued through the eighth day for each rat.
3. At the end of the eighth day's run the sated rat was put in the grid cage with the grid electrified at 100 microamps. It was not removed until it had reached the platform and remained on it for 30 sec. This procedure was repeated for 20 trials on this day.
4. From the ninth through the thirteenth day the rat was run for 20 irregularly mixed trials with electrified and non-electrified grid.

On the shock trials the rat had to learn to go to the platform and to remain on it for 30 sec. The time it took the rat to make the appropriate response, i.e., going to the platform, was recorded.

On the food trials the grid was not electrified and the rat was left in the apparatus for 100 sec. If by the end of this time it had not made the appropriate response (feeding) it was removed from the apparatus and given the next trial. The time it took the rat to commence feeding was recorded and if the rat did feed it was allowed to do so for 30 sec. so long as it started sometime within this 100-sec. interval.

By the end of the thirteenth day of the preliminary training the rats had mastered this discrimination. When dropped on a cold grid the rat immediately went to the food-crock and fed; when dropped on a hot grid the rat immediately went to the platform and remained on it for 30 sec. It made this discrimination in less than a second or before the experimenter could get to his stop-watch to start timing the rat.

Thus at the conclusion of the preliminary training each animal had learned to escape from shock or to feed in the apparatus depending upon the situation and it had learned this individually and in isolation.

The Experimental Trials

At the end of the preliminary training the rats were divided into pairs of as nearly equal weight as possible.

In the experimental trials *two* rats were put into the cage with the grid electrified and the food-crock in place. They remained in the experimental cage for one trial of 120 sec. duration. They were run 12 trials a day, a total of 1,440 sec., which, on the basis of preliminary experimentation, was adequate for the hunger satiation of both animals. The trial was timed by an electric stop-clock and the individual feeding times by a manually operated stop-watch.

Throughout the experimental trials the apparatus operated automatically. With one or both rats on the platform the shock was off. With the grid not charged, a rat could feed at the food-crock. Thus at least *one* rat had *always* to be on the platform if the other was to get to the food-crock. Occasionally both rats would leave the platform and attempt to feed and take shock simultaneously. If this behavior persisted for five consecutive times the shock was increased 50 microamps, and this "double feeding" stopped.

The rats were fed pulverized purina dog chow mixed with water in the ratio of 5:6 respectively.

At no time did the rats receive food other than that obtained in the experimental situation. This procedure was continued for 40 days at which time it appeared that the rats were doing as well as they ever would.

RESULTS

The most significant fact in the data is that the rats exchanged positions from food-crock to platform and from platform to food-crock. Many of these exchanges were accompanied by shock and many shocks were administered in between these exchanges. It will be remembered that whenever there was any shock both rats received it, but it was administered only by one rat (the platform rat) stepping off the platform and thus electrifying the entire grid. As the experiment progressed more and more of the position shifts were accomplished without shock. Also fewer and fewer shocks were administered which did not result in an alternation. This data for six pairs of rats are presented in Table 1. Since we are most interested in the final stage of this behavior the data are given in terms of the mean performances for the last 5 days of the experiment as compared with the mean performances for the first 5 days of the experiment.

Notice that with the exception of pairs 1–2 and 9–10 the critical ratios indicate that there is a marked and statistically significant decrease in the number of shocks not resulting in an alternation. The shock seems to have been quite effectively reduced.

The nature of the alternations is also important. Let us call one of these rats A and the other B. Now if rat A is feeding at the food-crock he may return to the platform and then again return immediately to the food-crock. We shall call this exchange in position an "individual" alternation since it is accomplished only by one rat. When rat B exhibits this behavior we shall also call this

TABLE 1

Pair	Mean Shifts for for Last 5 Days	Per Cent of Total Shifts Without Shock	C.R.
1– 2	18	14	−1.33
3– 4	15	68	4.64
5– 6	92	89	8.92
7– 8	14	92	6.04
9–10	25	89	1.29
11–12	95	93	8.89

TABLE 2

	Pairs					
	1–2	3–4	5–6	7–8	9–10	11–12
Total number of alternations	947	863	3748	957	1201	3271
Percentage which the mutual alternations are of the total	97	94	97	94	95	99

exchange in position an individual alternation. When rat A is at the food-crock and returns to the platform and rat B comes off the platform and goes to the food-crock we refer to this kind of a shift as a "mutual" exchange in position. The question is, then, what percentage of the total exchanges in position is mutual and what percentage is individual? These data are presented in Table 2.

These data support the conclusion that for the entire group the rats alternate in a *mutual* manner in at least 94 per cent of the total alternations. We can say then that they are "taking turns" 94 to 99 per cent of the time.

Our next question is, how well do these rats get fed while they are eliminating shock and exchanging positions in the experimental situation? Table 3 gives the mean weights for the first 5 days and the last 5 days of the experiment.

It is clear from this table that every rat gained weight during the experiment and these gains ranged from 24 to 140 grams. This, along with the fact of their general healthy and vigorous appearance, further support the conclusion that rats were adequately fed throughout the experiment.

Another factor of importance is the extent to which the animals use the total available time in the apparatus. Means of this data for the entire

experiment are given in Table 4. Since they were run for 12 2-min. trials a day the total available feeding time for each pair of rats is 1,440 sec. a day. Occasionally both rats would go to the platform and remain on it together. This is considered time wasted in as much as it is time during which food was available for one or the other animal but was taken by neither. We can see that the rats used practically all of the available feeding time. This speaks well for our final choice of time interval and also indicates that the rats were actively working on the problem set by the experimental situation practically all of the time that they were in the apparatus.

DISCUSSION AND INTERPRETATION

First of all let us re-emphasize the fact that this was a double motive situation. Our original intention was to arrange these motives in an experimental situation in such a manner that neither of them could be satisfied without the co-ordinated efforts of *both* animals.

Rather than put the organism into an experimental situation and observe if it exhibits "co-operative behavior" we attempt to put the animal through a procedure which will train it to be cooperative. Next we put it into a situation

TABLE 3

Rat	Grams Eaten per Day	Weight Before the Experiment	Weight After the Experiment	Weight Gained
1	21	96	153	57
2	21	113	253	140
3	25	109	144	35
4	20	121	179	58
5	25	118	205	87
6	27	126	189	24
7	27	127	170	43
8	23	140	211	71
9	30	121	160	39
10	21	158	230	72
11	32	182	250	68
12	31	161	231	70

TABLE 4

Pair	Mean Feeding Time	Mean Time Together on the Platform
		sec.
1–2	1425	15
3–4	1410	30
5–6	1358	82
7–8	1407	33
9–10	1407	33
11–12	1397	43

which is a cooperative one and quantify the extent to which it exhibits the behavior. In short, we train the behavior into rather than draw it out of, the animal.

At the end of the preliminary training the rats have mastered a discrimination basic to the satisfactory solution of the problem presented in the test situation. First they shocked each other a great deal as they exchanged positions from food-crock to platform. They received many shocks and they did not get as much food as they did later. But very soon they shifted more frequently and received more food. They shocked each other less and less in between the shifts in position. They accomplished a great and greater number of shifts without getting any shock. In doing this they fed more and escaped more shock. They satisfied both of the motives in the situation.

There are, finally, several observations that throw light on the nature of the behavior of the animals. As the experiment progressed the rats directed their behavior more and more towards each other rather than towards the food-crock or the platform. The rat on the platform would reach off holding the platform down with only one foot, and nudge the feeding animal. It would sometimes crawl up on the latter's back and paw it. This frequently resulted in the feeding rat's return to the platform. Sometimes it would hoist the feeding animal up on its shoulders. It might even bite and pull on the feeding rat's tail. These are overt responses directed towards the other animal.

It would seem, then, that the platform rat's leaving the platform is conditional and dependent upon the movements of the other animal. The rat keeps at least one foot on the platform until the other animal is on the platform, or until the feeding rat goes by the platform rat towards the platform, or at any rate until the feeding animal has completed its return. In short, the rat has apparently learned that it or the other rat must

be on the platform if it is to escape shock. The feeding rat, which has returned to the platform, remains there as the platform rat leaves. Then it exhibits the behavior typical of the platform rat. They "take turns" and thus they both get adequately fed and they both eliminate shock.

In all the other animal experiments on co-operation both animals do the same thing; i.e., they pull ropes, punch stimulus cards, operate levers, etc. In this experiment the cooperative aspect of the situation rests on the animals' doing distinctly *different things*; i.e., one feeds and the other turns off the shock. They synchronize their activity on *two* different tasks. The products of this solution are mutually shared. There is no simultaneous sharing of the goal achieved and thus there is no chance for competition. There is also an element of inhibition in the shape of delaying of responses since one animal waits on the platform while the other eats and they take turns doing this.

Since the animals do respond in the experimental situation in a manner consistent with the preliminary training, since they do distinctly different things in the cooperation testing situation, and since some of their behavior has the characteristics of synchronization, restraint and differentiation of response, it would seem that their behavior may fairly be called cooperative.

SUMMARY AND CONCLUSIONS

To investigate the development of cooperative behavior in rats, six pairs of rats were put into a double motive problem situation (feeding and avoiding shock) requiring the co-ordinated efforts of both animals for its adequate solution.

Each rat was individually trained to feed when the grid floor was not electrified, and when it was charged to go to a platform which shorted out the grid floor when a rat stepped on it. The rats were then paired, and the problem was to discover if cooperative behavior would be obtained when two rats were put into a double motive situation in which the satisfaction of both of these motives is contingent upon the behavior of both animals. One rat of a pair had to run to a platform which shorted out the electrified floor grid of a feeding box in order that a second rat might feed.

From the data obtained in this situation we might draw the following conclusions:

1. The rats learned to exchange positions in this situation and at the same time allow sufficient feeding time for each rat to become adequately fed in the course of the experimental session.

2. They showed marked improvement in alternating without shock and in eliminating the shocks which did not result in an alternation.

3. They learned to take turns at the food-crock and platform so that by the end of the experiment they spent almost all of the available time in the apparatus working on the problem and very little time together on the platform.

4. And finally, in this situation, cooperative behavior has been apparently established. In a food-shock situation both animals exchange positions so that both are adequately fed. Furthermore, they exchange positions with sufficient care and speed that they avoid shock. They satisfy both conditions of the experiment in a situation in which the satisfaction of both conditions was contingent upon the behavior of both animals.

REFERENCES

1. BERNE, E. VAN C. An experimental investigation of social behavior patterns in young children. Univ. Ia. Stud. Child Welf., 1930, 4, 61 pp.
2. CRAWFORD, M. P. and NISSEN, H. W. Gestures used by chimpanzees in cooperative problem solving. (Silent film.) New York: Instructional Films, Inc., 30 Rockefeller Plaza, 1937.
3. CRAWFORD, M. P. Cooperative behavior in chimpanzee. Psychol. Bull., 1935, 32, 714.
4. CRAWFORD, M. P. Cooperative solution by chimpanzees of a problem requiring serial responses to color cues. Psychol. Bull., 1938, 35, 705.
5. CRAWFORD, M. P. Further study of cooperative behavior in chimpanzee. Psychol. Bull., 1936, 33, 809.
6. LEWIN, K. and LIPPITT, R. An experimental approach to the study of autocracy and democracy: a preliminary note. Sociometry, 1938, 1, 292–300.
7. MOORE, E. S. The development of mental health in a group of young children: An analysis of factors in purposeful activity. Univ. Ia. Stud. Child Welf., 1931, 4, 128 pp.
8. WINSLOW, C. N. A study of experimentally induced competitive behavior in the white rat. Comp. Psychol. Monogr., 1940, 15, 35 pp.
9. WOLFLE, D. L. and WOLFLE, H. M. The development of cooperative behavior in monkeys and young children. J. Genet. Psychol., 1939, 55, 137–175.

COMMUNICATION OF AFFECT IN "COOPERATIVE CONDITIONING" OF RHESUS MONKEYS*

Robert E. Miller, James H. Banks, Jr., and Nobuya Ogawa

AN approach to the investigation of variables in the communication of affects was made feasible by the development of techniques for conditioning monkeys to react to each other in response to experimentally induced affective states (Murphy, Miller and Mirsky, 1955). It was found that these affective experiences had profound effects on subsequent social relationships between the animals (Miller, Murphy and Mirsky, 1955; Murphy and Miller, 1956). The behavior of these animals implied that very subtle emotional expressions were detected and interpreted in the constant and complex interplay in the experimental social situation. In an attempt to measure this communication of affect, a new method was developed (Mirsky, Miller and Murphy, 1958) and some of the relevant expressive cues were identified (Miller, Murphy and Mirsky, 1959a, 1959b).

The present report deals with a new and more direct approach to the study of the phenomena of communication of affects. While the former

* From the Journal of Abnormal and Social Psychology, 1962, 64, 343–348. Copyright 1962 by the American Psychological Association. This investigation was supported by research grants from the National Institute of Mental Health of the National Institutes of Health, United States Public Health Service, the Foundations' Fund for Research in Psychiatry, and the Commonwealth of Pennsylvania.

method measured the effect of such communication on an existing conditioned response, it did not require the animals to communicate and, therefore, was subject to extinction upon repeated testing. The present method, however, requires successful communication between the animals for mutual solution of an avoidance problem.

METHOD

SUBJECTS. Three postadolescent male rhesus monkeys were used in this experiment. Each of them had been subjects in a previous investigation involving discriminated avoidance to visual form stimuli.

APPARATUS. Standard equipment consisting of primate chairs, response bars, relay and timing panels, programers, and cumulative recorders were used in these experiments. The only modification which was made was the insertion of a milk glass panel in front of the lights in a visual stimulus panel. This was done to reduce the intensity and diffuse the lights so that glare or reflections from the plastic and/or metal fittings of the primate chairs would not occur. The final result was effective in that the experimenter could not detect any reflection from in front of the chair using a violet stimulus light which was readily apparent from the monkey's position.

The animals were housed and tested in an air-conditioned isolation room. Two one-way vision screens were mounted in the walls enabling the experimenters to make observations of the animals during a test session without distracting them. The programing and recording equipment was located in another room some distance away so that relay clicks, etc. could not serve as conditioning cues. A clicker was presented via a loud-speaker in the isolation room throughout the daily test sessions to mask any extraneous noises from the corridors or other test rooms.

PRELIMINARY TRAINING. The animals were placed two at a time into individual primate chairs where they remained for the duration of the experiment. They were then given a 5-day period to become accustomed to this rather restrictive environment. On the sixth day avoidance conditioning was begun for each of the animals. A response lever was attached to the primate chair and a visual stimulus panel was located above and slightly to one side of the monkey's head. A flexible metal bracelet through which a shock stimulus could be delivered was placed upon the animal's right ankle. The other contact for shock was attached to the metal chair upon which the

monkey was sitting. The two chairs were positioned so that the monkeys were back to back and a wooden screen was placed between the chairs so that the animals could not see each other.

The stimulus panel and lever were attached to the chair of one of the animals. A 15-min. period then elapsed to permit the monkey to adapt to these changes and to allow the experimenter to obtain a measure of the operant level for bar pressing. The masking clicker was then turned on for the duration of the experimental session. Two minutes later a tape programer was started which automatically presented the 20 daily trials. Initially a tape was programed for a 30-min. session with trials randomly sequenced at 60, 90, or 120 sec. with an average intertrial interval of 90 sec. However, during the conditioning of the first two subjects another tape was introduced spacing trials at an average of 2.25 min. between the 20 trials for a 45-min. test session. This program, which was retained for the remainder of the experiments, presented trials at randomly sequenced intervals of 1.0, 1.5, 2.0, 2.5, and 3.0 min.

Each trial consisted of the presentation of the violet stimulus light for 6 sec. before the introduction of a shock delivered through the ankle bracelet and chair. A bar press terminated both the shock and the stimulus light. If the instrumental response occurred within 6 sec. of stimulus onset, the light was terminated and the shock was avoided.

When the conditioning session for the first animal was completed, the second monkey was then given a similar conditioning period. The first two animals (Nos. 120 and 132) received 10 such conditioning sessions for a total of 200 trials.

COMMUNICATION OF AFFECT. Following the avoidance training in which each animal had learned to perform a bar pressing response to a violet stimulus light in order to avoid shock, the primate chairs were turned so that the animals faced each other and the wooden screen between them was removed. The ankle bracelets and metal chairs were wired in parallel so that the two monkeys would receive simultaneous shocks.

On the eleventh test day the stimulus panel was attached to Monkey 120's chair but no response bar was provided him. No. 132 was equipped with the response lever but did not have the stimulus panel. Test trials were presented on the same schedule as during the conditioning sessions and the shock unconditioned stimulus continued to be presented for failure to press the bar within 6 sec. of the stimulus light.

The task confronting the animals is obvious.

The monkey with the light had no means of performing the instrumental response which would avoid the noxious shock stimulus. The second monkey could perform the response but had no stimulus to inform him when a response was appropriate. However, if the animal with the stimulus was able to communicate to his partner by means of expressive cues the information that the conditioned stimulus (CS) was being presented, the second monkey could then make the appropriate instrumental response which would enable both animals to avoid the shock.

Monkey No. 120 was given the stimulus panel and No. 132 the response bar for 22 consecutive test sessions. Following this, 4 days of individual reconditioning trials were given and, then, the roles of the two animals were reversed, No. 120 being given the bar and 132 the stimulus light, for ten more testing sessions. On several sessions a wooden screen was placed between the animals which masked off all but facial cues from the monkey having the CS.

After this pair of animals had been tested in both directions, i.e., first 120 with the stimulus and then 132 with the stimulus, Monkey 132 was removed from his chair and replaced by No. 129. This animal had also been avoidance conditioned to visual stimuli in a primate chair several months prior to this experiment. However, he had not been exposed to the CS of a violet light. No preliminary training was administered to Monkey 129. After a 5-day period of adaptation to the chair, he was placed facing No. 120 in the communication of affect situation. The stimulus panel was mounted on 129's chair and the response lever was attached to 120's chair. Twenty trials per day were administered for 10 days. Then the wooden screen masking all but 129's head was introduced for two test sessions and, finally, a wooden barrier completely obstructing the animals' view of each other was placed between the chairs for two additional test sessions.

The positions of response lever and stimulus panel were again switched: 129 now having the response bar while 120 had the stimulus light. Ten more test sessions were administered to this combination with the partial screen on days 6, 7, and 8 and the complete screen on days 9 and 10.

RESULTS

The initial adaptation to the confinement of the primate chairs and the establishment of a conditioned avoidance response in the first pair of animals were accomplished without incident.

these animals had prior experience in a similar situation, learning was rapid and uneventful. Both animals performed avoidance responses on better than 95 per cent of the trials on the last three conditioning days.

The communication of affect data had to be analyzed in terms of two response parameters. The first of these, of course, was the incidence of avoidances, i.e., trials on which a bar press occurred within 6 sec. of presentation of the CS to the stimulus animal. The second was the numbers of "spontaneous" or operant responses which occurred in the intervals between trials, i.e., in the absence of the conditioned stimulus. If the level of spontaneous responses was high, the probabilities of obtaining fortuitous or spurious avoidance responses would also be high. Therefore, in order to assess the level of authentic avoidances, the number of responses occurring in the absence of the CS was divided by 430. This divisor was derived by dividing the total intertrial time (nonstimulus periods) into 6-sec. intervals. This calculation gave the number of spontaneous responses (SRs) expected per 6-sec. interval assuming a rectilinear distribution of such responses. The data indicated that such an assumption was, in general, justified although there was some tendency for SRs to diminish in frequency during the session and for SRs to occur with somewhat greater frequency immediately after a trial as a part of a chain of responses triggered by the CS. These departures from rectilinearity tended to bias the data against the hypothesis of communication of affect. There was never any suggestion of a temporal conditioning effect or a pattern of responses resembling fixed ratio or fixed interval operant conditioning.

The number of avoidance responses which would be expected on the basis of chance was derived by multiplying the number of SRs per 6-sec. interval by 20, the number of such intervals during which the CS was actually present. Since the number of spontaneous responses varied from day to day, a separate analysis was made for each testing session.

The data for the first pair of animals (120 and 132) is presented in Table 1. As the footnote indicates, the experimenters discovered after 22 testing days that the monkey with the stimulus (No. 120) had been improperly connected to the shock apparatus so that he had received no shock during the entire communication of affect test series. The p values were determined from the Mainland tables (Mainland, Herrera and Sutcliffe, 1956) with the number of trials (20) as N. These data indicate that the animals performed a discriminated avoidance response in the social communication situation. The number of SRs was initially fairly high but diminished rapidly as the avoidance behavior became more discreet. The animal with the bar learned to attend very carefully to his partner with the CS and responded rapidly with bursts of bar presses when the stimulus monkey reacted to the CS. In this connection the behavior of Monkey 120 during those trials when he did not receive shock while No. 132 did was particularly interesting. From observations made through the one-way vision windows

it was impossible to detect the absence of shock from 120's behavior. He responded as 132 received the shock with a pronounced startle and leg withdrawal. It may be that this communication of affect (in the direction opposite from that being tested) served as reinforcement to maintain 120's affective response to the CS.

The data also indicate that visual cues alone were not the sole basis for solution of the mutual avoidance problem. The placing of either partial or complete screens between the animals did not eliminate successful avoidances. It was apparent

TABLE 1. *Cooperative Avoidance Conditioning: First Pair of Animals*

Test Day	Monkey With Light	Monkey With Bar	Number of Avoidances	Estimated Chance Avoidances	p
1	120	132	3	9.8	ns
2	120	132	17	20	ns
3	120	132	17	14.1	ns
4	120	132	19	17.1	ns
5	120	132	18	18.6	ns
6	120	132	19	3.8	< .01
7	120	132	14	6.2	ns
8	120	132	16	10.8	ns
9	120	132	19	6.6	< .01
10	120	132	19	10.2	< .01
11	120	132	19	5.4	< .01
12	120	132	19	7.2	< .01
13	120	132	18	8.4	< .01
14	120	132	18	3.1	< .01
15	120	132	15	4.3	< .01
16	120	132	15	2.6	< .01[a]
17	120	132	16	0.94	< .01[a]
18	120	132	15	0.74	< .01[a]
19	120	132	20	14.6	ns[a, c]
20	120	132	18	15.9	ns[a]
21	120	132	15	1.30	< .01[a, d]
22	120	132	16	1.07	< .01[a, e]
23	132	120	19	2.32	< .01
24	132	120	19	1.63	< .01
25	132	120	18	0.74	< .01
26	132	120	19	0.51	< .01
27	132	120	18	0.56	< .01[a]
28	132	120	20	0.18	< .01[a]
29	132	120	18	0.60	< .01[a]
30	132	120	20	0.56	< .01[a]
31	132	120	20	0.00	< .01[b]
32	132	120	19	2.14	< .01[b]

[a] Half screen between animals.
[b] Full screen between animals.
[c] Experimenter accidentally reduced bar tension to the point that chair shaking, etc. gave multiple bar presses.
[d] Bar tension increased.
[e] Discovery that misconnection of shock lead head prevented shock to Number 120 for previous 22 days.

TABLE 2. *Cooperative Avoidance Conditioning: Second Pair of Animals*

Test Day	Monkey With Light	Monkey With Bar	Number of Avoidances	Estimated Chance Avoidances	p
1	129	120	15	20.0	*ns*
2	129	120	20	16.5	*ns*
3	129	120	17	20.0	*ns*
4	129	120	19	8.7	< .01
5	129	120	20	6.6	< .01
6	129	120	19	6.5	< .01
7	129	120	20	3.8	< .01
8	129	120	19	2.2	< .01
9	129	120	19	5.8	< .01
10	129	120	20	4.7	< .01
11	129	120	19	9.6	< .01[a]
12	129	120	19	9.3	< .01[a]
13	129	120	20	1.9	< .01[b]
14	129	120	19	0.7	< .01[b]
15	120	129	7	1.4	*ns*
16	120	129	20	7.9	< .01
17	120	129	20	2.0	< .01
18	120	129	18	0.24	< .01
19	120	129	20	1.3	< .01
20	120	129	20	7.4	< .01[a]
21	120	129	20	3.8	< .01[a]
22	120	129	20	0.8	< .01[a]
23	120	129	18	11.8	< .01[b]
24	120	129	20	8.6	< .01[b]

[a] Half screen between animals.   [b] Full screen between animals.

from observation that auditory cues were also a factor in the communication process. The animal with the stimulus panel frequently squirmed and slapped at his chair when the stimulus was being presented. It was not possible to determine whether vocalization may also have occurred to the CS. That the results of the experiment were not attributable to spurious clicks, etc. from the apparatus itself was demonstrated by tilting the stimulus panel so that neither animal could see it and presenting a series of stimuli with the shock turned off. On the two occasions when this control procedure was run not a single bar press occurred while the CS was present.

The cooperative avoidance data from the second pair of animals, Nos. 120 and 129, are presented in Table 2. It may be recalled that Monkey 129 had no preliminary avoidance training to the stimulus employed in this investigation. As indicated in Table 2, 120 responded to the new situation by making large numbers of intertrial lever responses for the first three test days. The number of SRs was high enough so that merely by chance alone avoidance would be

expected to occur on the majority of test trials. As Table 2 indicates the two animals actually received only eight shocks during these first three days. Nevertheless, this relatively few pairings of the CS and UCS was sufficient to condition 129 and from Day 4 the discrimination between CS presentations and intertrial intervals became significant.

When this pair of animals was switched so that 129 had the response mechanism while 120 had the stimulus light, problem solution was very rapid. Monkey 129 which had never had the bar in this test situation attended very closely to the behaviors of No. 120 after the first test day and responded only rarely between trials (see Fig. 1).

The insertion of screens between the animals in the second pairing seemed to have a more disrupting effect on the discrimination than in the initial pairing. This suggests that visual cues may have played a more important role in the communication of affect in the second pair of animals than in the first.

The data for all of the combinations of animals which were tested for mutual avoidance

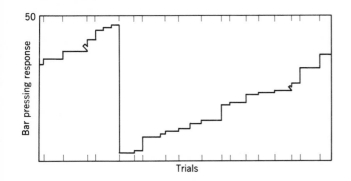

FIGURE 1. Mutual avoidance conditioning. (Monkey 120 with stimulus light and 129 with bar. Trials 1 through 18 of test day 18 are shown. Diagonal jogs on trials 3 and 16 indicate that avoidance was not successful and shock was delivered to both animals. Vertical pips on the abscissa indicate trials delivered at variable intervals by a tape programmer, from trial 1 on the extreme left to trial 18 on the extreme right. The 15-min. pretrial control period is not shown.)

behavior revealed that when a discrimination had been established, the avoidance responses occurred as a short burst of bar presses. The first such response was, by definition, the avoidance since only a single response was required to terminate the CS and preclude the UCS. However, the number of responses occurring within the 6-sec. interval beginning with stimulus presentation was much greater than chance expectation. On several occasions the mean number of bar presses occurring within stimulus intervals was over 100 times greater than the frequency of such responses expected purely by chance.

DISCUSSION

The phenomenon of intraspecies communication has been prominently mentioned in the experimental literature on the infrahuman primates. Carpenter (1952) has stated, "Each known genus of primates has a repertoire of gestures which are employed consistently and which stimulate consistent reactions." The studies of cooperative problem solving (Crawford, 1941) and food sharing (Nissen and Crawford, 1936) in the chimpanzee indicate the importance and effectiveness of communication in that species. It was found that the chimpanzee could "solicit" food from a partner or direct its behavior in a cooperative serial problem solving task.

While the communication between primates has been recognized and described, there have been no systematic attempts to investigate the phenomenon experimentally. The present experiment represents an attempt to develop a methodology through which the communication processes may be more intensively studied. The data suggest that the "cooperative conditioning" technique is an exceptionally efficient and sensitive tool for the investigation of nonverbal

communication. The method would appear to lend itself to the isolation and identification of the discrete expressive cues which convey affects between animals.

Some question arose before these animals were tested as to how the monkey with the response mechanism might be expected to behave in the social situation. While we hoped that a communication of affect would be utilized in the solution of the problem, there was a reasonable possibility that a second but less efficient solution would be learned. From the responding animals' viewpoint, an adequate avoidance could be achieved by setting up a rate of bar pressing which would insure that no more than 6 sec. elapsed between successive responses. In other words, the problem might have been structured by the subject in the same fashion that monkeys and other species learn to avoid in the Sidman type of problem. There is no doubt that monkeys can learn to maintain a rather steady, high rate of responding in such a problem situation (Sidman, 1958). However, steady state responding at a high rate is a very costly adaptive mechanism from the subject's point of view if it must be maintained for any extended period of time. Porter, Brady, Conrad, Mason, Galambos, and Rioch (1958) demonstrated that severe, indeed, terminal gastrointestinal lesions develop in a remarkably short period of time in monkeys required to lever press every 20 sec. for a 6-hour test period. The Porter et al. study, incidentally, resembles in some ways the present experiment in that two animals were hooked together in a "yoked chair" setup. The experimental animal could prevent shock not only to himself but to his paired control by pressing the lever. The control animal, however, had no direct role in the avoidance behavior in that experiment while both animals in the present study played indispensable roles in problem solution.

It might also be pointed out that the present

method can be viewed from several different viewpoints. While we were oriented primarily in terms of nonverbal communication of affect, it seems quite appropriate to consider these experiments as studies of cooperative behavior. By making some changes in initial training it would appear to be feasible to attempt to isolate some parameters of cooperation-competition in this experimental situation. Likewise, since the two animals are mutually dependent for a satisfactory solution to their joint problem, it might be possible to structure the experiment in such a way as to study the development of dependency relationships.

At the present time more intensive studies of the nonverbal communication of affect are being conducted. Improved control of the sensory modalities involved in the communication has been achieved so that the two animals may be isolated from each other, thus eliminating extraneous noises from twisting or shaking in the chairs plus vocal and olfactory cues. Further, joint reward conditioning, and combined reward-avoidance schedules are being tested to determine the cues utilized in the communication of affects other than fear.

SUMMARY

Rhesus monkeys in primate chairs were conditioned to bar press within 6 sec. of presentation of a light in order to avoid electric shock. Following acquisition of this avoidance response two animals were placed facing each other and the bar was removed from the chair of one monkey and the stimulus light from the chair of the other. In order for either monkey to avoid shock a communication was necessary since neither animal had access to all elements of the problem. The results indicated that through nonverbal communication of affect an efficient mutual avoidance was performed. It was concluded that this paradigm is an exceptionally efficient and sensitive method for investigations of nonverbal communication.

REFERENCES

CARPENTER, C. R. Social behavior of non-human primates. Structure et physiologie des sociétés animals. *Colloq. int. Cent. nat. Rech. scient.*, 1952, **34**, 227–246.

CRAWFORD, M. P. The cooperative solving by chimpanzees of problems requiring serial responses to color cues. *J. soc. Psychol.*, 1941, **13**, 259–280.

MAINLAND, D., HERRERA, T. and SUTCLIFFE, M. *Statistical tables for use with binomial samples.* New York: New York Univer. Press, 1956.

MILLER, R. E., MURPHY, J. V. and MIRSKY, I. A. The modification of social dominance in a group of monkeys by interanimal conditioning. *J. comp. physiol. Psychol.*, 1955, **48**, 392–396.

MILLER, R. E., MURPHY, J. V. and MIRSKY, I. A. Non-verbal communication of affect. *J. clin. Psychol.*, 1959, **15**, 155–158 (a).

MILLER, R. E., MURPHY, J. V. and MIRSKY, I. A. The relevance of facial expression and posture as cues in the communication of affect between monkeys. *Arch. gen. Psychiat.*, 1959, **1**, 480–488 (b).

MIRSKY, I. A., MILLER, R. E. and MURPHY, J. V. The communication of affect in rhesus monkeys: I. An experimental method. *J. Amer. Psychoanal. Ass.*, 1958, **6**, 433–441.

MURPHY, J. V., MILLER, R. E. and MIRSKY, I. A. Inter-animal conditioning in the monkey. *J. comp. physiol. Psychol.*, 1955, **48**, 211–214.

MURPHY, J. V. and MILLER, R. E. The manipulation of dominance in monkeys with conditioned fear. *J. abnorm. soc. Psychol.*, 1956, **53**, 244–248.

NISSEN, H. W. and CRAWFORD, M. P. A preliminary study of food-sharing behavior in young chimpanzees. *J. comp. Psychol.*, 1936, **22**, 383–419.

PORTER, R. W., BRADY, J. V., CONRAD, D., MASON, J. W., GALAMBOS, R. and RIOCH, D. Some experimental observations on gastrointestinal lesions in behaviorally conditioned monkeys. *Psychosom. Med.*, 1958, **20**, 379–394.

SIDMAN, M. By-products of aversive control. *J. exp. Anal. Behav.*, 1958, **1**, 265–280.

"ALTRUISM" IN THE ALBINO RAT*

George E. Rice and Priscilla Gainer

COOPERATION has to date been demonstrated repeatedly in experiments with animals. Tsai (1950) and Daniel (1942, 1943) found this in rats, Crawford (1936, 1938) and Nissen and Crawford (1936) in chimpanzees, and Winslow (1944) in cats. It seems fairly clear that cooperation can exist at the infrahuman level, and, if so, it is conceivable (albeit a long jump) that behavior homologous to altruism in humans might be exhibited in animals. Webster (1941) defines altruism as "regard for and devotion to the interests of others." The authors felt that altruism could be operationally defined as "behavior of one animal that relieves another animal's 'distress'." Further, if an animal exhibited visual and auditory signs of discomfort such as squealing and convulsive wriggling, the animal was assumed to be "distressed."

It was felt that this concept could be tested using albino rats as Ss if one "distressed" rat was suspended by a harness so that it hung free of the floor in full view and hearing of a second rat. The second or operating rat would be in a compartment with an accessible bar that, if depressed, would automatically lower the "distressed" or hanging rat to the floor, thus presumably relieving the "distress."

METHOD

SUBJECTS. The Ss were 40 experimentally naive albino rats of the Wistar strain, half male and half female. All were about 12 weeks old at the onset of training.

APPARATUS. The apparatus was a two-compartment plywood and Lucite box 21 in. long, 7 in. wide, and 8 in. high. The larger of the two compartments was 13 in. long, and one of the 13-in. sides was made of Lucite to facilitate observation of the rat. This compartment was equipped with a 3-in.-long depressable bar located below a signal light. Both were at the end near another Lucite panel separating the two compartments. This compartment was covered to prevent the occupant's escape. The smaller compartment was bare except for a hoist mechanism which lowered the "distressed" rat to the floor and subsequently returned this animal to a suspended position. A grid floor wired for shock was beneath both compartments although shock was administered only to the experimental animal in the larger compartment. A geared Erector set motor powered the hoist, and raising or lowering of the hoist was accomplished by shifting gears. An aluminum tray for wood shavings was inserted below both compartments.

PROCEDURE. From the age of 2 to 12 weeks all Ss were handled and petted 10 min. daily. Following this a Styrofoam block 2 in. by 2 in. by 5 in. was suspended from the hoist while 1 of the 10 male and 10 female Ss to be trained was placed in the larger compartment. Ten seconds after placement in the test chamber the signal light was turned on, followed 5 sec. later by a mild electric shock to the S which continued until the bar was pressed. Pressing the bar before the 5-sec. time limit prevented the onset of shock. Upon bar depression, the signal light went off and the hoist mechanism lowered the block to the floor, where it remained for 15 sec. At the end of that time, the block was again raised, and 10 sec. after the block reached its zenith, the signal light came on, thus starting the entire training cycle again. Each experimental S remained in the training situation for 10 min. at a time and for five separate training periods approximately equally spaced over a 3-day period. All Ss that had not reached the minimum criterion of either pressing the bar before the onset of shock or within 3 sec. following the administration of shock were discarded, and others were trained to bring the total of trained Ss to 20. All trained animals were then subjected to CR extinction procedures by being placed in the same situation with the exception of shock. This training continued until the

* From the *Journal of Comparative and Physiological Psychology*, 1962, **55**, 123–125. Copyright 1962 by the American Psychological Association.

bar-pressing response disappeared, which usually occurred within another 3-day period.

The trained rats were then randomly distributed between Experimental Group 1 and Control Group 1 with five males and five females in each group. The Ss from Experimental Group 1 were placed individually in the test chamber with a "distressed" rat suspended from the hoist by means of a "corset" sewn from a 2-in. elastic band which allowed the legs to hang free through apertures in the harness. This animal typically squealed and wriggled satisfactorily while suspended, and if it did not, it was prodded with a sharp pencil until it exhibited signs of discomfort. When the hoist lowered this rat to the floor, it was able to stand on its own four feet and signs of discomfort ceased. Each experimental S remained in the chamber for five 10-min. trials distributed equally over a 3-day period. In this and in all subsequent conditions, the total number of bar presses and the general behavior of the rat being tested were noted.

The Ss from trained Control Group 1 were subjected to the same conditions except that the Styrofoam block was on the hoist in place of the "distressed" rat.

Experimental Group 2 consisted of ten rats with no prior training. Each S was placed in the experimental situation identical to that of Experimental Group 1 with a "distressed" rat in the suspension chamber.

Control Group 2, ten untrained rats, was tested with the suspended Styrofoam block as was Control Group 1.

RESULTS

Experimental Group 1 pressed the bar a mean of 14.7 times per rat, Control Group 1 pressed the bar a mean of 0.8 times, Experimental Group 2 achieved a mean of 17.6 bar presses, and Control Group 2 pressed the bar 5.4 times per S.

It is to be noted that in every case the difference between the bar presses of the experimental Ss confronted with a "distressed" rat and the control Ss faced only with a suspended block was beyond the 0.01 level of chance expectancy.

The critical ratio between the two experimental groups, however, was not a significant or noteworthy difference.

The Ss of the control groups typically either did not press the bar at all or pressed the bar with any frequency only during the early trials and subsequently not at all, while it was typical of animals in the experimental groups to increase in bar-pressing rate throughout the trials. Another noteworthy result is that the untrained Ss of Experimental Group 2 actually pressed the bar more frequently and in addition exhibited more signs of interest in the suspended rat than did the conditioned rats of Experimental Group 1. This result was evidenced by the fact that the Ss of Experimental Group 2 remained at the end near the suspended rat most of the time and usually remained oriented toward that rat.

DISCUSSION

The outstanding fact of these findings was that the rats confronted with a rat exhibiting auditory and visual signs of distress acted in a manner very different from those closeted with a block. The behavior of the former resulted in what might be termed relief of the distress signs, and it is suggested that this behavior might be homologous to altruism.

It does not seem that the increased bar pressing in the experimental groups could be due to conditioning since this response was extinguished in the trained animals and the Ss that had not been conditioned exhibited the highest number of bar presses; nor does it seem that the difference was due to the S's curiosity about the bar because of the significant difference between the experimental and the control groups. If curiosity had been the prime motive for bar pressing per se, the Ss lowering a block might have been expected to press the bar as often as those lowering a rat. However, it is worth noting that, not only did the groups differ in their bar pressing, but that often when the suspended animal had been quiet and then squeaked, the experimental rat promptly pressed the bar. Also, often the experimental S would approach the Lucite barrier between chambers and would then press the bar.

One possible difficulty with the interpretation of this behavior as altruistic is that the experimental rat may not have been aware of any "distress" although it is hard to see what benefit the experimental animal received from its response. Of course, it is possible that distress calls may generate distress in other members of the species. It also may be that the presence of another rat was sufficient cause for increased activity and subsequent bar pressing. A follow-up control group of untrained animals placed in the apparatus with a nonsuspended rat in the distress chamber is planned for a future study.

SUMMARY

Forty albino rats were placed individually in a compartment equipped with a bar which, when pressed, lowered either a "distressed" rat or a plastic block to the floor of an adjoining compartment, thus apparently relieving the suspended animal's "distress." Half the Ss learned to press the bar by avoidance conditioning, followed by extinction procedures. Half of these were confronted with the suspended rat and half with the block, and the former pressed the bar significantly more often than the latter. Another 20 untrained animals, similarly divided, performed according to the same pattern. It is suggested that this behavior might operationally be termed altruistic.

REFERENCES

CRAWFORD, M. P. Further study of cooperative behavior in chimpanzees. *Psychol. Bull.*, 1936, **33**, 809.

CRAWFORD, M. P. Cooperative solution by chimpanzees of a problem requiring serial responses to color cues. *Psychol. Bull.*, 1938, **35**, 705.

DANIEL, W. J. Cooperative problem solving in rats. *J. comp. Psychol.*, 1942, **34**, 361–368.

DANIEL, W. J. Higher order cooperative problem solving in rats. *J. comp. Psychol.*, 1943, **35**, 297–305.

NISSEN, H. W. and CRAWFORD, M. P. A preliminary study of food-sharing behavior in young chimpanzees. *J. comp. Psychol.*, 1936, **22**, 383–419.

TSAI, L. S. Rivalry and cooperation in white rats. *Amer. Psychologist*, 1950, **5**, 262. (Abstract.)

Webster's Collegiate Dictionary. (5th ed., Abridged) Springfield, Mass.: Merriam, 1941, p. 32.

WINSLOW, C. N. The social behavior of cats: I. Competitive and aggressive behavior in an experimental runway situation. *J. comp. Psychol.*, 1944, **37**, 297–313.

ALTRUISM OR AROUSAL IN THE RAT?*

J. J. Lavery and P. J. Foley

RICE and Gainer (1) have described behavior in the albino rat which, they suggest, might be homologous to altruism. Their rats would press a bar to lower a suspended animal showing obvious signs of distress, such as "squealing and convulsive wriggling." Since the operating rats in this situation pressed the bar significantly more often than control animals, their behavior was interpreted as being altruistic.

This conclusion does not follow unless it can be demonstrated that the bar is pressed specifically to relieve the distress of the other animal. Indeed, there could be various reasons for the behavior described by Rice and Gainer. The extra stimulation in the experimental, as opposed to the control situation, may raise the general activity level of the organism and hence increase the number of bar presses. This could be considered an arousal phenomenon. For the purposes of the present study, the question is simply: is there some specific component in the squeal of a distressed rat which triggers, in a listening rat, behavior calculated to relieve the distress, or does the increased stimulation result in increased activity?

To resolve this problem an experiment was carried out with ten male albino rats of the Wistar strain, all experimentally naive. From the age of 5 weeks they were handled and weighed daily. At 6 weeks they were introduced to an experimental box constructed according to the specifications of Rice and Gainer (1), in two groups of five, chosen at random. Each group remained in the box for 10 min. For the next 7 days, each rat spent 10 min. alone in the experimental box.

Two weeks later each rat of Group I was placed in the box, individually, and immediately the white noise was switched on by the experimenter. Whenever the subject touched the bar, the noise was turned off for 15 sec. If the subject touched the bar again during the 15-sec. quiet period, an

* From *Science*, 1963, **140**, 172–173. Copyright 1963 by the American Association for the Advancement of Science.

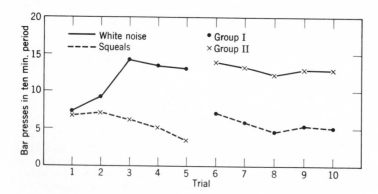

FIGURE 1. Average number of bar presses in a 10-min. period for two groups of rats, one group being exposed to white noise on five successive days followed by exposure to "distress" squeals on another five successive days, the other group being exposed to "distress" squeals, followed by white noise.

additional 15-sec. period of silence was allowed before the noise was switched on. Group II was subjected to exactly the same treatment, with the exception that it was exposed to the recorded squeals. Each rat remained in the box for 10 min., and the number of times it touched the bar was recorded.

Both groups were tested under the conditions described above for 5 consecutive days. After 2 days of rest, they were tested for another 5 days, with the conditions interchanged, that is, Group I was exposed to squeals and Group II to white noise.

To obtain "squeals," two other rats were placed in the box and subjected to electric shock. The squeals were recorded on Ampex Instrumentation tape, with an Ampex tape recorder, type 311-2. A loop was made from this recording and analyzed with the Bruel and Kjaer noise-measuring system, consisting of a condenser microphone cartridge type 4131, a cathode follower (type 2613), an audio spectrum analyzer (type 2109), and a level recorder, type 2304. The tape loop could be played back, through a power amplifier, and an Altec 604 speaker. A white-noise generator, Grason Stradler No. 455.3, could be switched into the system instead of the tape recorder, the output being 80 dB relative to 0.002 dyne/cm². The speaker enclosure was mounted in such a way that the experimental box could be placed directly under the speaker.

The results of the experiment are shown in Fig. 1. The average number of times the bar was touched for white noise is roughly the same for both groups after the first two trials, and is consistently higher than that for squeals, which is also roughly the same for both groups. Group I, exposed initially to white noise, takes three trials to reach asymptote. The analysis of variance (Table 1) shows a significant difference between the "noise" and "squeal" treatments.

TABLE 1. *Analysis of Variance on Number of Bar Presses Under the Two Treatments With Trials Pooled*

Source	df	MS	F
Between subjects	9	315.1	3.63
Between sessions	1	57.8	
Between treatments	1	5313.8	61.3*
Remainder	8	86.7	

* $p < 0.001$.

The results show clearly that there is no specific component in the squeal of a distressed rat which evokes what might be called altruistic behavior, when this behavior is defined as pressing a bar to stop the squeal. On the other hand, when the sound of white noise can be stopped by pressing a bar, rats learn to do this very quickly, and maintain a comparatively high level of responding. Therefore, the squeals and the white noise must be regarded simply as two sources of auditory stimulation, the latter giving rise to more behavioral activity than the former.

Further, the fact that the group exposed to white noise on the first five trials reaches asymptote only on the third trial suggests that this increased activity is directed rather than undirected. This might indicate that the stimuli are noxious in differing degrees. However, comparison between data from the present study and that of Rice and Gainer shows that the number of bar presses reported by the latter authors lies between those of the recorded squeals and the white noise. This adds support to the activation explanation, since increased arousal could be expected in the Rice and Gainer situation where squeals were provided by a wriggling rat (on a hoist), prodded by an experimenter, in a compartment very close to the bar. On the other hand, one would not expect the squeals to be more noxious simply because the experimenter and a live rat are present. Furthermore, in the Rice and Gainer situation, bar pressing did not suppress the presence of either the rat or the experimenter. The comparison is also more difficult to explain if the bar-pressing behavior is interpreted as altruistic, since, in that case, both situations with squeals—either live or recorded—should yield more frequent bar pressing than the white noise.

REFERENCES AND NOTES

1. RICE, G. E. and GAINER, P. *J. Comp. Physiol. Psychol.*, **55**, 123–125 (1962).

2. Defence Research Medical Laboratories Project No. 242, DRML Report No. 242-6, PCC No. D50-89-01-01, H.R. No. 261.

CHAPTER EIGHT

COMPETITION

LIKE cooperation, competition is a form of behavioral interdependence in which the reinforcement of two interacting individuals is mutually contingent. The difference between cooperation and competition, however, is that in cooperation the response of one individual leads to either positive or negative reinforcement of both coactors, while in competition the reinforcement is at the same time positive for one and negative for the other.

The experimental interest in competitive behavior among animals has focused primarily on the question of how their performance is affected in the competitive situation. In this sense competition is simply an extension of the problem of social facilitation with the difference that here there is apparently some sort of invidious comparison by the competing individuals of their performances, or at least, that the performance of one individual becomes a source of cues for the quality of the performance of the other. In an attempt to see whether rats respond to each other's performance in this manner, Lepley observed rats competing for food. His results are of interest because they seem to contradict our intuitive expectations about the effects of competition on performance and also to contradict most of the research on competition using human subjects, which seems to show that performance improves under competitive conditions. Here we note that, the performance of the winner is unaffected, while the performance of the loser is characterized by, anthropomorphically speaking, "giving up." Scott and McCray show a similar effect for dogs, but to a lesser extent. The experiment by Church, on the other hand, demonstrates that there exist conditions under which the performance of a rat can be enhanced by the competitive situation; and we note from the study by Bayroff that among some of the conditions which enhance performance under competition is the presence of threat. When the mutual contingency of reinforcement applies to avoidance responses, a clear enhancement of performance is obtained. It is of interest that Bayroff failed to find any effects of early socialization on competition. It will be recalled that these effects were quite pronounced in imitative behavior as well as in aggression. The final selection by Warren and Maroney demonstrates the consistency with which individuals win contests over food.

COMPETITIVE BEHAVIOR IN THE ALBINO RAT*

William M. Lepley

INTRODUCTION

THE simple experiment to be reported was performed in an attempt to answer the question: can rats be induced to exhibit behavior analogous to that which is loosely called competitive with reference to human behavior?

The most casual observation of a group of rats at feeding time reveals that the animals forcefully oppose one another in the press about the food receptacle. Harlow (2) has described this behavior as competitive. In this study, Harlow concludes that: "The essential condition for the occurrence of social facilitation is the presence of rats unrestrained and actively competing with each other for food." He demonstrated that the food consumption was greater in these competitive circumstances. Several animal studies have shown this phenomenon of social facilitation and that it is intimately related to the sort of behavior ordinarily called competitive. Harlow's study of monkeys (3) provides a further example. In this case, competition, although probably the most important factor, does not seem to be essential to the appearance of social facilitation. This seems to be also true in Bayer's study of the feeding of hens (1).

In the experiment reported below, an attempt was made to arrange the competitive circumstances so as to require of the animal: (1) a discrimination between a pre-goal winning situation and a pre-goal losing situation; and (2) sustained, appropriate anticipatory activity in order to attain the goal for which it is competing.

THE EXPERIMENT

APPARATUS. The apparatus consisted of a straight unobstructed alley 6 in. wide, 5 in. high and 30 ft. long (inside dimensions). This alley was constructed of wood, painted a dull black and provided with a top of wire mesh. The starting box was divided into two compartments by a longitudinal partition and was closed at the forward end by a vertically sliding gate. This gate was so wired that the opening thereof closed the circuit through a magnetic signal marker arranged to record upon a constant speed polygraph (paper speed = 7.86 in. per min.). The goal box was also divided into two compartments and was closed by a fixed gate over which the animals were required to climb or jump. This goal box was situated 6 in. beyond an arbitrarily established "finish line," which was plainly marked upon the floor of the alley. The distance between starting gate and "finish line" was exactly 30 ft. An animal's run was recorded as complete when any portion of the body crossed the "finish line." The time of crossing was recorded on the polygraph by a magnetic signal marker which was manually regulated by the experimenter. The apparatus as a whole was operated behind a screen placed at the goal end of the alley. Running time was obtained by reading the polygraph record with an appropriately graduated scale.

SUBJECTS. The experimental animals used were albino rats. The group consisted of 12 litter-mates, 6 males and 6 females. These rats were 125 days old on the first experimental day.

GENERAL PROCEDURE. 1. Throughout the experiment the animals were run after a starvation period of approximately 47 hours. The food reward consisted of pulverized, 100 per cent whole wheat bread mixed with whole milk. Each animal was given four trials per day.

2. After each day's runs the animals were fed to satiation upon this same diet.

3. All experimentation was scheduled between the hours of 5.30 and 6.30 P.M., at which time extraneous noise was at a minimum.

4. The ample, artificial lighting of the apparatus was held constant.

5. Throughout the experiment the positions of the animals in the double starting box were varied according to a chance order.

SPECIFIC PROCEDURE. The successive experimental procedures were as follows:

1. Each animal was given 20 manually guided (if necessary), untimed, rewarded runs in isolation.

2. Each animal was given 80 unguided, untimed, rewarded runs in isolation.

3. Each animal was given 40 unguided, timed,

* From the *Journal of Experimental Psychology*, 1937, **21**, 194–201.

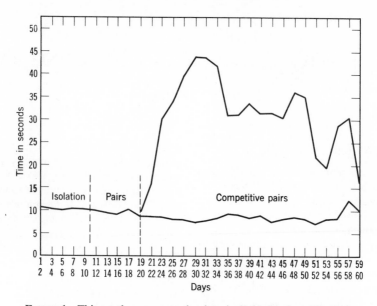

FIGURE 1. This graph represents the data in Tables 1, 2 and 3. Each point on the graph from day 1 to day 20 inclusive represents the mean of forty-eight measurements. Each point of the graphs from day 21 to day 60 inclusive represents the mean of twenty-four measurements. The upper line in this phase represents the performance of the losers. The lower line represents the performance of the winners.

rewarded runs in isolation. The mean time scores for these 10 successive experimental days are presented in Table I and represented graphically in Figure 1. Each point in this phase of the graph is based upon the mean of 48 measurements.

4. Upon the basis of the mean time scores obtained in the foregoing procedure the animals were then matched and paired within each sex group. The mean score for each animal was based upon 40 measurements. The range of these mean scores was from 14.2 sec. to 8.5 sec. These pairs were then given 40 unguided, timed, rewarded runs *in pairs*. In this procedure and in the later procedures one animal was placed in each compartment of the starting box and the order of placement was varied according to chance. Food was placed in each of the two goal compartments. The mean time scores for these 10 days appear in Table 2 and are represented as days 11 to 20 in the graph. Here again each point represents the mean of 48 measurements. Though the animals ran in pairs they were timed independently. In this procedure the order of crossing the finish line within each pair varied inconsistently from trial to trial which indicated that the pairs were well matched.

5. These pairs were then given 160 additional unguided, timed runs. In these 160 runs only the winning rat in each run, that is to say, only the

animal first to cross the finish line, was rewarded. This was accomplished by placing food in only one of the goal-box compartments and by requiring the losing rat to enter the unbaited side. In the early stages of this procedure it became necessary to establish a criterion of refusal to run because the losing animals rapidly slowed down and occasionally refused to leave the starting box. A run was scored as a refusal if at the end of 60 sec. the animal had not crossed the finish line. These scores were included in the computations. The mean time scores obtained by this procedure appear in Table 3 and are represented on the graph as days 21 to 60. In this phase of the graph the scores of the winners and the scores of the losers are plotted separately. The upper line represents the losers and the lower line the winners. Each point in each graph represents the mean of 24 measurements. *This method of plotting is made legitimate by the fact that from the twenty-third day to the fifty-eighth day inclusive the personnel of the losing group was always the same.* Obviously then the personnel of the winning group was likewise constant. In other words, 6 of the 12 contestants ran for 36 days or 144 trials without food reward. On the fifty-ninth day one female loser began winning. On the sixtieth day a second loser, also a female, began winning. At this point the experiment was discontinued.

TABLE 1. *The Speed of Locomotion of Rats Running in Isolation. Each Value in this Table Represents the Mean of Four Measurements. Time in Seconds Tabulated by Days*

Animal	1	2	3	4	5	6	7	8	9	10
Male 1	14.1	13.8	14.1	15.6	12.8	11.0	15.6	14.7	18.4	11.9
Male 2	16.2	10.0	8.4	8.8	11.1	8.4	10.5	10.5	9.7	9.7
Male 3	9.5	8.2	8.7	14.0	10.7	9.6	9.2	8.9	7.7	9.7
Male 4	11.0	10.2	10.0	10.7	8.6	12.0	10.6	10.4	15.2	12.2
Male 5	10.7	11.9	12.6	11.5	12.6	14.8	14.4	13.0	11.2	10.2
Male 6	10.0	12.0	7.5	11.8	8.8	8.6	8.2	11.1	9.0	9.7
Female 1	8.0	7.4	7.4	8.9	6.9	10.7	10.0	8.2	9.4	7.5
Female 2	7.4	10.4	7.3	9.5	8.0	8.7	8.3	10.3	11.9	8.5
Female 3	9.5	8.3	9.7	10.2	8.6	8.2	9.3	8.8	6.7	7.1
Female 4	9.9	13.3	8.9	11.5	8.8	9.9	8.6	10.1	9.4	11.7
Female 5	12.0	9.0	10.1	8.3	7.5	7.8	8.0	11.0	9.7	9.1
Female 6	10.6	11.1	10.0	8.1	7.3	12.4	8.9	8.4	8.3	7.2
Mean	10.7	10.5	9.6	10.7	9.3	10.2	10.1	10.5	10.6	9.5

[1] The complete records from which the summary tables in this study were compiled are on file at the Department of Psychology of The Pennsylvania State College.

TABLE 2. *The Speed of Locomotion of Rats Running in Pairs. Each Value in this Table Represents the Mean of Four Measurements. Time in Seconds Tabulated by Days*

Animal	11	12	13	14	15	16	17	18	19	20
Male 1	11.9	13.1	12.4	10.6	11.1	11.2	12.5	10.8	9.7	9.5
Male 2	10.7	12.7	9.9	8.9	10.9	9.0	9.6	12.0	8.6	11.9
Male 3	9.6	8.6	8.7	8.9	7.4	7.2	9.4	9.5	9.6	9.4
Male 4	11.4	8.3	10.1	8.8	8.7	9.0	10.3	12.0	8.9	10.6
Male 5	11.1	13.9	13.4	10.3	11.0	9.7	8.5	10.1	8.1	8.0
Male 6	8.8	7.6	7.8	8.1	6.8	6.9	9.1	9.7	8.9	8.8
Female 1	11.2	7.4	9.3	6.6	7.5	9.3	7.4	8.0	6.9	7.9
Female 2	7.6	8.2	9.5	9.2	8.0	8.3	7.8	9.2	6.9	7.0
Female 3	8.6	7.7	11.5	8.4	8.1	7.4	7.6	9.8	7.8	8.8
Female 4	11.0	12.0	8.9	9.6	9.2	11.5	12.7	12.8	8.2	9.7
Female 5	8.9	10.1	7.7	8.1	10.0	10.5	12.5	10.8	8.6	8.9
Female 6	7.1	8.1	9.8	9.5	7.6	8.6	9.0	11.0	7.8	6.8
Mean	9.8	9.8	9.9	8.9	8.9	9.1	9.7	10.5	8.3	8.9

RESULTS AND INTERPRETATIONS

1. The results from the last 10 days of runs in isolation appear to show that the group had been trained to a speed of locomotion plateau. A comparison of the mean scores for the first 5 of these days with the mean scores for the last 5 days reveals an obtained difference of 0.0.

2. The data from the first 10 days of trials in pairs, in which both winner and loser were rewarded, appear to show that the interstimu-

TABLE 3. *The Speed of Locomotion of Rats Running in 'Competitive' Pairs. Each Value in this Table Represents the Mean of Four Measurements. Time in Seconds Tabulated by Days*

Animal	21	22	23	24	25	26	27	28	29	30
Male Winner 2	10.2	8.9	7.9	10.0	9.3	6.7	7.2	7.0	6.3	8.5
Male Winner 5	9.4	7.3	6.9	6.5	8.5	7.2	8.3	7.6	6.9	7.2
Male Winner 6	9.2	9.4	9.8	8.6	7.8	10.3	8.9	7.3	6.7	7.4
Female Winner 5	8.9	8.7	7.5	9.9	8.6	8.8	9.7	11.0	9.5	9.3
Female Winner 2	6.4	7.4	6.8	7.1	7.6	7.4	7.5	6.8	6.4	7.2
Female Winner 1	7.9	7.3	8.8	11.3	5.9	6.6	6.5	6.2	6.1	6.9
Mean	8.7	8.2	8.0	8.9	8.0	7.8	8.0	7.7	7.0	7.8
Male Loser 4	10.4	11.8	32.0	31.3	34.6	32.0	45.8	48.7	49.4	50.3
Male Loser 1	9.5	9.6	27.0	38.3	30.0	38.6	51.4	25.5	48.6	40.2
Male Loser 3	9.9	10.7	12.7	23.2	33.6	25.4	48.7	36.7	31.8	44.0
Female Loser 4	25.9	47.3	46.2	47.9	49.8	48.3	48.2	49.1	58.5	49.0
Female Loser 6	7.5	7.9	12.4	14.6	16.6	39.0	27.2	17.2	34.0	31.7
Female Loser 3	11.8	18.9	36.6	34.5	33.8	20.1	47.6	33.6	48.9	37.4
Mean	12.5	17.7	27.8	31.6	33.1	33.9	44.8	35.1	45.2	42.1

Animal	31	32	33	34	35	36	37	38	39	40
Male Winner 2	6.6	9.1	7.3	8.9	8.4	7.6	7.5	17.4	9.4	8.3
Male Winner 5	7.0	7.9	7.6	9.1	7.8	7.1	7.0	8.0	8.1	7.7
Male Winner 6	7.2	11.1	8.1	8.1	7.4	8.5	8.1	8.0	7.0	7.4
Female Winner 5	8.5	8.1	8.8	10.8	11.3	14.3	10.9	14.3	13.4	12.4
Female Winner 2	7.0	7.6	7.0	9.1	16.1	7.8	8.2	6.7	6.3	6.7
Female Winner 1	5.8	6.8	6.4	6.6	7.0	8.0	7.6	6.4	6.1	7.5
Mean	7.0	8.4	7.5	8.8	9.7	8.9	8.2	10.1	8.4	8.3
Male Loser 4	47.1	45.6	50.8	47.5	50.5	29.8	48.6	25.6	24.5	34.7
Male Loser 1	39.7	47.5	32.3	42.1	35.2	11.4	28.2	17.3	15.9	45.8
Male Loser 3	45.4	42.8	44.7	24.9	35.3	12.2	18.1	14.5	10.3	25.2
Female Loser 4	49.5	38.4	48.6	50.7	51.1	29.4	26.6	39.9	33.7	38.6
Female Loser 6	46.3	44.3	43.7	42.3	19.8	35.3	35.2	47.8	48.1	60.0
Female Loser 3	36.0	39.3	35.3	35.9	40.5	16.6	35.0	31.3	24.1	41.2
Mean	44.0	43.0	42.6	40.6	38.7	22.5	32.0	29.4	26.1	40.9

TABLE 3—*Continued*

Animal	41	42	43	44	45	46	47	48	49	50
Male Winner 2	14.8	11.5	7.4	8.1	9.4	9.6	10.0	15.1	6.8	6.7
Male Winner 5	11.7	7.9	6.5	6.9	6.8	7.0	7.3	7.6	7.3	6.4
Male Winner 6	7.8	6.8	7.2	7.6	7.6	7.7	9.5	6.5	13.3	7.6
Female Winner 5	8.4	9.7	8.7	10.9	9.0	8.4	8.3	8.5	10.3	10.0
Female Winner 2	6.8	7.0	7.0	6.0	7.5	8.7	6.6	7.3	6.3	9.8
Female Winner 1	6.5	6.3	6.9	5.9	7.8	6.4	7.7	7.1	6.7	5.7
Mean	9.3	8.2	7.3	7.6	8.0	8.0	8.2	8.7	8.5	7.7
Male Loser 4	16.2	24.1	22.4	17.3	22.3	12.6	32.6	21.7	13.8	19.0
Male Loser 1	29.1	34.0	27.1	27.2	33.0	14.5	40.3	29.4	35.6	37.0
Male Loser 3	11.6	26.1	15.1	20.7	14.9	20.0	25.1	25.8	22.0	15.6
Female Loser 4	48.9	51.1	56.8	54.3	52.7	54.3	53.6	46.5	57.1	60.0
Female Loser 6	45.3	48.2	59.5	48.0	49.0	46.0	52.1	44.3	58.1	60.0
Female Loser 3	25.2	13.7	14.5	12.0	27.2	15.2	23.4	34.9	16.4	22.7
Mean	29.4	32.9	32.6	29.9	33.2	27.1	37.9	33.8	33.8	35.7

Animal	51	52	53	54	55	56	57	58	59	60
Male Winner 2	6.8	7.0	10.6	7.5	7.5	8.1	8.7	8.3	9.4	10.3
Male Winner 5	6.3	6.1	6.1	6.0	5.7	6.9	7.1	26.1	9.2	10.0
Male Winner 6	6.9	6.5	6.4	5.3	6.3	6.6	9.9	8.8	12.2	9.0
Female Winner 5	7.7	10.3	7.8	6.9	21.1	9.2	15.3	9.8	11.6	10.5
Female Winner 2	7.8	6.3	6.3	6.2	6.7	6.5	8.0	6.3	6.0	6.9
Female Winner 1	4.7	7.0	5.7	20.1	5.4	6.8	9.9	26.7	18.4	7.7
Mean	6.7	7.2	7.2	8.7	8.8	7.4	9.8	14.3	11.1	9.1
Male Loser 4	13.9	12.0	13.1	9.9	10.7	17.6	14.4	13.1	11.9	11.4
Male Loser 1	15.3	16.6	14.2	20.9	23.7	31.3	24.8	26.9	11.7	24.6
Male Loser 3	9.2	9.9	7.7	14.6	26.3	47.2	60.0	33.3	25.7	20.0
Female Loser 4	42.9	38.9	29.6	60.0	49.2	52.5	52.1	58.1	32.7	23.3
Female Loser 6	48.1	22.0	23.1	18.9	30.6	39.7	37.1	33.5	10.9	10.2
Female Loser 3	12.2	19.9	9.9	9.6	5.6	5.7	5.5	6.1	7.0	6.0
Mean	23.6	19.9	16.3	22.3	24.4	32.3	32.3	28.5	16.7	15.9

lation resulting from running in pairs was adequate to produce an increase in the speed of locomotion. A comparison of the mean scores for these 10 days with the mean scores for the preceding 10 days of isolated trials reveals a difference of 0.8 sec. which is 3.08 times as large as the standard error of the difference. This interpretation must be made with caution because

it is possible that this small increase in speed of locomotion may represent nothing more than a later phase of a learning process. This caution is offered in spite of the apparent plateau that precedes.[1]

3. The results from the 40 days of racing were unexpected and baffling. It was not anticipated that winners and losers would be segregated so

---

[1] An experiment now in progress shows promise of contradicting the suggested conclusion that the rat's

locomotor behavior is susceptible to social facilitation.

quickly and consistently. As before stated, from the twenty-third day to the fifty-eighth day inclusive, the personnel of each group remained unchanged. The lower line which represents the performance of the winners in this phase of the graph reveals nothing notable except a further increase in speed and an apparently increased variability of performance at those points where the losers increased their speed. The upper line representing the performance of the losers rises rapidly to a maximum at about the thirtieth day and supposedly represents the effect of frustration resulting from the removal of the reward. From this point on, the graph falls gradually and very irregularly, indicating a slow increase in the running speed of the losers. It may be worthwhile to mention that during this period the percentage of refusals decreased from 50 per cent to about 15 per cent and on some days to zero. Casual observation of the losing animals during this phase of the experiment seemed to reveal first an increase and then a decrease in agitated behavior, as described by Hull (4) and Miller (5). These changes appeared to parallel the changes in running speed. As before stated, two female losers began winning during the last 2 days of the experiment. The author is at a loss to account for this latter phase of the loser's performance. It is indicated that some significant change has taken place among the factors determining the speed of locomotion. This change will not be identified without further experimentation but the author wishes to advance one highly speculative interpretation. It is not wholly unreasonable that the internal inhibitions generated by frustration have raised or released to a dominant status some unidentified motive, such as that of social facilitation or of escape.

SUMMARY

It may be said in conclusion:

1. That if competitive behavior has been demonstrated it has not been identified.

2. That there is some evidence that the speed of locomotion in the white rat is susceptible to influence through social facilitation.

REFERENCES

1. BAYER, E. Beiträge zur zweikomponenten Theorie des Hungers, *Zeit. f. Psychol.*, 1929, **112**, 1–54.
2. HARLOW, H. F. Social facilitation of feeding in the albino rat. *J. Genet. Psychol.*, 1932, **41**, 211–221.
3. HARLOW, H. F. and YUDIN, H. C. Social behavior of primates. I. Social facilitation of feeding in the monkey and its relation to attitudes of ascendance and submission. *J. Comp. Psychol.*, 1933, **16**, 171–185.
4. HULL, C. L. The rat's speed-of-locomotion gradient in the approach to food. *J. Comp. Psychol.*, 1934, **17**, 393–422.
5. MILLER, N. E. and STEVENSON, S. S. Agitated behavior of rats during experimental extinction and a curve of spontaneous recovery. *J. Comp. Psychol.*, 1936, **21**, 205–331.

ALLELOMIMETIC BEHAVIOR IN DOGS: NEGATIVE EFFECTS OF COMPETITION ON SOCIAL FACILITATION*

J. P. Scott and Curt McCray

ALLELOMIMETIC behavior, observationally defined as doing what another animal does, with some degree of mutual stimulation, is commonly seen in packs of dogs, herds of hoofed mammals,

* From the *Journal of Comparative and Physiological Psychology*, 1967, **63**, 316–319. Copyright 1967 by the American Psychological Association. This research was supported by Grant CRT 5013 from the National Institutes of Health and performed at the Jackson Laboratory.

troops of primates, flocks of birds, and schools of fishes. Synonyms are "contagious behavior" and "schooling behavior" (in fishes). Previous experimental work with dogs (Vogel, Scott and Marston, 1950) yielded three findings. (*a*) When *S*s ran through the same course, singly and together, running times under paired conditions showed greater correspondence than those under individual conditions, thus demonstrating the reality of allelomimetic behavior. (*b*) Running with the same *S* each day produced social facilitation amounting to 18 per cent, chiefly shown by the slow *S* of each pair, whereas running with a different animal each day lessened the degree of mutual mimicry and produced a decrement in running time of 73 per cent. (*c*) Running with the same *S* each day produced the same amount of social facilitation whether or not *S*s were previously familiar with each other.

The present experiment introduces the element of competition into the situation. It would be predicted, on the basis of reinforcement theory and previous results, that competition, defined as a situation in which only the winning *S* is rewarded, would cause the winning *S* to continue to run at top speed, whereas the losing *S* would tend to gradually run more slowly, showing the phenomenon of extinction. If competition has no effect and motivation connected with allelomimetic behavior is the important factor, *S*s should run just as fast whether rewarded or unrewarded.

METHOD

SUBJECTS. Thirty-two adult basenji dogs, 13 males and 19 females, were paired, with one exception, into like-sexed pairs and, with two exceptions, with unfamiliar animals, in order to avoid any effects of previous experience. Basenjis were chosen because they readily compete over food. These *S*s had the additional advantage of an annual breeding cycle, beginning near the autumnal equinox; since the experiment was done from July 18 to August 11, no *S*s were in estrus during this period.

APPARATUS. The apparatus consisted of a runway 200 ft. long and 4 ft. wide with a 90° turn at the halfway point. The outer wall of the runway was a 7-ft.-high solid board fence, and the inner wall was a 4-ft.-high strand of poultry netting. The start box was the same width as the runway and consisted of two identical chambers with a common guillotine door. The goal box was a large area enclosed by a solid board fence. Once the

*S*s entered, a guillotine door could be lowered to prevent retracing.

PROCEDURE. After 3 days of preliminary individual training, each *S* was given one single and one paired trial each day, alternating the order on successive days. In competitive trials the loser was unrewarded and led back to a retraining cage. The winner was given a food reward of 1 tsp. of canned fish, petted, and carried back to the holding cage. In non-competitive double runs *S*s were met by two *E*s, both fed similar amounts of fish, and both petted and carried back to the holding cage. In single runs each *S* was rewarded as above. Running times were taken from the time the start-box gate was raised until *S* entered the goal box. If an *S* did not finish the course by the end of 2 min. it was led or called into the goal area and there rewarded or not, as described above.

The research design included four blocks of 5 days each. The *S*s were tested Monday through Friday on 4 successive weeks. In Group 1, consisting of eight pairs of *S*s, the first ten trials were competitive and the second ten trials noncompetitive. For the eight pairs of *S*s in Group 2 this treatment was reversed. Thus competition was introduced in both an early and an advanced stage of the learning process.

RESULTS

RUNNING TIME. The principal effects of the experimental variables upon individual performance are summarized in Table 1. The *S*s ran faster under paired conditions than when alone in approximately two-thirds of the trials. Competition had no facilitative effect on running speed, and, if anything, had a detrimental effect. Finally, *S*s ran faster in the second half of the experiment than in the first, showing somewhat more improvement under paired conditions than singly. This nonparametric analysis shows no difference between Groups 1 and 2, the data being remarkably similar, except that Group 1 showed a somewhat greater tendency to improve on single runs.

The majority of *S*s learned the task quickly and performed satisfactorily throughout the experiment, but in two pairs in each group, competitive behavior produced a definitely detrimental effect, culminating in fights in two pairs, and causing avoidance behavior and irregular running times in the others. Both of the latter pairs were the exceptional familiar *S*s (paired littermates) mentioned above. As might be

TABLE 1. *Relative Performance of Individual Ss Under Different Conditions of Running*

Condition	Number of Trials			Number of Ss With Majority of Trials Faster
	Faster	Slower	Equal	
Paired vs. single runs in same day	407	188	29	24**
Competitive paired runs vs. non-competitive paired runs in correspondingly ordered trials	141	155	8	13
Paired runs in second half vs. paired runs in first half in correspondingly ordered trials	215	81	8	25**
Single runs in second half vs. single runs in first half in correspondingly ordered trials	192	92	20	22*

* $p < 0.05$, Wilcoxon matched-pairs signed-ranks test.
** $p < 0.01$, Wilcoxon matched pairs signed ranks test.

expected, these Ss showed little tendency to improve while paired, but most improved under single conditions. Removing their data has no effect except to reduce the ratio of Ss improving under single conditions to a nonsignificant level. After data from these four pairs were removed, an analysis of variance was performed on the rest, based on the means of each block of five trials and involving two variables, weeks, and single *vs.* paired conditions. Running time varied from week to week ($p < 0.05$), as the Ss generally improved their performance. There were also significant differences between single and paired runs ($p < 0.01$), indicating that the Ss ran faster in pairs than singly.

A second two-way analysis of variance was based on comparisons between the groups run under competitive and noncompetitive conditions, and between weeks. Weeks again gave significant F value ($p < 0.01$), but the F ratio for the competitive and noncompetitive comparison was very small and nonsignificant.

In order to further explore these results, data from one additional pair (in which one S consistently refused to run in the single trials) was removed from Group 1 and further analysis of variance done using only Trial Blocks 1–5 and 16–20. Means of these data are shown in Table 2. The apparent differences between Group 1 and 2 in single runs are nonsignificant by either parametric or nonparametric tests.

Running time in these Ss was significantly faster on the last block of trials only in paired runs ($p < 0.01$), indicating that the principal improvement in motivation came from this source.

Paired runs were again significantly faster than single runs ($p < 0.01$). Finally, there was a significant ($p < 0.01$) Groups × Trial Blocks interaction, indicating a possible effect of competition. Since running time under competition was slower in both Blocks 1–5 and 16–20, it may be concluded that competition had, if anything, a negative effect. This relationship was obviously less clear-cut if the intermediate trial blocks were included and would therefore seem to hold only in the early and late stages of the learning process, if it is indeed a real effect.

It may be concluded that Ss ran significantly faster in paired runs than alone, that their performance improved with time more consistently in paired than single trials, and that competition had only a negative effect on running speed.

TABLE 2. *Mean Running Time (in Sec.)*

Group	Trial Blocks			
	1–5	6–10	11–15	16–20
	Single trials			
1	12.7	11.9	11.0	11.3
2	15.1	13.1	13.7	12.8
	Paired trials			
1	11.9[a]	11.1[a]	10.3[b]	10.2[b]
2	11.6[b]	11.2[b]	10.4[a]	10.7[a]

[a] Competitive trials.   [b] Noncompetitive trials.

SOCIAL RELATIONSHIPS. A second dependent variable is the relationship between the two Ss composing each pair. This relationship can be measured in two ways: the win-loss or leadership ratio (the number of firsts compared to the number of seconds in a block of ten trials) and the degree of correspondence between paired runs as compared to single runs. Competition apparently had a slight effect upon win-loss ratios, as there were seven pairs with ratios of 9:1 or 10:0 when the Ss ran under noncompetitive conditions, as opposed to four pairs under competitive conditions. A Mann-Whitney U test of the entire range of ratios yielded a p value of 0.05. Thus, competition resulted in less decisive win-loss ratios.

On the other hand, the relative degree of correspondence between running times of paired Ss was unaffected, except in the four pairs of Ss which showed excessive competition and fighting. In the rest of the Ss, the correspondence between paired trials was closer than that between single runs in 84 per cent of the cases. If all Ss were included, this figure was reduced to 75 per cent, but, even so, 15 of the 16 pairs showed a majority of trials in which the correspondence was closer in paired trials. It may be concluded that excessive competition has a disruptive effect on this relationship, causing Ss to avoid each other, but that mild competition has no effect.

DISCUSSION

We may think of the performance of an S running alone for a food reward as a baseline which can be modified by experimentally varied conditions. In the present experiment, this performance was modified in time as the animals improved, but much more consistently under paired than single conditions. It was strongly modified by the condition of paired running itself, but making the paired runs competitive had no additional effect except a possible negative one. In the pairs where excessive competition arose, the result was definitely to lengthen running time rather than lessen it. The improved performance observed under the experimental conditions is therefore not the result of competition but the result of two factors, learning plus paired running.

The data thus confirm the existence of a basic source of motivation associated with allelomimetic behavior, independent of competition and additional to the process of learning. The essential nature of the motivation is a tendency to maintain close contact with other individuals, and social facilitation or increment of performance is a by-product which results chiefly from the fact that one member of a pair may be more highly motivated than another toward an external goal. Under certain conditions, allelomimetic behavior could theoretically produce decrements in performance as well as increments.

Other experimental data confirm the reality of this type of motivation in dogs and throw some light on its physiological nature. A long series of experiments (Bacon and Stanley, 1963; Stanley and Elliot, 1962; Waller and Fuller, 1961) indicates that puppies will readily learn to run to a goal in which the sole reward is contact with a passive human being, and that this will take place in Ss which have never been given food rewards by people. The response, in this case, is a tendency to make contact with people rather than other dogs, but we can assume that it results from the same basic mechanism.

A second line of evidence comes from experiments with social isolation in puppies (reviewed by Scott and Bronson, 1964). Beginning at about 3 weeks of age, a puppy shows distress vocalization when isolated in a familiar place, and the rate is greatly increased if isolation takes place in a strange situation. The symptoms can be alleviated by restoring the puppy to its normal social group and home pen.

The effect of this emotional response to isolation and strange places is normally to keep the puppy close to familiar surroundings and individuals. The puppy must rapidly learn that being away from familiar individuals and places is unpleasant and that this emotion can be relieved by returning to the group. The precise nature of the physiological mechanism underlying this emotional response is yet to be defined, but behavioral evidence indicates the existence of a physiological response system little more complicated than that of a simple reflex whose effects are magnified by the process of learning and reinforcement. Among dogs, at least, the resulting desire for companionship, whether canine or human, is a major source of motivation.

REFERENCES

BACON, W. E. and STANLEY, W. C. Effect of deprivation level in puppies on performance maintained
by a passive person reinforcer. *J. comp. physiol.
Psychol.*, 1963, **56**, 783–785.

SCOTT, J. P. and BRONSON, F. H. Experimental
exploration of the et-epimeletic or care-soliciting
behavioral system. In P. H. Leiderman and
D. Shapiro (Eds.), *Psychobiological approaches to
behavior*. Stanford. Stanford University Press,
1964. Pp. 174–193.

STANLEY, W. C. and ELLIOT, O. Differential human
handling as reinforcing events and as treatments
influencing later social behavior in basenji puppies. *Psychol. Rep.*, 1962, **10**, 775–788.

VOGEL, H. H., SCOTT, J. P. and MARSTON, M. V.
Social facilitation and allelomimetic behavior in
dogs. *Behaviour*, 1950, **2**, 120–143.

WALLER, M. B. and FULLER, J. L. Preliminary observations on early experience as related to social
behavior. *Amer. J. Orthopsychiat.*, 1961, **31**, 254–
266.

EFFECTS OF A COMPETITIVE SITUATION ON THE SPEED OF RESPONSE*

Russell M. Church

SEVERAL experiments have been reported in which pairs of animals were reinforced on a competitive basis. Lepley (1937) trained rats individually to run down a 30-ft. straight alley for food, and when they had reached an apparent asymptote in their running speed, he matched pairs of Ss for speed. The pairs were given 160 trials in which only the first S to reach the goal was reinforced. In this situation there was a separation of winners and losers within one or two trials. The losers radically decreased their running speed (extinction), and the winners did not show any increase in their running speed.

The effect of a competitive situation on the running speed of rats in a straight alley was also investigated by Winslow (1940). As in Lepley's experiment, the Ss were trained to an apparent asymptote in running speed. Then Ss were run in pairs, reinforcement given to the first S to reach the goal. (Each S was given a number of paired trials with many of the other Ss.) The difference between the speed of Ss alone and the speed of competitive pairs did not approach statistical significance. Winslow also found no facilitation in response speed in the case of pairs of rats in a multiple-unit T maze.

Winslow (1944a) also investigated the effect of a competitive situation on the running speed of cats in a straight alley. The method was similar to his study of rats, but the results indicated that the competitive pairs ran significantly more *slowly* than the individuals. In a pair of problem boxes, the mean time for escape to food did not differ significantly for cats tested in competitive pairs and cats tested individually (Winslow, 1944b).

Thus, there is no evidence that the speed of response of animals is facilitated in a competitive situation. Yet, on theoretical grounds, such a facilitation might be expected. In a competitive situation the available reinforcements are allo-cated among two or more individuals as a function of some characteristic of their behavior relative to one another. For example, in the experiments of Lepley and Winslow cited above, on each trial the available reinforcement was awarded to the faster of two animals. Unless one S receives all the reinforcements, a competitive situation must produce a differential reinforce-ment of some response characteristic. It has been shown that differential reinforcement of a re-sponse characteristic may produce an increase in that characteristic (Ferster and Skinner, 1957).

* From the *Journal of Comparative and Physiological Psychology*, 1961, **54**, 162–166. Copyright 1961
by the American Psychological Association. This

investigation was supported in part by Research
Grant M-2903 from the National Institute of Mental
Health, United States Public Health Service.

Therefore, a competitive situation might be expected to result in a facilitation of the response characteristic that is chosen as the basis for the competition.

The present experiment is an attempt to determine a set of conditions under which the speed of response will be increased in a competitive situation.

METHOD

Apparatus

The apparatus consisted of two lever boxes. Each box was $7\frac{1}{2}$ in. by 8 in., 9 in. high. The front and back were stainless steel; the sides and top were $\frac{1}{4}$-in. transparent Lucite. The boxes were located 1 in. apart, with the Lucite walls parallel to each other. The floors were composed of 16 stainless-steel bars; the levers were made of stainless steel rounded and smoothly finished, $\frac{1}{2}$ in. thick and 2 in. wide. By pressing the lever in its compartment, each S advanced its own stepping relay. The stepping relay that had most recently "passed" the other activated a holding relay. This indicated which of the two Ss had made the greater number of responses during each 15-sec. interval.

Measures were taken of the number of responses made by each S during each 15-sec. interval and of which of the two Ss was reinforced at the end of each 15-sec. interval.

Subjects and Procedure

The Ss were 52 naive albino rats of Sprague-Dawley strain, about 90 days old at the beginning of the experiment. They were caged individually and were fed wet mash, 8 gm dry food mixed with 25 cc water once a day about 22 hours before each experimental session. Water was accessible at all times in the home cages. Lab rat food pellets (45 mg) from the P. J. Noyes Co., Lancaster, N.H., were used as reinforcement for lever pressing. The Ss were handled and on a feeding schedule for 10 days. They were given magazine training and shaping procedures and allowed to press the lever 50 times for 100 per cent reinforcement. They were then given 10-min. sessions every 24 hours on a 30-sec. variable-interval schedule. The 28 Ss of experiment 1 were given 10 sessions, and the 24 Ss of experiment 2 were given 12 sessions.

Within each experiment Ss were partitioned into groups of four selected to minimize the difference within groups in the median number of responses of the Ss during the last 5 days of training. Two Ss in each partition were assigned to the experimental (competitive) group and two to the control (noncompetitive) group. (As a restriction, all Ss were kept in the same compartment in which they had been originally trained.)

On the next 18 days Ss were given one 10-min. session each day. In the case of an experimental pair of Experiment 1, the S that made the *greater* number of responses during the 15-sec. period was reinforced immediately following its next response. In the case of an experimental pair fo Experiment 2, the S that made the *fewer* number of responses durin. the 15-sec. period was reinforced immediately following its next response. Each experimental S was matched with a control S that received the same number and temporal distribution of reinforcements, each of which immediately followed a response. Presumably, any difference between the experimental and control group would be due to the fact that Ss in the former group were in a competitive situation, i.e., their reinforcements were contingent upon their response rate relative to that of another S.

RESULTS

The Effect of the Competitive Situation on Response Rate

During variable-interval training, Ss in Experiment 1 increased their response rate to a mean of 20.5 responses per minute on the sixth session (Fig. 1). During the last 5 days of training this rate was maintained. During the 18 days of competition the experimental Ss *increased* their rate of response relative to Ss that were in the control situation. The experimental Ss had a median increase of 5.7 responses per minute on the last 5 days of competition compared with the last 5 days of training; the control Ss had a median decrease of 1.0 response per minute during the corresponding periods ($p = 0.01$, Wilcoxon's matched-pairs test). Thus under the conditions of this experiment, Ss in a competitive situation clearly responded more rapidly than Ss in a control situation.

Not all Ss in the competitive group increased their rates of response. In some cases one S responded substantially faster than its opponent and thus came to obtain virtually all the reinforcements. For example, one S increased from about 24 to about 60 responses per minute and obtained all 40 reinforcements on the next 5 successive days. This S's opponent, meanwhile,

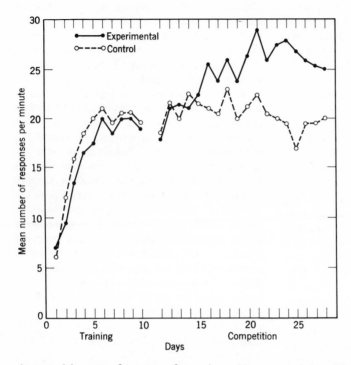

FIGURE 1. The effect of the competitive situation on response rate, Experiment 1. The number of responses per minute during training (days 1 to 10) and competition (days 11 to 28) for *S*s in the experimental (competitive) and control (noncompetitive) groups.

decreased its rate of response from about 20 to about 3 responses per minute (extinction). It is interesting to note that the original *S* decreased its rate from about 60 to about 40 responses per minute during the same 5 days when it was no longer subjected to the competitive pressure.

During variable-interval training, *S*s in Experiment 2 increased their response rate to about a mean of 16.0 responses per minute by the eighth session (Fig. 2). During the last 5 days of training this rate was maintained.

During the 18 days of competition experimental *S*s *decreased* their rate of response relative to *S*s in the control situation. The experimental *S*s had a median decrease of 4.0 responses per minute in rate of response on the last 5 days of competition compared with the last 5 days of training; the control *S*s had a median increase of 3.9 responses per minute during the corresponding periods ($p < 0.01$, Wilcoxon's matched-pairs test). Thus under the conditions of this experiment, *S*s in a competitive situation clearly responded more slowly than *S*s in a control situation.

The Effect of the Competitive Situation on Individual Differences

In Experiment 1 the degree of individual differences (variance between individuals) in the experimental group was greater during the last 5 days of competition than the last 5 days of training ($F = 7.1$, $p < 0.01$), but there was no corresponding increase in the individual differences of the control group ($F = 1.1$, $p > 0.05$). The data suggest that, relative to their matched controls, *S*s in the experimental group with faster initial rates of response showed a greater increase in their response rates than *S*s with slower initial rates of response. Although in the experimental group there was no significant rank-order correlation (rho = 0.09) between initial rate (median of last 5 days of training) and amount of gain in the competitive situation (increase of the last 5 days of competition over the initial rate), there was a significant negative correlation between these measures in the control group (rho = -0.63, $p < 0.05$).

In Experiment 2 the degree of individual differences (variance between individuals) in the experimental group was less during the last 5 days of competition than during the last 5 days of training ($F = 2.3$); but the individual differences in the control groups were greater during the last 5 days of competition than during the last 5 days of training ($F = 2.1$). Although these differences cannot be recognized as statistically significant, the individual differences within the control group were significantly greater than the individual differences within the experimental group on the last 5 competition days ($F = 3.3$, $p < 0.05$). The data indicate that, relative to their matched controls, *S*s in the experimental group with faster initial rates of response showed a greater decrease in their response rates than *S*s with slower initial rates of response. In the experimental group the rank-order correlation between the initial rate (median of the last 5 days of training) and the amount of gain in the competitive situation (decrease of the last 5 days of competition over the initial rate was -0.83, $p < 0.01$); there was an insignificant -0.30 correlation between these measures in the control group.

The Effect of the Competitive Situation on Response Variability

In Experiment 1 the degree of within-subject variability (variance of number of responses in each 15-sec. interval during the 10-min. session) did not change significantly as a function of days, and it was not significantly different in the experimental and control groups.

In Experiment 2 the degree of within-subject variability (variance of number of responses in each 15-sec. interval during the 10-min. session) decreased in the experimental group and increased in the control group. Eleven of 12 experimental *S*s had less variability on the final day of competition than on the first day of competition; 10 of 12 control *S*s had greater variability on the final day of competition than on the first day of

competition ($p < 0.01$, Wilcoxon matched-pairs test).

Differential Reinforcement in the Competitive Situation

One measure of the amount of differential reinforcement is the correlation between the number of responses made by an *S* during a session and the number of reinforcements that it received at that session. In the experimental group of Experiment 1, the median rank-order correlation between the number of responses and the number of reinforcements was 0.50; the median correlation in the control group was 0.22. This measure of the amount of differential reinforcement was significantly higher in the experimental group than in the control group ($p < 0.01$, Wilcoxon's matched-pairs test). In the experimental group of Experiment 2 the median rank-order correlation between the number of responses and the number of reinforcements was -0.38; the median in the control group was $+0.17$. This correlation was significantly lower in the experimental group than in the control group ($p = 0.01$, Wilcoxon's matched-pairs test), demonstrating differential reinforcement of low response rates in this group.

In the experimental group of Experiment 1, the

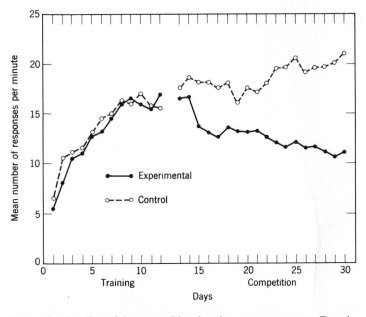

FIGURE 2. The effect of the competitive situation on response rate, Experiment 2. The number of responses per minute during training (days 1 to 12) and competition (days 13 to 30) for *S*s in the experimental (competitive) and control (noncompetitive) groups.

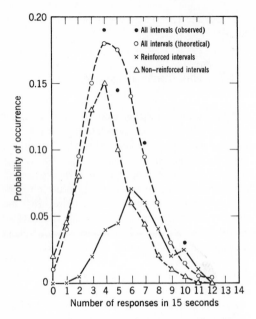

FIGURE 3. The observed and theoretical probability of X responses in 15 sec., and the probability of reinforcement and of nonreinforcement as a function of X. Data are from one experimental S of Experiment 1 during eighteen 10-min. sessions (i.e. 720 points).

that a competitive situation may result in facilitation of the response characteristic that is chosen as the basis for the competition. In the first experiment, in which the reinforcement was delivered to the faster of two Ss in each 15-sec. interval, there was an increase in the rate of response (relative to a control group); in the second experiment, in which the reinforcement was delivered to the slower of two Ss in each 15-sec. interval, there was a decrease in the rate of response (relative to a control group). A competitive situation need not result in an increase in the speed of response. Apparently it is possible to increase or to decrease the speed of response in a competitive situation. Presumably, it would be possible to develop an accurately timed response (by reinforcing the individual most closely approximating X responses per minute at each 15-sec. interval). It might further be supposed that the particular characteristic of the behavior that is chosen as the basis of competition is arbitrary. It should be possible to select as the basis of competition the accuracy, duration, magnitude, or other characteristic of the response.

rank-order correlation between the number of responses made by an S during a session and the number of reinforcements that it received at that session increased gradually as a function of sessions from 0.46 to 0.81. There was no corresponding increase in the control group. In the experimental group of Experiment 2, the rank-order correlation between the number of responses made by an S during a session and the number of reinforcements that it received at that session decreased gradually as a function of sessions from -0.20 to -0.78. There was no corresponding decrease in the control group.

Figure 3 illustrates the extent of differential reinforcement in the case of one experimental S of Experiment 1. It might be noted that the observed distribution of the number of responses in each 15-sec. interval is similar to a Poisson distribution $(a^x e^{-a}/x!)$, with a mean equal to the observed mean. The major point of the figure, however, is the displacement of the frequency distribution for reinforced intervals from that of the non-reinforced intervals.

DISCUSSION

The two experiments provide clear evidence

SUMMARY

The two experiments indicate clearly that a competitive situation may influence the response rate. The Ss, 52 rats, were trained two at a time in adjacent boxes to press a lever for food reinforcement. After this initial training a competitive situation was established. Every 15 sec. the S that had made the greater number of responses (first experiment) or the S that had made the lesser number of responses (second experiment) was reinforced with a food pellet. In comparison with control Ss (that received an identical number and temporal distribution of reinforcements) experimental Ss in the first experiment increased their rate of response ($p = 0.01$), and in the second experiment decreased their rate of response ($p < 0.01$). Thus, under these conditions, competition facilitated the response characteristic that was chosen.

The most reasonable theoretical explanation for this competitive facilitation is the following: under the competitive conditions the frequency of reinforcement was related to S's behavior (in this case, its speed of response) but under the control conditions the frequency of reinforcement was not related to S's behavior. Thus, under the competitive conditions, but not under the control conditions, there was a differential reinforcement of some response characteristic.

REFERENCES

FERSTER, C. B. and SKINNER, B. F. *Schedules of reinforcement*. New York: Appleton-Century-Crofts, 1957.

LEPLEY, W. M. Competitive behavior in the albino rat. *J. exp. Psychol.*, 1937, **21**, 194–201.

WINSLOW, C. N. A study of experimentally induced competitive behavior in the white rate. *Comp. psychol. Monogr.*, 1940, **16** (Whole No. 78).

WINSLOW, C. N. The social behavior of cats: I. Competitive and aggressive behavior in an experimental runway situation. *J. comp. physiol. Psychol.*, 1944, **37**, 297–314 (a).

WINSLOW, C. N. The social behavior of cats: II. Competitive aggressive, and food-sharing behavior when both competitors have access to the goal. *J. comp. physiol. Psychol.*, 1944, **37**, 315–326 (b).

THE EXPERIMENTAL SOCIAL BEHAVIOR OF ANIMALS II. THE EFFECT OF EARLY ISOLATION OF WHITE RATS ON THEIR COMPETITION IN SWIMMING*

A. G. Bayroff

INTRODUCTION

A PREVIOUS study (1) of the relationship between the nature of the early life of white rats and their later social behavior ("gregariousness") yielded a conclusion that the early life is not the only factor which determines the nature of adult behavior.

This conclusion followed from the fact that in a test situation which permitted *but did not require* the animals to react to each other, the result was that most of the animals reacted to the spatial rather than the social characteristics of the situation. That is to say, in a situation which permitted the establishment of position habits as adequate adjustments only a few animals failed to develop them. Concretely, if a life of solitude *makes* an animal seek or avoid the company of other animals, then this seeking or avoiding should appear on those occasions when other animals are present, regardless of the other features of the situation. Since such consistent seeking or avoiding of other animals did not appear in the test situations, it was concluded (1) that there is no assurance that the nature of the early life is the sole or most important factor in determining the nature of the adult *social* behavior, and (2) that an early life of a special social nature gives no assurance that the social behavior will appear unless the situation *requires* it. *In brief, the special early life did not produce an urge which drove the animals to exhibit one form oɟ social behavior or another* [see also (2), and (3): pp. 430–431].

The study of the relationship between the nature of the early life and adult social behavior was continued in July 1937. In the present paper will be presented results showing the effect of an early life of solitude on competition in swimming. The test situation was such that an animal's success depended upon the behavior of a competitor. Other forms of behavior may have been permitted but only the *social* behavior, it would appear, resulted in a satisfactory adjustment. It was on this characteristic of the test situation that the present experiment differed from the previous experiment.

PROBLEM

The general question involved was: which kind of white rat is more likely to be the victor in a situation involving serious competition, the one reared in solitude or the one reared in a group? As studied in this experiment, the question was:

* From the *Journal of Comparative Psychology*, 1940, **29**, 293–306. Copyright 1940 by Williams & Wilkins Co., Baltimore. Portions of this paper were read before the meeting of the Southern Society for Philosophy and Psychology, April 7, 1939, at Durham, N.C.

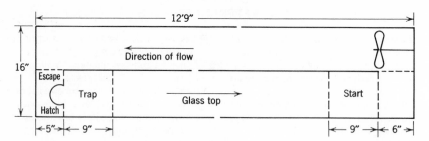

FIGURE 1.

which of two equally fast animals would win over the other in escaping first from under water, the animal reared in solitude or the animal reared in a group?

To emphasize the competition, an animal's escape was made conditional upon its arrival at the escape hatch before the second animal arrived. The animal arriving second was not allowed to escape immediately but was trapped under water. Competition as used in this paper is thus defined by these conditions of the experiment.

APPARATUS[1]

Figure 1 is the plan of the swimming tank. The tank contained two channels. A wire-mesh cage lowered the animals to the tank's bottom. The only way to escape from under water was by climbing the wire-mesh ladder inside the hatch. The under-water entrance to the hatch could be closed by a spring door. Nine inches in front of the hatch was a spring trap door so that an animal could be trapped under water, if so desired, near the escape hatch. A small light over the hatch illuminated the opening under water and provided heat for the wet animals.

The second channel communicated with the first at the two ends. The three-bladed propellor

in this channel produced the flow of water. A four-step pulley on a jackshaft and a variable-diameter pulley on the motor permitted variation of the speed of the propellor and consequently a variation in the velocity at which the water circulated.

At both ends of the tank were located single-pole double-throw snap switches so that the electric timers could be operated from either end.

PROCEDURE

The general plan of operations is outlined in Table 1.

1. Birth to Weaning

The mothers were obtained from two sources. Those giving birth to the first seven litters were obtained from the outbred colony stock. The mothers of the last seven litters were obtained from an animal dealer.

When the young rats were approximately 21 days old, their mothers were removed. One random half of each litter was distributed among individual living cages. The other random half was allowed to remain together.

TABLE 1. *General Procedure*

Step	Duration	Operation
1	3 weeks	With mother, birth to weaning
2	$4\frac{1}{2}$–8 months	Early life: solitary or social
3	$1\frac{1}{2}$ months	Preliminary training in swimming tank
4	$1\frac{1}{2}$ months	Timing trials for individual swimmers
5	2 months	Competitive trials for paired swimmers

Approximate ages at end of experiment, 10 to 14 months.

[1] Thanks are due Professors G. B. Dimmick and Edward Newbury of the University of Kentucky for the benefit of their experience with swimming problems in rats.

2. The Early Life

The animals lived in their cages from $4\frac{1}{2}$ months to 8 months before being started in the preliminary training. During this period the solitary animals lived by themselves, played by themselves, had no sexual experiences with other animals, and because of the wood partitions between the cells, saw no other rats. The animals reared in groups had the usual feeding, play, sex, etc., experiences common to group life. Both sets of animals were fed the same diet, namely, Purina dog chow, supplemented with lettuce, carrots, apples, citrus fruits, etc., and water.

3. The Preliminary Training

The general plan of preliminary training involved the gradual increase of the depth of water, its temperature, and its velocity. At first the animals could wade through the still water. Later, they could swim on the surface of the water. Beginning with the twelfth day, the animals were forced to swim completely under water and against the slight current created by the propellor revolving at the rate of 270 r.p.m. From the twentieth day to the forty-fourth and last day (i.e., on the last 25 days), the animals were swimming under the standard conditions: completely submerged, with the temperature of the water at $31.75°C \pm 0.25°C$, and the propellor turning over at the rate of 600 r.p.m.

All animals received one trial a day for the 44 days.

4. Timing Trials for Individual Swimmers

For the next 30 consecutive days, the animals were required once a day to swim under the standard conditions.

The speeds shown by the individual swimmers during this timing period were the basis for the composition of the pairs. Sample distributions of the speeds revealed their resemblance to normal distributions. Consequently it was assumed that the differences between the mean times of two animals were not significant if the standard deviation of one animal reached or exceeded the mean of the second animal. Each pair thus contained two animals whose means were equal, with the standard deviation for one overlapping the mean of the other.

5. Competitive Trials for Paired Swimmers

Each pair contained one animal which had been reared in solitude and one which had been reared with other animals. Twenty-eight pairs were arranged in which a standard deviation was such that it overlapped the mean of the other member of the pair. Table 2 contains the pairings (columns 1 and 2) and the times in seconds (columns 3 and 4).

Each pair was made to swim under the standard conditions which, to repeat, were complete undersurface swimming in the water at $31.75 \pm 0.25°C$, with the water being moved by the propellor revolving 600 times a minute.

Each pair was brought to the tank in a carrying box and gently pushed into the lift drop. The drop was lowered and at the moment it touched bottom, both timers, one for each animal, were started. As soon as the nose of one animal reached the hatch, one timer was stopped. That animal was allowed to escape. When the second animal reached the hatch, the second timer was stopped and the hatch door was snapped shut so that escape was impossible. At the same time the spring trap door was released behind the animal. For a period of 20 sec. this animal was kept under water. At the expiration of this time (which necessarily varied on occasions) the hatch was opened, permitting the second animal to escape to air.

In brief, then, two equally fast animals were forced to swim under water and against a current to escape to air. The first animal to reach the escape hatch was permitted to escape. The animal reaching the hatch second was trapped under water for a short time before it was allowed to reach air.

Each pair received one trial a day for 60 consecutive days.

THE DATA[2]

In Table 2 are the summaries and comparisons of the essential data for the matched animals.

1. The Relation Between Victory and the Early Life
(Columns 9, 10, and 11)

(a) Of the total of 28 pairs in which the animals were well matched for speed on the timing trials (i.e., equal means and standard deviations or similar means with standard deviations sufficiently great to overlap the means), there were 22 pairs in which one member won significantly more often

---

[2] Thanks are due Miss Doris Lindsay for her assistance with the computations.

TABLE 2. *Summary of Principal Results*

Pairs, Animal Numbers		Means and Standard Deviations of Timing Trials 1–30		Means and Standard Deviations of Competition Trials 1–60		Critical Ratio* of Time on Timing Trials 1–30 to Time on Competition Trials 1–60		Number of Victories in 60 Competition Trials		Critical Ratios of Differences in Number of Victories on Competition Trials 1–60
S	N	S animal	N animal	S animal	N animal	S animal	N animal	S animal	N animal	
				Matched animals—complete						
4	14	20.1±3.4	20.4±6.7	15.2±2.9	10.6±3.6	−6.7	−7.5	4	56	
6	30	17.3±3.1	17.0±2.3	12.4±1.2	13.7±2.7	−8.3	−6.2	43	17	
16	7	14.2±2.2	14.1±2.3	9.1±2.3	11.7±2.3	−10.4	−4.7	54	6	
24	31	12.2±4.1	12.2±3.1	9.2±2.0	8.2±1.0	−3.8	−6.8	14	46	
26	53	9.9±1.8	10.0±1.6	6.7±3.0	13.5±4.4	−6.2	+5.4	59	1	
27	43	10.2±1.3	10.2±1.1	8.3±1.4	9.9±3.0	−6.4	−0.7	46	14	
30	45	12.9±2.4	13.2±1.8	10.7±1.2	9.2±1.9	−4.6	−9.8	17	43	≧3.0
31	6	14.4±5.6	14.5±2.8	10.0±1.6	12.4±2.4	−4.2	−3.6	52	8	
35	47	13.7±1.6	13.5±1.5	13.1±2.3	11.7±1.2	−1.4	−5.7	18	42	
37	23	17.8±2.6	18.8±3.9	13.9±1.3	9.5±2.2	−7.8	−12.2	7	53	
45	2	11.7±1.5	10.8±3.7	10.1±2.1	7.9±4.9	−4.1	−3.2	8	52	
52	3	11.1±1.9	11.3±1.6	9.5±1.4	11.9±5.8	−4.1	+0.8	42	18	
54	55	11.8±2.1	10.4±1.5	10.1±1.8	9.3±1.4	−4.0	−3.5	18	42	
55	32	7.7±1.8	8.5±3.2	6.6±1.1	8.7±2.5	−3.0	+0.3	57	3	
17	25	11.7±4.4	10.4±2.5	8.1±1.3	7.7±1.6	−4.4	−5.6	19	41	
34	44	11.9±1.9	12.0±2.3	10.1±1.2	9.4±2.1	−4.5	−5.1	19	41	
41	40	11.4±1.5	10.4±1.2	9.7±1.3	8.9±1.0	−2.1	−5.9	19	41	=2.5–3.0
56	21	14.8±3.0	20.4±6.7	10.4±2.0	7.6±1.2	−7.2	−5.6	40	20	

TABLE 2. *Summary of Principal Results*—cont.

Pairs, Animal Numbers		Means and Standard Deviations of Timing Trials 1–30		Means and Standard Deviations of Competition Trials 1–60		Critical Ratio* of Time on Timing Trials 1–30 to Time on Competition Trials 1–60		Number of Victories in 60 Competition Trials		Critical Ratios of Differences in Number of Victories on Competition Trials 1–60
S	N	S animal	N animal	S animal	N animal	S animal	N animal	S animal	N animal	
				Matched animals—incomplete						
14	*X15*	15.6±2.8	16.4±2.5	12.7±2.3	11.2±2.3	−5.1	−9.6	12	39	
X22	*1*	12.8±1.0	13.4±1.7	9.4±2.7	11.8±2.1	−9.0	−3.8	39	3	$\geqq 3.0$
X23	27	11.1±1.4	11.1±1.1	8.4±1.6	12.3±3.1	−8.0	+1.7	32	4	
X1	*17*	13.5±2.1	13.5±2.7	13.0±1.1	11.2±1.8	−8.5	−4.4	16	34	
				Matched animals—complete						
21	20	9.4±1.2	9.6±1.4	8.1±1.1	7.6±1.2	−4.9	−6.9	27	33	
33	16	10.6±1.0	10.5±1.5	9.2±1.3	10.8±2.8	−5.6	+0.8	21	39	<2.5
36	33	10.9±1.9	11.0±1.2	8.7±2.4	8.8±5.1	−9.6	−3.1	33	27	
44	54	12.3±1.7	12.3±1.3	11.2±3.1	11.3±1.6	−2.2	−3.3	34	26	
47	42	9.1±1.5	9.0±1.3	8.5±1.5	8.9±1.5	−1.9	−0.4	32	28	
				Matched animals—incomplete						
X3	46	12.3±2.1	12.3±1.5	10.0±1.8	9.5±2.1	−5.2	−7.2	16	25	<2.5
1	2	3	4	5	6	7	8	9	10	11

Victor in italics.   S = special animals (solitarily reared); N = normal animals (socially reared).   X died.
* CR is negative if mean time for the competition trials is *less* than mean time for the timing trials.

TABLE 3. *Mortalities During Swimming Trials*

Pair		Dead Animal		Victor up to Time of Death	
S	N	S	N	S	N
Matched					
1	17	1		?*	17
3	46	3			
22	1	22		22	
23	27	23		23	
26	53	26†		26	
14	15		15		15
Unmatched					
25	34	25		25	
Unpaired					
11		11			
42		42			
53		53			
Totals					
11	8	10	1	4	2
1	2	3	4	5	6

* In the first 25 of the 50 completed competition trials, S1 had won not once. However, of the second 25 trials, S1 had won 16 before its death.

† S26 died 3 weeks after the conclusion of the last competition trial and is not recorded as a death in Table 2.

than did the other member as indicated by critical ratios greater than 2.5. (In 18 of these 22 pairs the critical ratios were at least equal to 3.)

(*b*) The total of 28 pairs included 5 pairs which, because of the death of one of the members, did not complete the entire series of 60 trials but are included in the data.

(*c*) Of the total 22 pairs in which one member won significantly more often than did the other, 10 of the victors had been reared in solitude and 12 in groups with other animals. There is thus no clear and simple relation between the nature of the early life and the success in the competition.

2. The Relation Between Victory and Increase in Swimming Speed
(Table 2, Columns 7 and 8)

(*a*) Of the 44 animals in the 22 pairs in which one member won consistently 37 (84 per cent) swam faster during the competition trials than they did during the timing trials (critical ratios greater than 3.0).

(*b*) Of the 12 animals in the 6 pairs in which one member won as often as did the other, 8 (67 per cent) swam faster during the competition (critical ratios greater than 3.0).

(*c*) Thus, most of the animals, whether reared in solitude or reared in groups, whether consistent victors or consistent losers, or equally successful competitors, swam faster during the competition.

(*d*) Of the 28 pairs, 27 *remained* matched for speed throughout the competition trials as indicated by the overlapping of the standard deviations. In 22 pairs, the standard deviation of one member overlapped the mean of the second member. Sample distributions showed no marked skewness in opposite directions for the two members of each pair examined. Hence, it was assumed that the overlappings indicated similarity of the speeds of the two members.

3. Mortalities

(*a*) In Table 3 are listed those animals which did not survive the experiment, excluding deaths occurring during the early life prior to the preliminary training. This table includes animals of an unmatched pair and animals which died before the pairings were made. Hence, these animals are not listed in Table 2. These animals are included here because their deaths may be of some significance.

(*b*) Of the total of 11 deaths throughout the entire experimental period, 10 were of animals which had been reared in solitude (columns 3 and 4).

(*c*) Of the total of 11 deaths, 6 occurred among the matched pairs. No one of these was a consistent loser. They were either consistent victors or won as often as did their competitors. The death in the unmatched pair was an animal which had been winning consistently.

(*d*) Small confidence can be put in the analysis of the mortalities because of the irregularities of procedure and the small number of cases involved. However, certain relationships are suggested.

(1) More deaths occurred among animals which were reared in solitude than among animals which were reared in groups.

(2) More deaths occurred among animals which were in *active* competition than among animals which were losing consistently.

The correctness of these observations cannot be determined with confidence from this experiment.

4. The Relation Between Victory and Incidental Factors

Dependable evaluations of the relation between victory and such factors as relative age, relative weight, sex, and litter membership are difficult to make because of the small number of cases. For the present it is sufficient to say that the available evidence does not indicate a dependence of success in the competition upon any *one* of these factors.

SUMMARY OF RESULTS

1. In the total of 28 pairs of swimmers, matched for speed and competing for escape from under water, were 22 pairs in which one member won consistently.

2. In 10 pairs the solitarily reared animal won consistently over its socially reared competitor; in 12 pairs the socially reared animal won consistently over its solitarily reared competitor.

3. Of the total of 56 animals in the 28 pairs, 45 animals (80.4 per cent) swam significantly faster during the competition trials than they did during the timing trials as indicated by critical ratios equal to or greater than 3.0. Thus, consistent winners, consistent losers, and animals which won as often as did their competitors increased their speeds. Similarly, solitarily reared and socially reared animals increased their

speeds. Furthermore, in general, the members of each pair increased their speeds equally.

4. Of the total of 11 deaths, 10 occurred among solitarily reared rats. Of the 6 deaths which occurred during the competition, none was a death of a consistent loser.

5. Victory did not seem to depend on any single factor such as relative age, relative weight, sex, or litter membership.

DISCUSSION

1. Adequacy of the Experimental Situation

The experimental situation seemed to be one requiring, primarily, social adjustments. As manipulated by the experimenter, an adequate adjustment was possible only if one animal reacted to its competitor. The animals were constantly under water and consequently appeared to be adequately motivated. The preliminary training had taught them the proper direction in which to swim. When the competition was introduced, the animals were likely, because of the motivation and preliminary training, to attend to the second animal. On the first trials of the competition, the animals may not have struggled for prior entrance into the escape hatch. However, in view of the continued motivation and the consistency of victories, it seems likely that the animals learned quickly to struggle for prior escape. The presence of a second animal, the animals learned, might result in a delay of the escape, and because of the continued motivation the animals learned this relationship quickly.

In brief, it appears that the only satisfactory adjustment possible was a social adjustment.

2. Significance of the Swimming Speeds

However, acceptance of the experimental situation as an adequate test of competition must await the determination of the swimming abilities of the animals. If one member of a pair of competitors swam much faster than the other member, the former would reach the escape hatch first, not because it succeeded in a competition, but because it could swim faster than the other. For this reason, the animals were matched for swimming speed. For this interpretation to be acceptable it is necessary to demonstrate that the speeds in the timing trials are safe measures of the abilities of the animals.

The fact that so many of the animals swam significantly faster during the competition trials is obvious evidence (Table 2) that the timing trials did not measure the maximum swimming speeds (abilities) of the animals. However, competing animals which swam *equally* fast before the competition began increased their speeds equally during the competition. In brief, then, the timing trials appear to be safe indicators of the relative swimming abilities even if they did not measure the maximum speeds. It seems unlikely that the winnings were spurious results determined by differences in individual abilities rather than by differences in competition-effect.

The explanation of the increases in speed during the competition trials must await further experimental work. Two explanations seem reasonable: (1) the increases in speed may have been the result of a social facilitation produced by the competition, and (2) the increases in speed may, on the other hand, have been the results of continued practice in swimming. An experiment is planned in which animals will swim for as long as have the animals in the present experiment but individually rather than in pairs. This experiment may reveal the maximal speeds resulting from continued practice in swimming. If these speeds are less than the speeds obtained from the competing animals of the present experiment, then it may be safe to conclude that the increases in speed in the present experiment are primarily social effects. Until such an experiment is performed there is no evidence for choosing between the two explanations.

3. Significance of the Mortalities

As previously pointed out, the mortalities must be dealt with cautiously. Since autopsies were not performed (it was only after several deaths that the relationships to nature of the early life and to victories became noticeable), it is not known what the physiological disorder was. Accordingly, no attempt will be made here to determine the significance of the mortalities.

4. The Relationship Between the Early Life and Victory

Since animals which lived in solitude since weaning were victors as often as were animals which lived in groups, it would follow that the nature of the post-weaning life is not the principal determiner of success in such competition as studied in this experiment.

(*a*) Victory could not be primarily a chance matter since these victories were consistent. There must be some factor or constellation of factors which accounts for such consistencies. The principal difference between the members of each

pair was the difference in the early life and according to the results, *this* factor is not the *one* which determined the victories.

(*b*) No one of such factors as relative age, relative weight, sex, and litter membership were found to bear a decided relationship to victory. Before accepting this conclusion it is necessary to determine to what extent the small number of cases influenced the lack of correlation. Further it is necessary to determine whether the victories might be determined not by a single one of these factors but by two or more acting together. That is, animals older than their competitors might not prove to be the victors but animals which were older and heavier, etc., than their competitors might be the victors.

(*c*) If the nature of the early life, age, weight, and sex prove to be unrelated to success in the competition, then it will appear that the related factor is in the heredity. To discover this relationship it will be necessary to breed selectively for succeeding generations. Only until genetically different strains are developed which differ also in their success in the competition ought it be concluded that the heredity is the principal determiner of success.

SUMMARY

1. Pairs of white rats were forced to swim under water and against a current to reach the exit. The rat to enter the exit first was allowed to escape to air. The second rat was trapped under water for a short time before it was permitted to escape.

2. Each pair consisted of one animal which had been reared in solitude and one which had been reared in a group with other animals. Each pair contained animals which previous tests of individual swimming proved to be of equal speed.

3. The number of solitarily reared victors in the competition was equal to the number of socially reared victors. In a small number of pairs neither member won consistently. It appears, then, that in this experiment, the nature of the early life is not the principal determiner of success in the competition.

4. Other factors such as age, weight, sex, and litter membership do not appear to be the principal individual determiners.

5. Significant and equal increases in speed during the competition trials occurred among most of the animals, whether solitarily reared or socially reared, whether victors or losers.

6. Attention is called to certain minimum cautions which must be observed in evaluating the procedure and results of this experiment.

REFERENCES

1. BAYROFF, A. G. The experimental social behavior of animals. I. The effect of early isolation of white rats on their later reactions to other white rats as measured by two periods of free choices. *J. Comp. Psychol.*, 1936, **21**, 67–81.

2. BAYROFF, A. G. Early environments and the social behavior of white rats (to appear).

3. CRAWFORD, M. P. The social psychology of the vertebrates. *Psychol. Bull.*, 1939, **36**, 407–445.

COMPETITIVE SOCIAL INTERACTION BETWEEN MONKEYS*

J. M. Warren and R. J. Maroney[1]

A. INTRODUCTION

THIS experiment was designed to determine the effects of several variables upon social interaction between monkeys in a competitive food-getting situation. The following problems were studied: (*a*) the effect of varying motivation and incentive conditions upon social dominance and submission; (*b*) the nature of the dominance-submission hierarchy in different groups of monkeys; (*c*) the relation between aggression and success in competition for food; (*d*) the relation of dominance status to weight, sex, and level of spontaneous activity.

B. METHOD

1. Subjects

The subjects were 18 prepubescent rhesus monkeys, nine males and nine females, which ranged in weight from $3\frac{3}{4}$ to $6\frac{1}{2}$ lb. at the beginning of the experiment. Six monkeys were assigned to three experimental groups, each consisting of three males and three females that had never been housed in the same cage. The composition of the groups, and the sex and median weights of individual monkeys during the two series of tests are given in Table 1. (One animal, No. 13, died after completing the first series.)

At the time this experiment began, the monkeys had been in the laboratory approximately 2

months and had been adapted to responding in the Wisconsin General Test Apparatus by 30 days' previous testing of lateral preference and of food preference.

2. Apparatus

The Wisconsin General Test Apparatus (Harlow, 1949) was used throughout the experiment; the monkeys occupy a large restraining cage which faces a table upon which is a movable test tray. On each trial, the experimenter placed the incentive upon the tray and, in full view of both monkeys, pushed it within reach.

3. Procedure—Series I

Within each of the groups of six monkeys, every animal was tested in competition with each of the other five monkeys every week for 4 weeks. Within a group there were 15 pairings of the six monkeys taken two at a time.

The effect of quality and quantity of incentive on social competition was studied by presenting nine incentive conditions: 1, 2, or 4 pieces of raisin, potato, or egg mash pellet. Each of the nine conditions was replicated six times in a test session, so that a total of 54 presentations was made in every test session. The position of the incentive on the test tray was varied from right to left according to a balanced irregular sequence from trial to trial.

* From the *Journal of Social Psychology*, 1958, **48**, 223–233. Copyright 1958 by The Journal Press, Provincetown, Mass.

[1] This investigation was supported (in part) by Research Grant M-835 from the National Institute of Mental Health, U.S. Public Health Service. The research was conducted at the University of Oregon.

TABLE 1. *Characteristics of Subjects*

Group	Monkey	Sex	Weight I	Weight II	Gain in Weight	Median Activity Score
I	4	F	$5\frac{1}{2}$	7	$1\frac{1}{2}$	201
	9	F	$5\frac{5}{8}$	$6\frac{3}{4}$	$1\frac{1}{8}$	384
	12	F	$4\frac{1}{2}$	$5\frac{1}{2}$	1	8
	13	M	$4\frac{3}{4}$	Dead		
	14	M	$4\frac{5}{8}$	6	$1\frac{3}{8}$	517
	18	M	$5\frac{3}{8}$	$6\frac{1}{2}$	$1\frac{1}{8}$	421
II	2	F	$5\frac{3}{8}$	$7\frac{1}{4}$	$1\frac{7}{8}$	96
	7	F	$4\frac{5}{8}$	6	$1\frac{3}{8}$	146
	8	F	$4\frac{1}{2}$	$5\frac{3}{4}$	$1\frac{1}{4}$	115
	10	M	$6\frac{1}{2}$	$7\frac{7}{8}$	$\frac{7}{8}$	327
	15	M	$4\frac{1}{2}$	$5\frac{5}{8}$	$1\frac{1}{8}$	109
	16	M	$4\frac{1}{4}$	$5\frac{1}{2}$	$1\frac{1}{4}$	46
III	1	F	$4\frac{1}{2}$	$6\frac{3}{4}$	$2\frac{1}{4}$	99
	3	F	$5\frac{1}{8}$	$5\frac{5}{8}$	$\frac{1}{2}$	97
	5	M	$5\frac{1}{8}$	$5\frac{3}{8}$	$\frac{1}{4}$	250
	6	M	$5\frac{1}{8}$	$6\frac{5}{8}$	$1\frac{1}{2}$	256
	11	F	$3\frac{3}{4}$	5	$1\frac{1}{4}$	39
	17	M	$4\frac{3}{8}$	$5\frac{1}{4}$	$\frac{7}{8}$	862

The results of the food preference experiment previously mentioned indicated that raisin, potato, and egg mash pellet were high, medium, and low preference foods, respectively, for these monkeys when tested individually.

In the second test period in this series, the dominant member of each pair in Groups I and II was fed his daily ration of chimcracker and fruit immediately before being tested with the subordinate of the first pairing. The subordinate member was tested after 23 hours of food deprivation, the condition which was standard for all monkeys throughout the remainder of the tests.

4. Series II

Approximately 6 months after the first four tests, all of the groups were retested twice. Two weeks intervened between the first and second of these retest pairings. In this series, 25 presentations of a single raisin were made in each testing session. At the time these tests were made, the monkeys' spontaneous locomotor activity was determined by testing each individual three times in a standard activity cage (French and Harlow, 1956).

Degree of dominance was measured by counting the number of trials on which each of the two competitors obtained the incentive. Occasionally, when several pieces of food were presented, both animals secured half of the food; each monkey was given credit for one-half success on these trials. In addition, the frequency of hitting, biting and mounting was recorded for Group *A*, on Series I, and for all groups on Series II.

C. RESULTS

1. Quality and Quantity of Incentive

The percentage frequency with which the subordinated monkey in each pair succeeded in securing food is plotted as a function of the quantity and quality of the incentive in Figure 1. Each of the points on these functions represents 3,240 individual observations. It is apparent that the quantity of the incentive has very little, if any, effect on the outcome of competition for food. The subordinated monkeys, however, obtained the low preference incentive, eggmash pellets, significantly more often than the more highly preferred incentives, potato and raisin ($X^2 = 4.63$, $P < 0.05$).

2. Characteristics of the Hierarchies

Since every monkey was tested in competition with every other member of his group in all of the testing periods, it was possible to obtain a composite measure of each animal's relative dominance in his group by computing the mean percentage of success in securing food, by the method of paired comparisons. The mean percentage of successful trials is plotted against successive test periods in Figures 2 to 4; the breaks in the abscissae between test periods 4 and 5 indicate the interval of 6 months between these tests. The most obvious similarity among the three groups is that each contained one or two monkeys of very low dominance status; the most striking inter-group differences are in the degree to which individuals attained consistent high dominance status. Clearly defined status differentiation emerged sooner in Group *A* than in the other groups.

The consistency of the dominance rankings as measured by rank correlations between successive

TABLE 2. *Correlations of Ranks on Successive Tests*

Test Periods	A	B	C
1 & 2	1.00**	.94**	.77*
2 & 3	.94**	1.00**	.71
3 & 4	.94**	.94**	.94**
4 & 5	1.00**	.37	.83*
5 & 6	.94**	.71	.77*

* Rho significant at 5 per cent level.
** Rho significant at 1 per cent level.

test periods is presented in Table 2. Twelve of the 15 correlations are significantly greater than zero, and in two of the groups high and significant correlations were obtained between tests after an interval of 6 months. The consistency of dominance rankings varied somewhat among groups, Group *A* showing almost perfect correspondence between tests, while Groups *B* and *C*'s behavior was less constant.

It should be noted that the correlations between tests 1 and 2, and tests 2 and 3, for Groups *A* and *B* are higher than the corresponding values for Group *C*, in spite of the fact that the dominant animal in each *A* and *B* pair was fed just before being tested in test period 2, while the *C* monkeys were tested under standard conditions.

Analysis of the outcomes of individual pairings provides additional information regarding the nature of the dominance hierarchy. The linearity of dominance relations may be inferred from the number of circular trials; i.e., relations of the type A > B, B > C, and C > A. Table 3 shows the number of such circular relations within each group on successive test periods; the maximum possible number of circular triads in a group of six is 8. The table shows that no such triads occurred in Group *A* after the first two periods, but that these departures from a transitive linear hierarchy were persistent in the other groups throughout test 5.

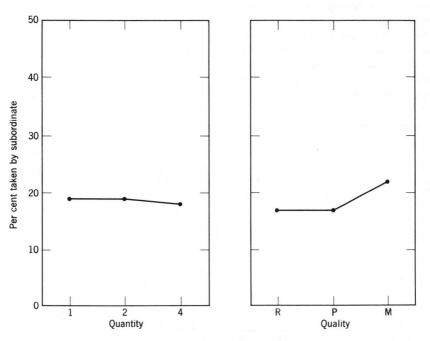

FIGURE 1. Frequency of success by subordinate monkeys in competing for food, as a function of quantity and quality of reinforcement.

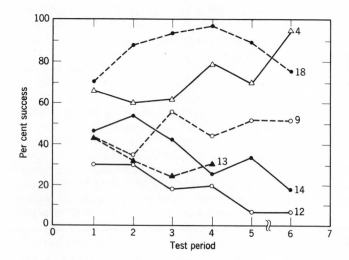

FIGURE 2. Mean dominance scores for individual monkeys in group *A*.

TABLE 3. *The Number of Circular Triads on Successive Test Periods*

Test	Group A	B	C	Sum
1	3	2	1	6
2	2	2	2	6
3	0	1	3	4
4	0	2	1	3
5	0	2	1	3
6	0	0	0	0

A second kind of information concerning the nature of dominance relations in young monkeys is derived from the results of individual pairings. The obtained frequency with which individual monkeys dominated competitors in their six encounters is tabulated in Table 4, and is compared with the expected frequency, assuming no

significant dominance, from the binomial expansion. The value of X^2 obtained (295.38, with 3 *df*) permits one to reject the null hypothesis at the 0.1 per cent level of confidence.

Another, and somewhat independent, measure of the stability of dominance-submission relationships is the frequency of reversals in dominance from one testing session to the next meeting of a given pair. One would infer a more stable change in dominance if, after being dominated in two tests, an animal was subsequently dominant in the remaining four tests, than if dominance shifted frequently between two monkeys. The frequency of reversals from one test period to the next is given for each group in Table 5; this table indicates, as do Tables 2 to 4, considerable variability between groups. There were proportionately fewer reversals between individuals in Group *A* than in Groups *B* and *C*. The total frequency of reversals over all groups remains

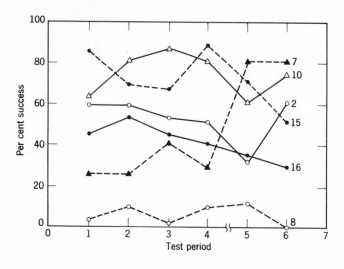

FIGURE 3. Mean dominance scores for individual monkeys in group *B*.

relatively constant; in about 20 per cent of the individual pairings the previous dominance-submission relation is reversed from one test period to the next.

TABLE 4. *Frequency With Which Monkeys Dominated Opponents in Six Testing Sessions*

Proportion	A*	B	C	Total Observed	Expected
6/6	6	8	6	20	1.25
5/6	3	1	5	9	7.50
4/6	1	5	3	9	18.75
3/6	0	1	1	2	12.50
Total	10	15	15	40	40

* Does not include pairings involving Monkey 13.

TABLE 5. *Number of Reversals in Successive Pairings*

Test Periods Compared	A	B	C	Sum
1 & 2	4/15	3/15	4/15	11/45
2 & 3	3/15	3/15	6/15	12/45
3 & 4	1/15	2/15	2/15	5/45
4 & 5	0/10	6/15	2/15	8/40
5 & 6	1/10	3/15	3/15	7/40
Total	9/65	17/75	17/75	43/215

3. The Influence of Weight, Activity, and Sex

In order to determine the effect of weight and level of spontaneous activity upon dominance

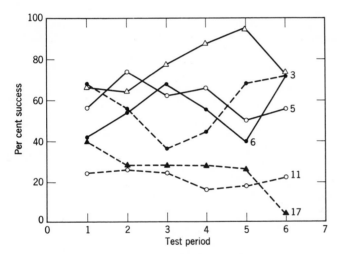

FIGURE 4. Mean dominance scores for individual monkeys in group C.

behavior, the average rank position in the dominance hierarchy for each monkey on Series I and II was correlated with his rank in weight within the group at the time of the two periods. Similarly, average dominance rank on Series II was correlated with median activity level. The results of these computations are presented in Table 6.

There is no indication of a significant relationship between activity level and dominance, nor is there any particularly striking evidence of a significant relationship between weight and dominance, although the correlations are all positive and those for the second series generally higher than for the first set of tests. One consideration, however, suggests that differences in weight are not ineffective in determining dominance: Group *A*, with the highest consistency in

dominance, was the only group with consistent individual differences in weight.

The influence of sex upon social dominance was determined by testing the significance of the difference between the mean ranks for males and females in all groups. The value of *t* obtained (0.61) was not statistically significant.

4. Aggression and Mounting

In an attempt to define the behavioral characteristics associated with dominant and submissive social behavior, detailed descriptive protocols were made for Group *A* on Series I and for all groups on Series II. The frequency of biting, striking, and mounting were recorded.

The records of aggressive behavior in Group *A* are summarized in Table 7, which gives the

TABLE 6. *Rank Correlations Between Dominance Status and Weight and Activity*

Variables Correlated	A	B	C
Weight I and dominance I	.48	.39	.89
Weight II and dominance II	.70	.66	.71
Activity and dominance II	.10	.54	− .14
Weight I and weight II	.90	.16	.37

frequency with which individual monkeys attacked others and were attacked by others on Series I and II. Note the great decrease in aggressive acts from Series I to II—the frequency of aggressive behavior was too low to permit any correlational analysis in the second series.

In order to determine the role of aggression in dominance among monkeys, three correlations were computed from the Series I data: (a) rho between rank in dominance and frequency of aggression is +0.77 and significant at the 5 per cent level of confidence; (b) rho between the frequency with which a monkey was attacked and

the number of his attacks upon others is −0.30; (c) rho between position in dominance hierarchy and number of times attacked is −0.47. These correlations suggest that, among these monkeys, aggression is an instrumental act to minimize frustration, rather than a response to frustration, since the lower an animal's position in the dominance hierarchy and the more frequently he is attacked by others, the *less* frequently he is aggressive toward others. This suggestion is compatible with the very low frequency of aggression in Series II. A clear-cut differentiation of status had been attained and little further

TABLE 7. *Incidence of Aggression by Individual Monkeys*

Group	Monkey	Bit Others	Struck Others	Was Bitten	Was Struck
AI	4	0	6	2	8
	9	3	7	1	21
	12	0	0	1	9
	13	16	5	1	10
	14	1	1	23	44
	18	9	73	1	0
Total		29	92	29	92
AII	4	3	0	0	0
	9	3	0	0	1
	12	0	0	1	0
	14	0	0	3	0
	18	1	1	3	0
Total		7	1	7	1
BII	2	0	0	0	1
	7	5	1	0	0
	8	0	0	1	1
	10	0	0	0	0
	15	1	1	2	0
	16	0	0	3	0
Total		6	2	6	2
CII	1	0	2	4	0
	3	1	0	0	0
	5	1	7	0	1
	6	3	1	0	0
	11	0	4	1	3
	17	0	0	0	10
Total		5	14	5	14

aggression was required to maintain it; this is supported by similarly low incidence of aggression in Groups *B* and *C*.

Mounting of like or opposite sexed subordinates by a dominant male or female is a frequently reported phenomenon in adult monkeys. Mounting was observed in 10 of the 80 pairings of animals in Group *A*, and only two males (18 and 14) exhibited this behavior; the lowest dominance animals were never mounted. No relation between mounting and social dominance is suggested by these observations.

D. DISCUSSION

The results of this study indicate that dominance behavior is not affected by prefeeding the dominant member of a pair of monkeys competing for food, nor is the competitive behavior influenced by varying the quantity of food incentive; subordinate monkeys obtain more low preference food than middle or high preference foods. These observations are compatible with the view that a separate dominance drive exists, which is essentially social and relatively independent of physiological, homeostatic drives in the primate (Maslow, 1936a).

Insofar as procedures and subjects are comparable, the results of this experiment are in very good agreement with those obtained in similar laboratory investigations with Macaca mulatta (Maslow, 1936b; Miller and Murphy, 1956). Just as Maslow reported, we found that a stable and linear dominance hierarchy was established within relatively few pairings; like Miller and Murphy, we found that the dominance hierarchies were reproducible after several months without testing. The present investigation is of value in confirming in general the results of the previous experiments, and in demonstrating that the dominance-submission behavior of prepuberal monkeys is very similar to that of adult animals.

The most striking difference between the results of this study and those of Maslow and of Miller and Murphy is in the much lower incidence of mounting observed in the present experiment. Such a discrepancy between experiments involving sexually mature and immature subjects is to be expected.

The behavior of cats in the same competitive social situation has been described in a previous publication (Baron, Stewart, and Warren, 1957). The most noteworthy difference between the species was with respect to the relation between aggression and dominance status. In cats, no well-defined relation was found; cats which are most successful in competing for food may be so nonaggressive as not to retaliate when attacked, while cats least successful in food getting may be highly aggressive. In monkeys, only high dominance animals exhibit much aggression, and only during the early tests. This difference suggests that dominance and submission may be more completely defined behaviorally in monkeys than in cats.

E. SUMMARY

Eighteen rhesus monkeys were divided into three subgroups consisting of three males and three females each. Within each group, every monkey was tested in competition for food with each of the other five members of the group on four occasions over 1 month (Series I). Two additional repetitions of the 15 paired comparisons with a group were made after an interval of approximately 6 months. The following results were obtained:

1. Variation in the quantity of incentive had no effect on the degree of dominance observed.

2. Subordinate animals were more successful in obtaining low preference incentives than highly or intermediately preferred foods.

3. Prefeeding the dominant animal did not affect dominant behavior appreciably.

4. A stable and eventually linear dominance hierarchy was obtained in each group.

5. Dominance behavior was not related to weight, sex, or level of spontaneous activity.

6. The correlation between aggression and success in getting food was $+0.77$.

REFERENCES

1. BARON, A., STEWART, C. N. and WARREN, J. M. Patterns of social interaction in cats (*Felis domestica*). *Behavior*, 1957, **10**, 56–66.

2. FRENCH, G. M. and HARLOW, H. F. Locomotor reaction decrement in normal and brain-damaged monkeys. *J. Comp. & Physiol. Psychol.*, 1955, **48**, 496–501.

3. HARLOW, H. F. The formation of learning sets. *Psychol. Rev.*, 1949, **56**, 51–65.

4. MASLOW, A. H. The rôle of dominance in the social and sexual behavior of infra-human primates: III. A theory of sexual behavior of infra-human primates. *J. Genet. Psychol.*, 1936, **48**, 310–338.

5. ——— (1936b). The rôle of dominance in the social and sexual behavior of infra-human primates: IV. The determination of hierarchies in pairs and in a group. *J. Genet. Psychol.*, 1936, **49**, 161–198.

6. MILLER, R. E. and MURPHY, J. V. (1956). Social interactions of rhesus monkeys: I. Food-getting dominance as a dependent variable. *J. Soc. Psychol.*, 1956, **44**, 249–255.

CHAPTER NINE

DOMINANCE AND SOCIAL ORGANIZATION

IN the last selection of the previous chapter Warren and Maroney studied the fate of competitive contests over a rather extended period of time. Repeated measures of competitive encounters revealed that monkeys are quite consistent in winning or losing a contest. But more important, Warren and Maroney also found that the differentiation between winners and losers increases over time. That is, after a series of trials one can observe a clear rank-ordering of animals in terms of their probabilities of success in winning an encounter. Over a period of time the initially successful animals become more successful and the initially unsuccessful animals sink even lower. One more observation should be made with respect to the Warren–Maroney data, and that is that while there was a good deal of aggression and fighting during the initial stages of competition, the fighting almost completely disappeared toward the end of the experiment. I have stated elsewhere (Zajonc, 1966) that the difference between conflict and competition lies in the control over the instrumental behavior that is directed toward the attainment of the desired incentives. In what is usually called "conflict" such instrumental behavior is primarily under the control of the internal states of the parties in conflict, their drives and emotions; and it is limited by their skills and fears. In competition, however, the critical instrumental responses also seem to be subject to *conventional* controls; that is, in addition to the above determinants they are also under the control of "rules" which the parties in competition recognize and observe. Very often these "rules" stand in opposition to internal impulses. The emergence and maintenance of a stable competitive rank-ordering, such as among Warren and Maroney's monkeys, and the elimination of physical strifes, implies the existence of some form of "rules" or conventions. To be sure, these conventions are of a very primitive character, and they consist, perhaps, simply of acknowledging the priority of each individual group member to some privileges coveted by all. Nevertheless, they have all the properties of other social norms, for they are evolved by the group, they are learned by the members of the group, they are shared by them, they are stable, they are arbitrary, they can be changed, and they are complied with by a great majority of the group members.

It may be argued that as a consequence of conflict and competition there emerges a stable social order, whose underlying basis is a dominance hierarchy. The existence of such a hierarchy defines for the individual members priorities of access to mating, to territory, to food, etc., and thus functions as a means of social integration. Wynne-Edwards (1962), who was concerned with the density of animal populations, suggested that the function of the hierarchy "*is always to identify the surplus individuals* whenever the population-density requires to be thinned out, and it has thus an extremely high survival value for the society as a whole" (p. 139).

The selections in this chapter deal with the establishment of dominance hierarchies, with the behavioral concomitants of such hierarchies, and with the antecedents and consequences of changes in dominance structures. The first article by Murchison describes an attempt to measure dominance structures in hens, and to relate the positions in the dominance structure to behavioral parameters, and to physical consequences for the animal. Wynne-Edwards' hypothesis receives some suggestive support from Ewing's study of the establishment of a hierarchy among cockroaches. His findings reveal that as was the case in the study of monkeys by Warren and Maroney, when hierarchy is established, fighting ceases. It is also consistent with the Wynne-Edwards theory that death, apparently from stress, is more likely to occur among the subordinate animals than among the dominant ones. The experiments by Smith and Hale and Leary and Slye deal with the modifications of dominance structures, the first employing avoidance conditioning by means of shock and the second through psychophysiological effects, using drugs that lower the general activation level of the animal. The first study used hens as subjects while the second observed monkeys. The psychophysiological correlates of dominance behavior among hens are extensively explored in the selection by Guhl. The nature of social organization and the intricate pattern of interrelationships within the hierarchy is seen when a dominant group member is removed. Bernstein's findings shed light on the strong behavioral interdependence among the animals in the dominance structure. Removing the dominant male clearly results in the increase of various social behaviors of the remaining animals, and his later reintroduction to the group causes a marked reduction in these behaviors. The role of the dominant male with respect to the external environment is illustrated in the final selection, also by Bernstein.

REFERENCES

WYNNE-EDWARDS, V. C. *Animal dispersion in relation to social behavior.* New York: Hafner Co., 1962.

ZAJONC, R. B. *Social psychology: An experimental approach.* Belmont, Calif.: Brooks/Cole, 1966.

THE EXPERIMENTAL MEASUREMENT OF A SOCIAL HIERARCHY IN *GALLUS DOMESTICUS:* IV. LOSS OF BODY WEIGHT UNDER CONDITIONS OF MILD STARVATION AS A FUNCTION OF SOCIAL DOMINANCE*

Carl Murchison

THREE previous papers (13, 14, 15), [are] highly important for the understanding of this paper. A brief abstract of the findings of those three papers is as follows:

> Beginning at 16 weeks of age, six young roosters are arranged in a hierarchy of dominance, the order being determined by the number of individuals in the group that each rooster is able to defeat in physical combat (Social Reflex No. 2). This order of ranking is revised at intervals of four weeks from the sixteenth to the thirty-sixth weeks. Beginning immediately after being taken from the incubator, these individuals had been tested at frequent intervals in the Social Reflex Runway. This test consisted simply of releasing two individuals simultaneously from opposite ends of the runway, and then observing the time spent and distance traversed by each in running to the other (Social Reflex No. 1). Various operations involving the concepts of physics were applied without great success to these data. Then simple measurements of time and space were applied. When plotted as a function of Social Reflex No. 2, it was found that Social Reflex No. 1, plotted in terms of space alone, was almost truly linear. A theoretical correction of the abcissa units, which agreed with the empirical data, satisfied the requirements of linear function.
>
> Social discrimination in *Gallus domesticus* is identified as it is measured in the Social Discrimination Cage. When the discriminations are plotted against Social Reflex No. 2, a relationship appears which approaches a linear form. When the discriminations are plotted against Social Reflex No. 1, true linearity is approximated at the thirty-sixth week. The analyses show that male discriminations for pairs of males are away from dominance, and that female discriminations for pairs of males are in

the direction of dominance. The constancy of this trait in males makes it possible to measure social discrimination in *Gallus domesticus* in units of space.

Social Reflex No. 3 (the sex reflex) is measured in terms of the total treadings in which each individual engages during a period of time. As so measured this reflex in male *Gallus domesticus* is a linear function of Social Reflex No. 1 and Social Reflex No. 2, while in female *Gallus domesticus* it is a function of Social Reflex No. 1. Observations seem to indicate that such subjective matters as sex favoritism, social insults, and social integration may eventually be exhibited as linear functions of such measurable quanta as Social Reflex No. 1, Social Reflex No. 2, and Social Reflex No. 3.

It is pointed out that these or some similar methods, built upon the presupposition that the methods operate on behavior quanta common to the social conduct of all social animals, can reach through the medium of covariable techniques to the eventual formulation of social law.

At this stage in these experiments it becomes desirable to examine some of the physiological relations of these social reflexes. These relations are to be exhibited without prejudice or theory as to causality. It is not entirely necessary that they be examined, as the social reflexes alone can be subjected to an infinite series of analytical experiments without any reference to any physiological basis. It is in the interest of a broader field of investigation that this examination is being made, and not because of any logical compulsion.

BRIEF HISTORY OF THE PROBLEM OF INSENSIBLE BODY LOSS

This problem was recognized as of scientific

---

* From the *Journal of General Psychology*, 1935, **12**, 296–312. Copyright 1935 by The Journal Press, Provincetown, Mass.

importance by Sanctorius in 1614 (16), who recorded the variations in his body weight over a period of 30 years. In 1740 Lining (10, 11) weighed himself each day for a year just after arising each morning and just before retiring each night. It was Lining who determined seasonal variations in body weight. Lavoisier in 1790 pointed out the necessity of making a distinction between cutaneous respiration and pulmonary respiration. Colin in 1862 (6) made a series of measurements on himself which indicated a great difference in body loss during repose and during activity. In 1869 von Leyden (17) applied the method to fever patients, and in 1873 Jürgensen (9) applied the method to normal men as a second variable in the measurement of temperature. Dennig (8) in 1898 studied the relation of water intake to body loss.

The use of an accurate balance scale in connection with a respiration calorimeter initiated the modern era in this field, and was the contribution of Atwater in 1896 (1). Lombard (12) in 1906 perfected an extraordinary balance that would determine losses during a period of a few minutes. Caspari (7) in 1910, while on an expedition, determined daily variations in body loss during variations in activity. In 1910 Benedict and Carpenter (4) made a large number of measure-

ments on subjects at rest and during work. Benedict (2) also applied the method in 1915 to a fasting subject. Benedict and Hendry (5) in 1921 applied the method to a group of girls, and Benedict (3) in 1923 continued the study with two other groups of girls. Benedict and Root (6) in 1926 determined an important relationship between body loss and metabolism, and concluded that accurate determinations of loss in body weight may be used clinically as a method of determining rate of metabolism.

SUBJECTS

The subjects used in this study are the six young roosters described as members of Group D in the previous papers of this series.

APPARATUS

The balance scale used in this study is a simple scale that can be purchased for about $40. The entire length of the scale arm represents 1 oz., and this arm is divided into 100 reading units. The individual isolation boxes were 18 in. square, each fitted with a wire-mesh floor underneath

TABLE 1. *Test Made During the 43rd and 44th Weeks*

	YY	Blue	Green	Yellow	Red	White
Weight in kilograms before 24 hours of isolation in darkness	4.00	3.64	3.49	3.92	3.75	3.42
Weight in kilograms after 24 hours of isolation in darkness	3.87	3.56	3.38	3.77	3.65	3.33
Gross loss in grams (a)	122.18	81.64	103.19	144.58	98.65	89.02
Weight in kilograms before 24 hours of social activity under light	4.09	3.75	3.59	3.95	3.85	3.40
Weight in kilograms after 24 hours of social activity under light	3.89	3.55	3.42	3.76	3.70	3.26
Gross loss in grams (a)	196.16	196.75	169.53	192.78	147.13	131.54
(a) and (b)	320.34	278.39	272.72	337.36	245.78	220.56
Weight in grams of feces	76.08	55.31	64.07	137.10	68.04	57.26
Net loss in grams during the 48 hours of starvation	244.26	223.08	208.65	200.26	177.74	163.30
Average weight in kilograms since birth	2.65	2.59	2.59	2.68	2.59	2.42
Percentage of average weight since birth lost during 48 hours of starvation	9.21	8.61	8.05	7.47	6.86	6.74

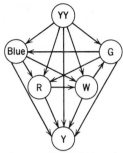

Male "dominance" hierarchy
16th week

FIGURE 1.

which a sheet of brown paper could be slipped into place. The Social Reflex Runway was used as the place for the social activity of the entire group.

PROCEDURE

At the end of the forty-third week the six roosters of Group D were carefully weighed and then placed in the six individual isolation boxes. These boxes were placed in a warm dark room. At intervals of 4 hours each chick was removed from his box, carefully weighed, returned to his box, the feces carefully weighed, and a new sheet of brown paper inserted. The chicks had no food or water during the experimental period, which lasted for 28 hours.

At the end of the forty-fourth week the six roosters of Group D were carefully weighed and placed in the Social Reflex Runway with the five pullets. The runway was in a warm room and was kept brilliantly illuminated during the entire

Male "dominance" hierarchy
20th week
24th week
28th week

FIGURE 2.

period of 28 hours. During this period none of the roosters had any food or water. The roosters were again carefully weighed at the end of 24 hours and 28 hours. It was assumed that the weight of the feces for each chick during this period would be the same as during the isolation period of the same length of time.

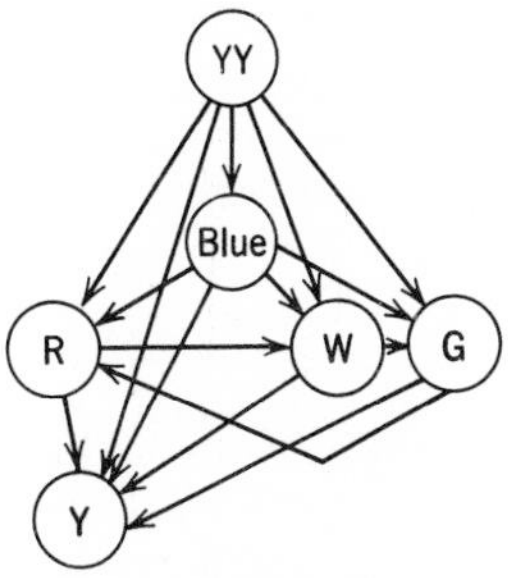

THE RAW DATA

The two series of measurements are given in all necessary detail in Table 1. The measures of body weight since birth to the thirty-sixth week were reported in [a previous] paper (14), while the weights for the fortieth week were reported in [a] second paper (15). The weights at the beginning of the test are in Table 1. The average weight since birth is the average of the monthly averages. It is necessary to assume familiarity with the raw data reported in the three papers already published.

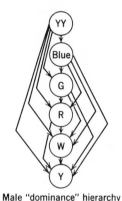

Male "dominance" hierarchy
32nd week

FIGURE 3.

Figures 1–4 contain the record to date of the male dominance hierarchy in Group D. The record stops at the fifty-first week, since "Y" has just died.

ANALYSES OF THE DATA

1. *In Terms of Net Loss of Weight Without Regard to Percentages.*

(a) *As a function of Social Reflex No. 2.* In Figure 5 the net loss of weight in grams during 48 hours of starvation at the forty-fourth week is plotted as a function of Social Reflex No. 1 at the

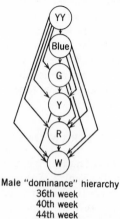

Male "dominance" hierarchy
36th week
40th week
44th week
48th week
51st week

FIGURE 4.

(c) *As a function of Social Reflex No. 3.* In Figure 7 the net loss of weight in grams during 48 hours of starvation at the forty-fourth week is plotted as a function of the total observed treadings of Social Reflex No. 3. The function is not linear in the three subjects, but the relation certainly approaches linearity. In Figure 8 the net loss of weight in grams during 48 hours of observation at the forty-fourth week is plotted as a function of the theoretical formulation of Social Reflex No. 3 in number of treadings based on the theoretical assumption that all six of the male members of Group D engage in treading with the total number of treadings remaining constant. In this case the function approaches true linearity.

thirty-sixth week. It is obvious that the function approaches linearity.

(b) *As a function of Social Reflex No. 1.* In Figure 6 the net loss of weight in grams during 48 hours of starvation at the forty-fourth week is plotted as a function of Social Reflex No. 1 at the thirty-sixth week. The function is almost truly linear.

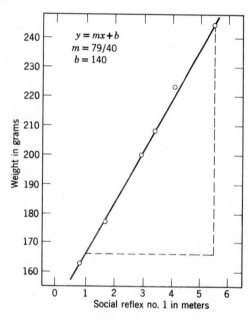

FIGURE 6. Loss of weight in grams during 48 hours of starvation at the forty-fourth week as a function of Social Reflex No. 1 at the thirty-sixth week. Male *Gallus domesticus*, group D.

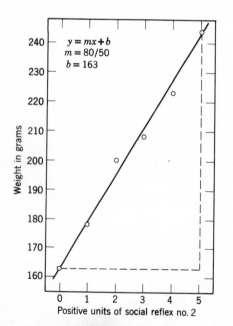

FIGURE 5. Loss of weight in grams during 48 hours of starvation at the forty-fourth week as a function of Social Reflex No. 2 at the thirty-sixth week. Male *Gallus domesticus*, group D.

(d) *As a function of social discrimination.* In Figure 9 the net loss of weight in grams during 48 hours of starvation at the forty-fourth week is plotted as a function of the percentages of times individuals receive negative social discriminations at the thirty-sixth week. The function is almost linear.

2. *In Terms of Net Loss of Weight Expressed in Terms of Percentages of the Average Weight Since Birth.*

(a) *As a function of Social Reflex No. 2.* In Figure 10 the net loss of body weight in grams during 48 hours of starvation at the forty-fourth week, expressed in terms of percentages of the average weight since birth, is plotted as a function of Social Reflex No. 2 at the fortieth week. With the exception of one experimental point, the function is truly linear.

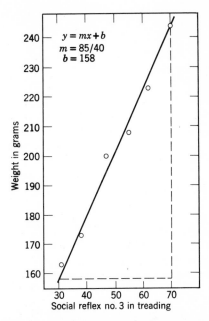

FIGURE 8. Loss of weight in grams during 48 hours of starvation at the forty-fourth week as a function of the theoretical formulation of Social Reflex No. 3 for all six roosters. Male *Gallus domesticus*, group D.

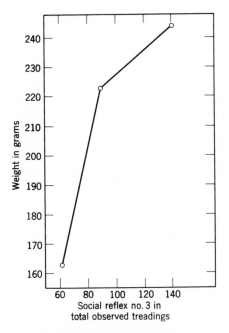

FIGURE 7. Loss of weight in grams during 48 hours of starvation at the forty-fourth week as a function of Social Reflex No. 3 in total observed treadings. Male *Gallus domesticus*, group D.

(b) *As a function of Social Reflex No. 1.* In Figure 11 the net loss of body weight in grams during 48 hours of starvation at the forty-fourth week, expressed in terms of percentages of the average weight since birth, is plotted as a function of Social Reflex No. 1 at the fortieth week. The function approaches true linearity.

(c) *As a Function of Social Reflex No. 3.* In Figure 12 the net loss of body weight in grams during 48 hours of starvation at the forty-fourth week, expressed in terms of percentages of the average weight since birth, is plotted in relation to Social Reflex No. 3 of the three treading roosters in Group D. The relation is not linear, but

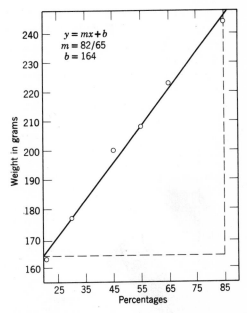

FIGURE 9. Loss of weight in grams during 48 hours of starvation at the forty-fourth week as a function of the percentages of times individuals receive negative social discrimination at the thirty-sixth week. Male *Gallus domesticus*, group D.

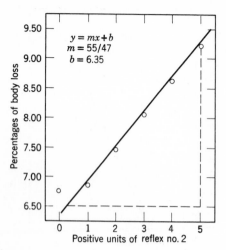

FIGURE 10. Net loss in body weight during 48 hours of starvation at the forty-fourth week. Expressed in terms of percentages of the average weight since birth, plotted as a function of Social Reflex No. 2 at the fortieth week. Male *Gallus domesticus*, group D.

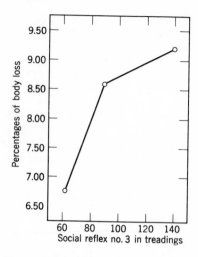

FIGURE 12. Net loss in body weight during 48 hours of starvation at the forty-fourth week. Expressed in terms of percentages of the average weight since birth, in relation to Social Reflex No. 3 of the three treading roosters in group D.

indicates the possibility of linearity in a larger hierarchy. In Figure 13 this latter possibility is realized theoretically in all but one experimental point.

(*d*) *As a function of social discrimination.* In Figure 14 the net loss of body weight in grams during 48 hours of starvation at the forty-fourth week, expressed in terms of percentages of the average weight since birth, is plotted as a function

of the percentages of times each individual receives negative social discriminations at the thirty-sixth week. The function is almost truly linear.

DISCUSSION

No assumption is made in this paper concerning the problem of metabolism except such as is given by the reference to the 1926 paper by Benedict and Root (6).

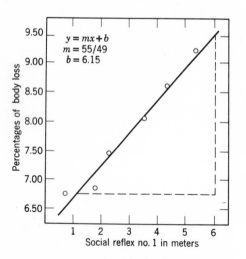

FIGURE 11. Net loss in body weight during 48 hours of starvation at the forty-fourth week. Expressed in terms of percentages of the average weight since birth, plotted as a function of Social Reflex No. 1 at the fortieth week. Male *Gallus domesticus*, group D.

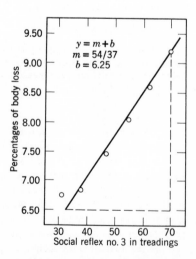

FIGURE 13. Net loss in body weight during 48 hours of starvation at the forty-fourth week. Expressed in terms of percentages of the average weight since birth, plotted as a function of the theoretical distribution of Social Reflex No. 3 if all six individuals participated. Male *Gallus domesticus*, group D.

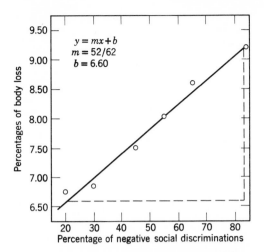

$y = mx + b$
$m = 52/62$
$b = 6.60$

FIGURE 14. Net loss in body weight during 48 hours of starvation at the forty-fourth week. Expressed in terms of percentages of the average weight since birth, plotted as a function of the percentages of times each individual receives negative social discrimination at the thirty-sixth week. Male *Gallus domesticus*, group D.

It is not accidental that no reference has been made thus far to the measurement of loss in body weight expressed as a percentage of the body weight just previous to the period of starvation. That has been the traditional method, but there is little theoretical support for such a method. The average weight since birth is a much more stable base from which to determine the percentage of loss. Obviously such a method can be applied only to animals that spend their entire lives in the laboratory.

It is pointed out once more that the raw data concerning the social reflexes and social discriminations have been published in previous papers and are not reproduced here (14, 15, 16).

A FURTHER THEORETICAL POINT

It has been pointed out that there is still a slight confusion concerning the reading of the units in the Social Reflex Runway. At the risk of boring those who are conversant with the theory of this method of measurement, the point will be cleared up. In Figure 15 is reproduced once more the runway as it has previously been reproduced in this series of papers. Following the method used by the Bureau of Standards in numbering the units of deflection from the theoretical zero of a balance scale, the theoretical zero of the scale of this runway could not be located at either end. The theoretical zero is located at the exact center of the runway, and the numbers from 1 to 5 are intended merely to indicate the point of origin of the series on the right half of the scale. Figure 16 gives the scale as it is actually used in the readings of experiments. Perhaps this improper method of printing the scale will clear up any confusion concerning how readings are taken from such a scale.

GENERAL SUMMARY

For male *Gallus domesticus* under the conditions of this experiment the net loss in body weight, over a divided period of 48 hours of starvation containing equal increments of inactivity and social activity, approximates true linear functions of Social Reflex No. 1, Social Reflex No. 2, Social Reflex No. 3, and of social discrimination. The linearity of these functions remains highly constant whether the net loss in body weight is expressed without reference to percentages or is expressed in terms of percentages

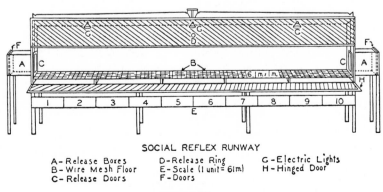

SOCIAL REFLEX RUNWAY

A – Release Boxes
B – Wire Mesh Floor
C – Release Doors
D – Release Ring
E – Scale (1 unit = 6 ml)
F – Doors
G – Electric Lights
H – Hinged Door

FIGURE 15.

of the average weight since birth. These findings support the validity of the identification and measurement of the three social reflexes, but raise no question as yet concerning the relation of the reflexes to the general biochemical problem of metabolism.

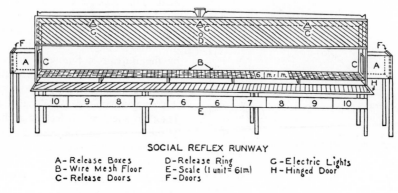

SOCIAL REFLEX RUNWAY

A - Release Boxes   D - Release Ring   G - Electric Lights
B - Wire Mesh Floor   E - Scale (1 unit = 61m)   H - Hinged Door
C - Release Doors   F - Doors

FIGURE 16.

REFERENCES

1. ATWATER, W. and BENEDICT, F. Experiments on the metabolism of matter and energy in the human body, 1898–1900. *U.S. Dept. Agric., Exp. Station Bull.* 109. Washington, D.C.: Govt. Print. Off., 1902. Pp. 147.

2. BENEDICT, F. A study of prolonged fasting. Washington, D.C.: Carnegie Instit. Wash., 1915. Pp. 84.

3. ———. The basal metabolism of young girls. *Boston. J. Med. & Surg.*, 1923, **188**, 127–138.

4. BENEDICT, F. and CARPENTER, T. The metabolism and energy transformations of healthy man during rest. Pub. 126. Washington, D.C.: Carnegie Instit., Wash., 1910. Pp. 255.

5. BENEDICT, F. and HENDRY, M. The energy requirements of girls from 12 to 17 years of age. *Boston J. Med. & Surg.*, 1921, **184**, 217; 257; 282; 297; 329.

6. BENEDICT, F. and ROOT, H. Insensible perspiration: Its relation to human physiology and pathology. *Arch. Int. Med.*, 1926, **38**, 1–35.

7. CASPARI, W. Ueber den Stoffwechselversuch in Alagna und über die Einwirkung kurzdauernden Aufenthaltes in grösseren Bergeshöhen auf den Stoffwechsel. *Denkschr. d. math. naturwissensch. Klasse d. kaiserl., Akad. d. Wissensch.*, 1910, **86**, 483.

8. DENNIG, A. Die Bedeutung der Wasserzufuhr für den Stoffwechsel und die Ernährung des Menschen. *Zsch. f. diätet u. Physik. Therap.*, 1898, **1**, 281–299.

9. JÜRGENSEN, T. Die Körperwärme des gesunden Menschen. Leipzig: Vogel, 1873. Pp. iv + 100.

10. LINING, J. Extracts of two letters to James Jurin. *Phil. Trans.*, 1742–43, **42**, 491–509.

11. ———. A letter to James Jurin. *Phil. Trans.*, 1744–45, **43**, 318–330.

12. LOMBARD, W. A method of recording changes in body weight which occur within short intervals of time. *J. Amer. Med. Asso.*, 1906, **47**, 1790–1793.

13. MURCHISON, C. The experimental measurement of a social hierarchy in *Gallus domesticus*: I. The direct identification and direct measurement of Social Reflex No. 1 and Social Reflex No. 2. *J. Gen. Psychol.*, 1935, **12**, 3–39.

14. ———. The experimental measurement of a social hierarchy in *Gallus domesticus*: II. The identification and inferential measurement of Social Reflex No. 1 and Social Reflex No. 2 by means of social discrimination. *J. Soc. Psychol.*, 1935, **6**, 3–30.

15. ———. The experimental measurement of a social hierarchy in *Gallus domesticus*: III. The direct and inferential measurement of Social Reflex No. 3. *J. Genet. Psychol.*, 1935, **46**, 76–102.

16. SANCTORIUS. Medicina statica. (Trans. by John Quincy.) London: Newton, 1720. Pp. viii + 344.

17. VON LEYDEN, E. Utersuchungen über das Fieber. *Dtsch. Arch. f. klin. Med.*, 1869, **5**, 308.

FIGHTING AND DEATH FROM STRESS IN A COCKROACH*

L. S. Ewing

INTRASPECIFIC fighting between males has been described for a number of insects including field crickets (1), cicadakiller wasps (2), ants where fighting occurs between colonies of the same species (3), and for a wood roach where the male defends a mating chamber against rivals (4). Although no published records have been found, fighting can be observed commonly between adult males of the cockroach *Nauphoeta cinerea*, where it appears to be associated with a loose territorial system. Fighting, which first appears on the second and third days after the imaginal molt, involves a complex sequence of events that eventually establishes a stable dominant-subordinate relationship between members of a fighting pair.

In an encounter between two more-or-less evenly matched males, a fairly consistent sequence of events can be observed. Both animals, with heads lowered, extend upwards the last three or four abdominal segments, simultaneously lifting the body high off the ground (Fig. 1). This posture may be assumed on sight, when one aggressive male crosses the path of another, or after brief but rapid antennal flagellation (fencing) between the two animals. This posture could be described as aggressive. It always precedes fighting but may cause a less aggressive male to flee. Following this display, two aggressive animals charge towards each other with their heads lowered and butt on contact. If one cockroach successfully engages its pronotum under that of its opponent, it may toss the rival in the air so that it falls on its back. Less frequently, males may grapple with their legs locked together and bite at each other as they roll over and over. A critical stage is usually reached within a few minutes, and one animal emerges superior.

The behavior of the loser is quite characteristic. After prolonged chasing by the dominant, it suddenly lies still and tucks its limbs under its body and its head under the shield of the prono-

tum. The antennae either lie flat on the ground, straight in front of the animal, or, less frequently, pointing backwards, parallel to the body, which is always pressed close to the ground (Fig. 2). Once this posture has been adopted, the subordinate animal no longer attacks the dominant; it runs underneath its superior and adopts the subordinate posture if subjected to attack itself.

It is possible to determine the dominant-subordinate relationship within a pair of males from their response to a test animal, a live adult male attached to a thin balsa-wood stick by means of a wire threaded through the pronotum. Amputation of its tibiae and tarsi restricts the movements of the test animal, thus facilitating manipulation. Antennal contact is allowed between the test animal and the subject, this generally being sufficient to elicit a response. Dominant animals typically adopt the aggressive posture and may attack the test animal, whereas subordinate animals do not respond aggressively and adopt the subordinate posture. Test animals can also be used in the determination of the earliest age at which the fighting response can be elicited in young adult males. Indeed, until fighting has appeared, dominant-subordinate relationships cannot be established.

TABLE 1. *The Relationship Between Fighting and Death, Single Males, Still Teneral, Were Left Untouched as Controls (Group 1); Pairs of Males Still Teneral, Were Left Untouched (Group 2); and Pairs of Males, Still Teneral, Were Confronted Once Daily With The Test Animal (Group 3)*

Group	Males (No.)	Deaths (No.)	Deaths in subordinates (No.)
1	30	3	
2	30	7	
3	30	15	12

FIGURE 1. The aggressive posture.

In the course of experiments to determine the time at which fighting first appears, it was noticed, incidentally, that, over the first few days after the adult molt, the death rate among paired males was rather high. I set up three groups of young adults to see if there was some connection between these deaths and fighting (Table 1).

The numbers of deaths occurring in the first two groups do not differ significantly ($\chi^2 = < 0.3$). The findings for the third group indicate that there is an increase in the number of deaths following repeated presentation with the test animal and that, where deaths do occur, they occur in subordinate animals in 80 per cent of the cases. In all 12 subordinate animals, death occurred on days 3 and 4, that is, after fighting had first appeared in the dominant. The increase in death rate is significantly different from that of the the control group ($\chi^2 = 0.01$) and probably significantly different from group 2 ($\chi^2 = 0.05$).

It would seem that there is some connection between deaths in subordinate animals and fighting. It is puzzling, however, that more deaths do not occur in group 2. Casual observation suggests that, if these animals fight at all, they fight considerably less than animals of group 3. It may be that cage-mates reared together from the time of emergence do not fight each other unless presented with a novel stimulus—in this case, the test animal. This phenomenon is not unknown and has been called passive inhibition. It occurs in young mice that normally fight at 35 days of age but do not fight cage-mates with which they have been reared (5).

To test whether deaths from fighting occurred in older cockroaches, I set up two more groups. Pairs of males were left together for 24 hours, then presented once with the test animal. As controls, single animals were treated in the same way. Both groups were observed for a period of 10 days, and the deaths were recorded.

No deaths occurred in the control group of 30 animals. Four of the 30 experimental males died, and all of these were subordinate animals. This offers some support for the suggestion that there is some connection between being subordinate and dying after fighting. It also suggests that older adults are less susceptible to death under these conditions.

Some mention must be made of the events leading up to deaths. Once a dominance relationship has been established, the frequency and intensity of fighting is usually reduced. Sometimes, however, the adoption of a subordinate posture on the part of the inferior animal fails to allay attack, and the dominant animal may continue to fight for long periods of up to 30 min. Towards the end of this time the movements of the submissive become sluggish and stiff; the righting reflex disappears; and a state of semi-paralysis, affecting the abdomen and limbs, sets in. Animals in this state apparently do not recover, and death soon follows. They die after an extended bout of fighting but characteristically show no signs of external damage. Some animals do survive prolonged attack over a period of several weeks. Such animals may have entire wings or limbs removed by their superiors.

The situation bears striking resemblance to the social stress found in mammals. Male rats in particular show a well-marked dominant-subordinate behavior. Prolonged aggression produces stress in the subordinate, and ultimately a disease state, characterized by the stress syndrome, leads to death, which cannot be attributed to external damage (6). Subordinate cockroaches may die from some internal changes comparable to those accompanying the stress syndrome in mammals.

Several problems arise from these observations; the corpus cardiacum may play some role in the development and maintenance of dominant-subordinate behavior and in death from stress. Indeed, it repeatedly shows severe depletion of neurosecretory material during artificially induced stress, such as forced hyperactivity and electrical shocks (7).

FIGURE 2. The subordinate posture.

REFERENCES AND NOTES

1. ALEXANDER, R., *Behaviour*, **17**, 131 (1961).
2. LIN, N. *ibid.* **20**, 115 (1962).
3. WALLIS, D. *Anim. Behav.*, **10**, 267 (1962).
4. RITTER, H. *Science* **143**, 1459 (1964).
5. SCOTT, J. P. *Aggression*, P. DeBruyn, Ed. (Univ. Chicago Press, Chicago, 1958.)
6. BARNETT, S. A. *Viewpoints Biol.*, **3**, 204 (1964).

7. HODGSON, E. S. and GELIDIAY, S. *Biol. Bull.*, **117**, 275 (1959).
8. The drawings were made from photographs.
9. I thank the Department of Zoology, Edinburgh for use of facilities, Dr A. Manning for criticism of the manuscript, and Dr L. Roth for the original culture of *Nauphoeta cinerea*.

MODIFICATION OF SOCIAL RANK IN THE DOMESTIC FOWL*

Wendell Smith and E. B. Hale

UNDER typical conditions of rearing, many species of birds have been reported to develop relatively stable social ranks in the flock of residence. Such peck orders, when studied in the domestic fowl, have been found to be related to a complex of factors including breed, weight, comb size, hormonal and neurological state, and past experience in fighting (Alee, 1950; Armstrong, 1947; Guhl, 1953; Guhl and Ortman, 1953; Hale, 1957; Thorpe, 1951).

The factor of past experience in fighting raises the possibility that under an appropriate arrangement of environmental conditions, social rank can be established through learning in contradiction to the influence of genetic and physiological factors. Recent studies by Miller, Murphy, and Mirsky (1955), Murphy, Miller, and Mirsky (1955), and by Radlow, Hale, and Smith (1958) indicate that interanimal conditioning in rhesus monkeys and chickens is feasible. Radlow *et al.* have shown that the most dominant bird in a small flock of cocks can be conditioned to become the most subordinate member of the flock. The purpose of the present study is to attempt a complete reversal of ranks in a flock of hens through avoidance conditioning. In a small flock of hens in which the individuals have been designated as A, B, C, and D, from greatest to least dominant, the goal is to change the order to D, C, B, A, which requires differential con-

ditioning of C and B, while A learns to submit to all other *S*s of the flock, and D presumably learns that all other flockmates avoid her.

METHOD

Three groups, each containing four single-comb white Leghorn hens, were used as *S*s. One group of hens was 14 months of age while the other two groups were 7 months of age at the beginning of the experiment. The *S*s were drawn from a flock containing 55 birds by retaining every fifth bird caught. This technique was employed to minimize any bias arising from using as *S*s the birds which were easiest to catch.

Each *S* was placed in an individual cage, measuring 17 in. by 19 in. by 17 in., 3 weeks before the beginning of the study to provide for adaptation to cage living and to being handled by *E*. Subjects were kept in the cages throughout the study except when training sessions were being carried on. No visual or physical contact between members of a group was possible.

Following the adaptation period, dominance relationships within each group were determined by staging contests between each pair of *S*s on two successive days. A contest consisted of placing a pair of *S*s in a small pen in the center of which a small dish of scratch grain had been

* From the *Journal of Comparative and Physiological Psychology*, 1959, **52**, 373–375. Copyright 1959 by the American Psychological Association. Publica-

tion No. 2247 in the Journal Series of the Pennsylvania Agricultural Experiment Station.

placed. No scratch grain was fed in the cages. All Ss were run on a 4-hour food- and water-deprivation schedule to insure their attention to food. Each contest lasted for 20 min. or until a fight ensued, whichever occurred first. Fights, pecks, threats, and amount of time spent in eating were recorded.

For each group of four Ss, six pair-relationships existed. Since each pair was tested twice, for the three groups a total of 36 contests were staged during the 2 days to determine the rank order of each S in its group. In two contests in each group, no fights, threats, or pecks occurred. For these six contests, the relative rank of each S was determined by assigning the higher rank to the member of the pair which had precedence to the food. In all six contests of this type, one S in each pair was clearly dominant in the food intake. For the other 30 contests, the social relationship between members of a pair was easily determined from fights, pecks, and threats. Since the same relative rank was maintained by each S in its group on the 2 days of contests, no further testing for rank in the group was necessary.

Seventy-two hours following the second set of contests, interanimal conditioning was begun. All Ss were trained under a 4-hour food- and water-deprivation schedule. A small dish of scratch grain in the pen served to increase the probability of contact between Ss and it served as positive reinforcement for the S serving as a CS. A 15-min. trial was given each pair in each group on alternate days until the originally dominant member of each pair failed to eat in the presence of, or direct aggression toward, its partner for three successive trials. On each trial, leads from an electric stimulator (Harvard Apparatus Co., Model 935B) were attached to the wings of the originally submissive member of the pair. A single-pulse a.c. shock of 120 V. was applied to the dominant member of the pair when it (a) tried to eat in the presence of its partner, (b) threatened, pecked, or attacked its partner, and (c) was pecked or attacked by its partner. Each S was given an opportunity to eat in the training pen on each trial with its partner absent to insure that the differential cue to which the animal was being trained was its partner, not simply approach to or avoidance of the scratch grain.

When the criterion for reversal of dominance was attained, a test session was run 72 hours following the last training session to insure that a stable peck order had been established. Testing for the stability of the peck order was done with Group 1 at 2, 4, and 6 weeks following the termination of training. Group 2 was tested at

3 and 5 weeks after the termination of training, while Group 3 was tested at 5 and 9 weeks following the termination of training. During these intervals, all Ss were housed in social isolation in individual cages.

For each test session, each pair of Ss in each group was placed in a pen without the leads being attached to the wings. The pair of birds were left together until the relationship between them was determined. Threats, pecks, fights, and precedence to food were recorded. Testing was done under the same deprivation schedule used during the training sessions.

The stability of the relationship in a pen different from the one in which the conditioning occurred was determined, also.

RESULTS

The criterion for the learning of a new dominance relationship was considered to be met when the originally dominant member of a pair failed to eat in the presence of, or direct aggression toward, its partner on three successive trials. Two of the groups attained this criterion by trial 13, while the third group required 12 trials. The mean number of shocks per trial was 12.9, 11.2, and 10.9 for the three groups.

TABLE 1. *Mean Number of Shocks per Trial Received by Each S*

Group	Subject			
	A	B	C	D
1	1.6	3.4	1.5	0
2	1.5	3.2	1.8	0
3	1.2	2.8	1.8	0

Averages are for shocks through the trial on which the criterion was met.

From Table 1, it is apparent that not all shifts of status were acquired with equal ease. In all three groups the highest number of shocks per training session was required by the bird (B) which had to shift downward one rank. The shift from the alpha to the omega rank and shift upward one rank required approximately the same amount of training for each pair relation reversed. The t tests for the differences among the groups of Table 1 indicate no statistical significance. The number of trials which each pair in each group required to attain the criterion of three successive trials without shock varied with the particular rank held by each S before training. The fewest trials were required to change the relationship between Ss C and A. The lowest mean number of trials required to reach the criterion occurred for S A, indicating that A learned to submit to former subordinates in fewer trials than did Ss B or C.

For groups, the number of shocks administered was closely related to the number of trials required to reach the criterion. The pairs D-A, C-B, and C-A received an average of 60.5 shocks while the remaining pairs received 78.5 shocks ($p = 0.01$).

The tests for retention of the new social ranks were all positive. In all three groups, the Ss retained the rank attained through conditioning. The relationship between pairs was maintained in a strange pen even after a period of isolation as long as 5 weeks.

DISCUSSION

In this study the steps which were followed were (a) to establish the social rank of 12 hens divided into three groups, (b) through avoidance conditioning, to change the social rank of each S, and (c) to determine the stability of the new social ranks.

The results indicate that a peck order can be modified through avoidance conditioning in a relatively few trials. Stable CRs were established with one hen serving as a CS for another. Since all Ss ate readily when alone in the training pens but only the "to be made dominant" Ss ate when another S was present, the originally social inferior was the CS for avoidance and perhaps for fear, since Miller et al. (1955) point out in their work on interanimal conditioning that a fear-reduction interpretation of avoidance conditioning best fits the behavior evoked by this type of learning situation.

What is each S required to learn? Descriptively, the alpha bird has only to learn to avoid all members of the flock. The omega bird, accustomed to being submissive to all flock members, has to learn that it is no longer threatened, pecked, or attacked by its flockmates. It has ready access to food without interference. Seemingly, changes in the behavior of flockmates toward it result in behavioral changes in the omega bird. It always receives positive reinforcement (food) on each trial. Since its flockmates are always shocked for aggressive responses toward it, the omega bird receives further positive

reinforcement through an escape from punishing pecks and attacks to which it formerly was subjected by its flockmates. After a few trials, further reinforcement is received from its success in defeating its formerly dominant flockmates.

True discrimination learning is required of the Ss which originally hold the second and third ranks in the flock, for they must respond differentially to each flockmate. Subject B has to learn to submit to C and D and to dominate A. Subject C has to learn to submit only to D and to dominate A and B. Since the physical appearance of all Ss presumably remains fairly constant during the conditioning period, the change in response of the stimulus animal to the previously dominant bird must clearly be based upon the behavioral characteristics of that bird.

The memory of the social rank of a chicken by its flockmates has been reported to be approximately 2 to 3 weeks (Guhl, 1953; Schjelderup-Ebbe, 1923). If these reports of retentivity of social ranks are accurate, the stability of the peck orders established in this study should be of short duration, i.e., 3 weeks or less, since all Ss were housed in isolation cages throughout the study. The finding that all the social ranks in all three groups were stable for periods ranging from 2 to 9 weeks raises doubts concerning the common practice of treating, as strangers, birds which have been isolated for 2 or 3 weeks unless it is assumed that peck orders developed through avoidance conditioning are more stable than those which develop in natural flocks.

SUMMARY

The social ranks in three groups of four hens were reversed completely in from 12 to 13 interanimal avoidance-conditioning trials per pair relationship. Shifting a bird from position 1 to position 4 in the flock was accomplished with fewer shocks and fewer trials than shifting a bird from position 2 to 3, which demanded more precise discrimination learning. Birds in subordinate positions responded to changes in the behavior of previously dominant hens by attacking or threatening them.

Acquired social ranks were not specific to the cage in which the CRs were acquired. Retention of acquired ranks during periods of separation extended beyond the 2- or 3-week maximum generally accepted for chickens.

REFERENCES

ALLEE, W. C. Dominance and hierarchy in societies of vertebrates. In Pierre P. Grassé (Ed.), *Structure et physiologie des societies animales*, Colloques internationaux du Center National de la Recherche Scientifique, 1952, **34**, 157–181.

ARMSTRONG, E. A. *Bird display and behaviour*. London: Lindsay Drummond, 1947.

GUHL, A. M. Social behavior in the domestic fowl. *Tech. Bull.* 73, Kans. Agr. Exp. Sta., 1953.

GUHL, A. M. and ORTMAN, L. L. Visual patterns in the recognition of individuals among chickens. *The Condor*, 1953, **55**, 287–298.

HALE, E. B. Breed recognition in the social interactions of domestic fowl. *Behaviour*, 1957, **10**, 240–254.

MILLER, R. E., MURPHY, J. V. and MIRSKY, I. A. The modification of social dominance in a group of monkeys by inter-animal conditioning. *J. comp. physiol. Psychol.*, 1955, **48**, 392–396.

MURPHY, J. V., MILLER, R. E. and MIRSKY, I. A. Inter-animal conditioning in the monkey. *J. comp. physiol. Psychol.*, 1955, **48**, 211–214.

RADLOW, R., HALE, E. B. and SMITH, W. I. A note on the role of conditioning in the modification of social dominance. *Psychol. Rep.*, 1958, **4**, 579–581.

SCHJELDERUP-EBBE, T. Weitere Beiträge zur Sozial und Individual-psychologie des Haushuhns. *Zeitsche. f. Psychol.*, 1923, **92**, 60–87.

THORPE, W. H. The learning abilities of birds. *Ibis*, 1951, **93**, 252–296.

DOMINANCE REVERSAL IN DRUGGED MONKEYS*

R. W. Leary and Donald Slye[1]

A. PROBLEM

SOCIAL dominance hierarchies of rhesus monkeys are readily established (12), and exhibit considerable stability in time (7). Hierarchies have remained constant despite variation in food deprivation (12) and despite the selective administration of failure or success experiences to individual animals (4, 5). Changes in dominance status have been achieved by shock conditioning a normally dominant monkey with sight of another monkey as conditioned stimulus (8, 9), by brain operation (11), and, very recently, by placing a relatively dominant animal in the home-cage of another animal (3).

It has been suggested that submissive behavior in the monkey is fear-determined, e.g., Maslow (6). The results of studies utilizing shock tend to support this hypothesis, but perhaps the most relevant information comes from observations of normal behavior. In social dominance tests at the University of Oregon rhesus monkeys do not ordinarily struggle for food; in fact, they frequently pass through tests without physical aggression, and when such aggression does occur it customarily lacks violence. It appears that some psychological process prevents the submissive animal, who may be clearly the strongest animal in a pair, from engaging in direct competition. The retreating, cringing, and unaggressive attitude of the submissive monkey suggests that this process involves fear.

The purpose of the present investigation is to study the test behavior of normally submissive monkeys, while their normally dominant pairmates are "tranquilized" by chlorpromazine at a dose level which reduces most activity, including food-getting activity. If fear in submissive monkeys is attendant upon stimuli produced by the usual behavior of dominant monkeys, then we should expect the drugging procedure to attenuate this fear. A further purpose is to find out how the test situation, especially the "failure" experience, affects the drugged animal. The behaviors seen during the use of chlorpromazine will be evaluated in light of tests performed before its use and after its withdrawal.

B. METHOD

1. Subjects

The subjects were eight rhesus monkeys,

[1] This study was supported in part by a grant (M-1664) from the United States Public Health Service.

weighing 5.7 to 7.5 lbs., who had participated in various learning experiments over a period of more than 2 years. Their previous experience in social dominance work has been described by Warren and Maroney (12), Maroney and Leary (5), Maroney (4), and Leary (3).

2. Test Procedure

A pair of animals was transported to a Wisconsin General Test Apparatus (*WGTA*) in a quiet room. One piece of food (raisin or peanut) was placed on a sliding tray which was then pushed up to the cage holding the animals. Two experimenters behind a one-way screen recorded for each monkey success (or failure) in getting the food, aggressive acts (hitting or biting), husking (picking up empty peanut husks from the cage floor), and wood-picking (picking at, chewing on, or biting the wooden parts of the cage).

This was the procedure for a single trial which required about 10 sec. Ten such trials constitute a test. Successive trials were separated by a few seconds during which observation continued.

3. Design of Preliminary Experiments

The first preliminary experiment, Pre-I, involved the last 2 days of an unpublished 10-day study of social dominance, in which all possible pairs of animals were tested once each day. The order of pairs was randomly varied from one day to the next. The first 8 days are excluded in order to minimize temporally correlated training effects.

The second preliminary experiment, D-I, was concerned with the discovery of a satisfactory drug dosage, one which would reduce competitive effort. The three highest ranking members of the group were given increasing amounts of chlorpromazine (0.3, 0.6, and 0.9 mg/kg) on 3 consecutive days. On each day a drugged monkey was tested with two of the remaining normal monkeys at times ranging from 30 min. to 2 hours 30 min. post-injection. Subsequently, one animal was tested twice while under the influence of a 1.2 mg/kg dose.

Finally, all monkeys were tested in all 28 possible pairings as a check against possible changes in the hierarchy. These non-drug tests, which were accomplished on a single day, constitute the third preliminary experiment, Pre-II.

4. Design of Tranquilization Experiments: D-II

Monkeys Philosopher and King who ranked first and third respectively in Pre-I and Pre-II were given a daily injection of 1.0 mg/kg for 10 days. From approximately 90 min. to 180 min. after injection each of these two monkeys was tested once with each of the six remaining monkeys. The two drugged animals were also tested once with each other. A different semi-random order of testing was applied to the 13 pairs every day.

5. Design of Post-tranquilization Experiments

After the completion of D-II 2 rest days were allowed. Then the 13 pairs were tested on 2 days. This experiment is designated Post-I. Next, the eight monkeys were tested in all possible pairs, and then in all possible trios. Only the first 2 days of the paired testing (Post-II) will be included in the present study since our only purpose is to provide a post-tranquilization standard which is comparable to the pre-tranquilization standard.

C. RESULTS

Complying with current psychological practice, we have referred to a monkey as dominant when, in a competitive test, he obtained more than half of the pieces of food; the monkey who obtained less than half we have termed submissive. Dominance ranking has been based on number of other monkeys dominated except where this method yielded a tie, whereupon further ranking was accomplished by consideration of the total pieces of food obtained by the tied subjects.[2] Thus, a monkey ranked 1 was dominant in more pairings than was any other monkey, and a monkey ranked 8 was submissive in more pairings than was any other monkey. On the other hand, in scoring and ranking aggression total instances of biting and hitting have been used. Similarly, wood-picking and husking measures are based on total acts.

Since we are mainly concerned with changes which are attendant upon drugging a normally dominant monkey we have tended to focus our attention on those 10 pairs in which Philosopher and King were dominant in Pre-I. Of course, in the case of dominance rankings, data from all 28 pairs have been employed.

---

[2] Actually there were two such ties in both preliminary and post-drug periods: Philosopher dominated Maud who dominated King who dominated Philosopher, and King dominated Becky who dominated Roman who dominated King, thus Philosopher and Maud were tied as were also Becky and Roman.

TABLE 1. *Mean Daily Scores for Aggression (A), Wood-Picking (W), and Husking (H)**

Pair**	Pre-I, Pre-II			Post-I, Post-II			All Nondrug Tests			D-II		
	A	W	H	A	W	H	A	W	H	A	W	H
P–M	0.0	2.3	0.0	0.0	1.0	1.3	0.0	1.6	0.7	0.2	0.1	1.2
P–B	1.3	0.0	2.3	2.0	0.0	0.0	1.6	0.0	1.2	0.0	0.0	0.1
P–R	0.0	0.0	0.3	0.0	0.6	0.0	0.0	0.2	0.2	0.4	0.0	0.0
P–C	1.0	1.3	2.0	1.0	2.7	1.3	1.0	2.0	1.7	1.1	0.3	0.9
P–I	0.0	0.0	0.3	0.6	0.0	0.0	0.3	0.0	0.2	1.6	0.0	0.0
P–A	2.3	0.7	4.0	1.0	0.0	0.3	1.6	0.3	2.2	0.8	0.0	0.3
K–B	0.0	3.3	0.3	2.0	1.3	1.3	1.0	2.3	0.8	0.4	0.0	0.0
K–C	1.0	3.3	1.7	2.0	3.3	2.0	1.5	3.3	1.3	0.6	0.2	1.1
K–I	0.3	0.3	0.0	2.7	0.0	0.3	1.5	0.2	0.2	0.0	0.0	0.0
K–A	0.0	0.7	1.7	3.3	0.0	0.0	1.6	0.3	0.8	0.2	0.6	0.1

* *A* scores are for animals *P* and *K*, while *W* and *H* scores are for the remaining animals.
** The first letter of each monkey's name is used for identification.

1. Nondrug Experiments
(Pre-I, Pre-II, Post-I, Post-II)

Table 1 shows the mean daily preliminary (Pre-I and Pre-II) and post-tranquilizer (Post-I and Post-II) scores for aggression, wood-picking, and husking. There is no evidence of significant variation in any of these three categories between the two sets of means. Nor was a significant difference found in any comparison of Pre-I with Pre-II, or of Post-I with Post-II. The following dominance rankings were obtained in Pre-I: 1, Philosopher; 2, Maud; 3, King; 4, Becky; 5, Roman; 6, Clown; 7, Arab; 8, Ingrid. In Pre-II the rankings were unchanged but in Post-II Clown gained a rank of 4, while Becky and Roman dropped to ranks of 5 and 6 respectively. In all the nondrug periods Philosopher was dominant in all pairings except the one involving King, and King was dominant in all pairings except those involving Roman and Maud.

These results indicate quite clearly that the limited drug experience of three animals in D-I, and the more extensive drug experience of two animals in D-II, had no effect on their subsequent (normal) behavior. Comparable stability is seen in the performance of undrugged animals which were paired previously with drugged animals.

2. Drug Experiments (D-I, D-II)

In D-I there was no evidence of a temporal (postinjection) effect, and therefore the scores for different test intervals were combined. At 0.3, 0.6, and 0.9 mg/kg the average percentages of food obtained by the drugged animals were 95, 90, and 60 respectively. The subject given 1.2 mg/kg

was successful on 13 per cent of the trials. For both members of pairs the amount of wood-picking activity was slight. On the other hand, aggression was somewhat increased, mainly due to attacks perpetrated by the drugged monkeys at dose levels where they still obtained the majority of the food.

In D-II obvious effects of injection were drastic reduction in activity, incoordination, awkwardness, and unusual rigid postures. Throughout most of the tests the two drugged monkeys appeared lethargic with their eyes often shut or half closed. It was our impression that Philosopher was considerably less affected than King.

If the dominance relations previously observed in Pre-I had been sustained, Philosopher would have been dominant in 60 pairings and King in 50 pairings. Actually, Philosopher was dominant 13 times and King six times, but this includes five times that King dominated Philosopher, and five times that Philosopher dominated King. In only 9 out of their 120 pairings with six normal animals were our two drugged subjects dominant: King once and Philosopher eight times.

As seen in Table 1, a nondrug standard of wood-picking, aggression, and husking has been calculated for one animal in each of the ten critical pairs by averaging the data from Pre-I, Pre-II, and Post-I, and Post-II. Assuming independence between pairs, these standards were compared with the mean scores for D-II by Wilcoxon's matched-pairs technique. A significant reduction in D-II was found for the wood-picking ($p = 0.01$) and husking ($p = 0.02$) of the undrugged animals. Both of these activities, which were at a near zero level in Philosopher and King during the nondrug experiments, were essentially unchanged in D-II. Statistically signifi-cant alteration of aggression was not found. Since aggression in the six normal animals was exceedingly low outside D-II, its almost complete disappearance in D-II did not represent much of a change. King's aggression declined considerably while Philosopher's aggression remained at approximately the normal level. It is noteworthy that despite the drug-induced submission of these two monkeys, they were responsible for almost all the aggressive acts which occurred in D-II. Out of 58 such acts they were responsible for 55, and of these, 27 took place in pairings in which they failed to secure half the pieces of food.

D. DISCUSSION

The picture of social relations among normal monkeys obtained here is a familiar one and requires no elaboration. Suffice it to say that the hierarchy was highly stable in time, that dominant monkeys rarely obtained less than 80 per cent of the food in a test, that aggression was the prerogative of dominant monkeys, and that husking and wood-picking behavior were typical of submissive monkeys.

The initial and delimited experiment with chlorpromazine, D-I, revealed individual differ-ences in sensitivity. However, an approximately correct threshold dosage for reducing com-petitive success was discovered. This dosage of about 1.0 mg/kg is considerably above the maximal level of 0.2 to 0.3 mg/kg we have found permissible for discrimination learning in the WGTA. It may be noted that a dose of 1.0 mg/kg was found by Pfeiffer et al. (10) to block a learned avoidance response. On the other hand, doses over 20 times as large have been used without eliminating an aggressive response to handling (1, 2).

We had hoped to find a dose level which impaired attention only slightly at the same time that food-getting was eliminated. This was not achieved. Apparently a dose that attains the latter goal will overshoot the former one. Perhaps only an injection which produces nearly complete sensory isolation will reduce the food-getting of a normally dominant animal to the zero level. Whether prehension and ingestion of food out-side of the test situation are more susceptible to drugging we do not know, but we suspect this to be the case.

Contrary to our intent, the undrugged and normally submissive animals did not lose their fear of King and Philosopher during D-II. At any rate, they did not attack them. These undrugged monkeys were usually dominant, but failed to display the "indignation" which usually occurs when a submissive animal tries to take food. The foregoing interpretation is strengthened, of course, by their subsequent failure to maintain any vestige of success in Post-I. That there was some reduction in their fear during D-II, however, is suggested both by their sitting at the front of the WGTA cage and by their decrease in husking and wood-picking. These last two activities may be thought of as forms of displacement. It should be noted that we do not know whether fear or frustration is the primary cause of such dis-placement, and therefore the change we observed might be due to a decrease in frustration rather than a decrease in fear.

A tentative analysis can be applied to D-II data with respect to the stimulus or informational change for the normally submissive animals. It is apparent that despite the reduction in com-municative activity and the production of abnormal test-cage postures in King and Philo-sopher, they were recognized by the other subjects. This recognition was shown by, and sufficed to, maintain an attitude of respect. The subnormal aggressiveness of King did not harm his status, his freedom from the aggression of the normally submissive animals, and his ability to regain dominance. From this result, as well as from non-drug data which show little if any aggression in certain pairs, we are led to hypothesize that the kind of aggression we measure is certainly not a necessary ingredient in the maintenance of hierarchical position in dominance tests.

In giving a further psychological interpretation to our results it should be pointed out that the word *fear* has been used, here and elsewhere, in a rather vague way to characterize a host of behavioral events as well as an underlying moti-vational condition. It might be wise to differen-tiate fear as a habit from fear as a motive or drive. Our results are consistent with the view that, normally, the behavior of a dominant animal creates fear in a submissive animal as a drive which then activates certain behavior. This behavior is a habit in the sense that it is a con-sistent and previously learned way of behaving to certain stimuli, the stimuli provided by one animal being linked to a somewhat different habit from the stimuli provided by another animal. When drugging the dominant animal in a pair changes his behavior so as to diminish motiva-tional fear in a normally submissive animal, there is not necessarily any influence exerted on the latter animal's fear habit.

Informational change was also present for the drugged subjects in D-II. Although chlorpromazine generally suppresses responsiveness to stimulation, it is likely that Philosopher and King were aware of their cage-mates during testing, and of the food-getting success of these animals. This did not impair their ability to regain dominance in D-II.

In the aggression of the drugged animals we have evidence that an emotional response can still occur. Furthermore, incidental observations were made of a fear response to leather gloves which had been used many months ago to hold the monkeys. Another pertinent fact is that when the drugged monkeys were paired with each other, all the food was consumed; hence, food motivation was not lacking. Yet in pairings with other animals, Philosopher and King had a relatively low rate of food-getting. We suspect that three factors may be involved: first, the drug does cause a lowering of food-getting motivation; second, the presence of another animal causes an increase in this motivation; third, the presence of another animal, especially an active one, acts as a kind of barrier against food-getting.

The evidence cited above gives us the picture of a more excitable drugged monkey than suggested by the results of Pfeiffer *et al.* (10). At the drug level we used, the central integrative mechanisms of fear, attack, and food-getting may be operative, but the level of stimulation which was formerly effective may now evoke subnormal activation. Since the food dominance of a previously submissive animal is probably an unusually potent stimulus for a normally dominant animal, we might anticipate that under the influence of drug the latter animal would show aggression (sometimes approaching or exceeding the normal level) whilst not showing food-getting.

Although we failed to upset our normal hierarchy by the drug tests involving the second and third ranking animals, we certainly did not place much stress on them, nor did we give the other animals much time for social exploration. The importance of these factors is suggested by comparing the results of Maroney and Leary (5) with those of Leary (3). It is possible that the drugging of one or two monkeys who are living in a colony rather than in separate cages would achieve more striking and persistent results.

E. SUMMARY

Two monkeys, ranking first and third respectively in the dominance hierarchy of eight young rhesus monkeys, were injected for 10 days with 1.0 mg/kg of chlorpromazine, prior to pairing with the remaining subjects for competitive food-getting tests in a *WGTA*. Tests preceding and following the drug phase showed nearly identical food-getting (dominance) rankings, and similar scores for aggression, wood-picking, and husking. Clearly, the drug phase had no persisting effects. In the drug phase the drugged animals generally failed to get the food and the normally submissive animals became successful. Furthermore, the wood-picking and husking of the undrugged monkeys were significantly reduced from the non-drug period standards, but significant changes were not obtained in these two categories for the drugged animals. Aggression, which was normally very low in the undrugged monkeys, remained low in the drug phase. The aggression of one drugged monkey decreased radically, while the other drugged monkey maintained a normal level of hitting and biting. However, in their pairings with the other animals, both of the two drugged subjects contributed almost all of the aggressive acts which occurred. It appears that fear as a drive or motive is reduced somewhat in normally submissive animals by drugging the animals who usually dominate them, but this procedure does not reduce fear as a learned response.

REFERENCES

1. Essig, C. F. and Carter, W. W. Convulsions and bizarre behavior in monkeys receiving chlorpromazine. *Proc. Soc. Exp. & Biol. Med.*, 1957, **95**, 726–729.

2. Hendley, C. D., Lynes, T. E. and Berger, F. M. Effects of meprobamate (Miltown), chlorpromazine, and reserpine on behavior in the monkey. *Federation Proc.*, 1956, **15**, 436.

3. Leary, R. W. The monkey as host and guest: Laboratory territoriality. *Amer. Psychol.*, 1958, **13**, 383–384. (Abstract.)

4. Maroney, R. J. The effects of success and failure experiences in altering dominance hierarchies of rhesus monkeys. (Unpublished M.S. Thesis, University of Oregon, 1957.)

5. Maroney, R. and Leary, R. A failure to condition submission in monkeys. *Psychol. Rep.*, 1957, **3**, 472.

6. MASLOW, A. H. Dominance-quality and social behavior in infra-human primates. *J. Soc. Psychol.*, 1940, **11**, 313–324.

7. MILLER, R. E. and MURPHY, J. V. Social interactions of rhesus monkeys: I. An examination of the stability of dominance relationships. *J. Soc. Psychol.*, 1956, **44**, 249–255.

8. MILLER, R. E., MURPHY, J. V. and MIRSKY, I. A. The modification of social dominance in a group of monkeys by interanimal conditioning. *J. Comp. & Physiol. Psychol.*, 1955, **48**, 392–396.

9. MURPHY, J. V. and MILLER, R. E. The manipulation of dominance in monkeys with conditioned fear. *J. Abn. & Soc. Psychol.*, 1956, **53**, 244–248.

10. PFEIFFER, C. C., RIOPELLE, A. J., SMITH, R. P., JENNEY, E. H. and WILLIAMS, H. L. Comparative study of the effects of meprobamate on the conditioned response, on strychnine and pentylenetetrazol thresholds, on the normal electroencephalogram, and on polysynaptic reflexes. *Ann. N.Y. Acad. Sci.*, 1957, **67**, 734–743.

11. ROSVOLD, H. E., MIRSKY, A. F. and PRIBRAM, K. H. Influence of amygdalectomy on social behavior in monkeys. *J. Comp. & Physiol. Psychol.*, 1954, **47**, 173–178.

12. WARREN, J. M. and MARONEY, R. J. Competitive social interaction between monkeys. *J. Soc. Psychol.*, 1958, **48**, 223–233.

PSYCHOPHYSIOLOGICAL INTERRELATIONS IN THE SOCIAL BEHAVIOR OF CHICKENS*

A. M. Guhl

IN the course of over 20 years of research in the social behavior of chickens there have been recurring evidences that the habits of domination and subordination have influenced activities regardless of the physiological conditions of the individual birds. These behavioral phenomena have not received the attention that they appear to merit. Beach (1958) has made the comment that psychologists and endocrinologists have worked independently, with the result that the interrelations of their disciplines have remained a matter of speculation. It is the aim of this review to bring some evidence from these two fields for further consideration.

The writer, a zoologist, has viewed behavior as the actions and reactions of the whole organism and as physiology outwardly expressed. Although conditioning, learning, and memory have a physiological basis, they are not well understood as physiological phenomena and therefore the term physiological has a useful connotation within the context of this review.

The observations and experiments cited herein have considered the following factors, among others, which may alter behavior: physical

factors such as extremes of temperature; changes in the thresholds of response caused by fatigue and drive reduction following repeated elicitation, and from stimulation that results in what ethologists call "overflow activities"; conditioning and learning; maturation and aging, which include experience; seasonal factors, such as internal rhythms and endocrine effects due to changing day lengths; genetic basis for strain differences and individuality; and dietary deficiencies or avitaminosis.

Social behavior has been defined variously although with somewhat similar denotations (Guhl, 1962b). According to Carpenter (1942) "social behavior refers to the reciprocal interactions of two or more animals and the resulting modification of the individual's action systems" [p. 248]. Others prefer certain limitations, such as occur within the same species. The type of social organization within a given species is determined by the behavior patterns characteristic of that species. Since the phenomena discussed here pertain to the common domestic fowl a brief description of the social hierarchy, or peck order, is required.

* From the *Psychological Bulletin*, 1964, **61**, 277–285. Copyright 1964 by the American Psychological Association. Contribution No. 321, Department of

Zoology, Kansas Agricultural Experiment Station, Kansas State University, Manhattan.

THE SOCIAL ORGANIZATION

Agonistic behavior, which includes attack and escape reactions, forms the basis for the social order in chickens (Guhl, 1962a). When individuals are marked for identification, one can record all social encounters involving the one pecking and the one pecked. When these observations are tabulated it becomes apparent that such domination is unidirectional; that is, peck rights are established, and that all the birds may be ranked according to the number of hens each pecks without any retaliation. This ranking has been called a dominance order, or peck order. In very small flocks the order may be a straight line hierarchy; however, pecking triangles are common and give the order of precedence a geometrical structure.

Repeated pecking reinforces the dominance relations between all pair combination in the flock. Thus special habits are established between the individuals. In time symbolic threats, or the mere raising or lowering of the head may suffice for the maintenance of pair-reaction patterns. This reduction in frequency of pecking or threatening in time is shown in Figure 1 (intensity of agonistic behavior is not illustrated, Siegel and Hurst, 1962). There are two points of particular interest: one, that the dominance order channels the flow of activity by way of precedence in competitive situations and thus precludes fighting and conserves energy; the other concerns the gradient of habits (or attitudes) from that of a high level of domination to a characteristic response of subordination at lowest levels. Social mobility (changes in rank) is unusual and the relative fixity of response patterns results in social inertia within the flock. The mutual interindividual adjustments promote toleration which

in turn facilitates the activities within the group (Guhl and Allee, 1944). It is the relation of social habits to the physiology of behavior which will be discussed.

GONADAL HORMONES AND SOCIAL INERTIA

ANDROGENIC TREATMENT. Since the early studies of the endocrines suggested a relationship between androgen and aggressiveness, Allee, Collias, and Lutherman (1939) injected testosterone propionate into low ranking hens within several small flocks to investigate its effects. Several of the low ranking treated birds staged successful revolts and rose in rank, even to top level. However, treated hens in one flock failed to show any mobility in the social order, for which no explanation could be given. Nevertheless, the significance of social inertia was recognized by the statement that "memory and habit reinforce, and may entirely replace the aggressive behavior which is so important in the origin of a social order" [p. 436]. During the postinjection period the hens that improved their rank tended to retain their new social status. This was attributed to psychological factors which retarded the rise but now helped to maintain the newly acquired rank. Similar results were obtained by Allee and Foreman (1955).

During the course of this experiment the treated birds were tested for aggressiveness in initial pair encounters. This test is similar to that used by Maslow (1934a, 1934b). Briefly, when two unacquainted hens are placed into a neutral area they readily establish dominance relations, and relative aggressiveness is indicated by the number of contests won. The treated hens were matched in pairs with normal controls and won most of these encounters, thus indicating that androgen may enhance aggressiveness in the absence of social inertia.

Subsequently some of these birds were used in an experiment on social discrimination. One hen which received a pellet of testosterone propionate showed no social mobility, although she threatened two males when tested in a discrimination cage (Guhl, 1942, p. 145). When she was placed into a pen of strange hens she drove them about the pen (the opposite of what was expected), but on return to her own flock she resumed the behavior associated with her rank (Guhl, Collias, and Allee, 1945, p. 380).

With a flock consisting of 16 individuals, four each of four breeds, Williams and McGibbon

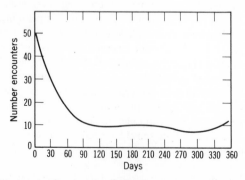

FIGURE 1. The decrease in domination after assembly of flocks of hens as shown by total social encounters as a function of time (data from Siegel & Hurst, 1962).

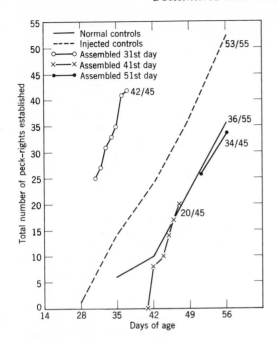

FIGURE 2. The total number of peck rights established at various ages among males reared as a group and others reared in spatial isolation and assembled at different ages. (The isolated chicks and one group reared together were injected with androgen. See Guhl, 1958.)

(1956) implanted a pellet of testosterone into one of each breed and at several levels in the social order. Observations over a 6-week period failed to show any social mobility. It was concluded that the hormonal effects were too gradual to overcome the already established behavior patterns (social inertia).

These results all suggest that recognition, memory, and habits underlie the social inertia that is responsible for a stable social order. Ill or injured birds may maintain their status provided that factors for recognition (e.g., combs) do not change abruptly. Successful revolts may occur among normal hens, but rarely in well-managed small flocks. To resolve these apparent discrepancies a study of the conditions under which revolts may be successful or unsuccessful was indicated.

Recent experiments (Guhl, unpublished) have used several flocks of ten females. Peck orders were allowed to develop to stability and testosterone propionate was then injected daily into certain individuals. The hens to be treated were selected according to the frequency at which they pecked or were pecked. Competitive interactions were stimulated during timed observation periods by presentation of scratch grain and wet mash. Relative social stress was measured in rates of pecking and threatening, which could be used to determine differences in social tension between flocks as well as between individuals within flocks. In competitive situations the injected bird would readily assume a threatening stance, with varying intensity, which in turn would evoke an attack by its ranking penmates. If the social tension was high, then the revolt was unsuccessful, whereas there was a greater chance of success when toleration (tendency toward extinction) was noteworthy. For example, in one set of experiments involving four flocks during several months of observations, there were 292 observed unsuccessful revolts of which 215 were initiated by treated hens. Of only 20 successful revolts, 16 were by treated birds.

It may be concluded that social inertia tends to stabilize the social order irrespective of changes in relative aggressiveness produced by treatment with androgen. Revolts may be successful if the subordinate individual is stimulated sufficiently in a competitive situation and the existing social relations are marked by toleration.

ESTROGENIC TREATMENT. The effects of an estrogen on social mobility have received little attention. Allee and Collias (1940) injected estradiol into several hens in five different flocks. Only two shifts in dominance relations were obtained. This was in contrast with the higher upward mobility which resulted when treated with androgen, and contrariwise the status displacement was downward with estrogenic treatment. Williams and McGibbon (1956) failed to obtain any social mobility by the implantation of pellets of diethylstilbestrol. The analysis of recent experiments with estrogen (Guhl, unpublished) have not been complete. No social mobility was recorded. Treated hens showed little change in frequency of pecking although some treated birds appeared to be less intense in their domination. This modification of behavior promoted toleration of subordinate penmates rather than the evocative threatening stance shown after injection with an androgen.

DEVELOPMENT OF SOCIAL ORGANIZATION IN CHICKS

In the course of an experiment on selective breeding for levels of aggressiveness (Guhl, Craig, and Mueller, 1960), it was deemed essential to examine the development of aggressiveness and the resultant peck order in chicks (Guhl, 1958). The age, in weeks, at which each chick established definite dominance relations was determined. Normal males began to set up a peck order at 5–7 weeks of age and completed the dominance order at about 8–10 weeks. Females initiated their social positions about 1 week later and required more time to complete the organization. Chicks kept in isolation from hatching onward were assembled when their penned controls formed a peck order and were able to establish dominance relations in a matter of hours. These results immediately raised the question of the relative effects of maturation and/or learning.

Unisexual groups of males and females were set up as controls and experimentals from the same hatch with all the chicks in the treated group receiving a gonadal hormone from the second or third day after hatching. Groups of each sex which received an androgen formed their social dominance earlier and in only slightly less time (about 1 week) than their respective controls. Males castrated (capons) at 10 days of age were slow in eliciting aggressive behavior and formed an order 4 weeks later than their controls. But capons treated with an androgen established pair reactions and their peck order earlier than the normal controls. Contrary to expectancy, treatment with an estrogen resulted in both male and female groups showing a similar advanced development of social organization. However, there was a marked difference in response behavior. Estrogen-treated chicks showed little pecking but avoided readily, with the consequence that the social system was essentially a consistent avoidance order. It was somewhat surprising that the hormonal treatments did not induce greater precocity. Could the lag have been due to some form of social inertia, even among immature birds?

To determine whether androgen treatment could induce an earlier formation of peck rights in the absence of learned social responses, five sets of males from the same hatch were used (Guhl, 1958). Two groups were penned from hatching, with one serving as a control and the other as an experimental control receiving androgen. Three sets were placed in spatial isolation and also treated with androgen. The isolates were assembled at 31, 41, and 51 days of age, and observed for the following 7 days to determine dominance relations.

As may be seen from Figure 2, the experimental controls developed peck rights earlier than the normal controls as expected, and formed 53 out of a possible 55 dominance relations by the end of the observations (56 days), whereas the uninjected chicks established only 36. Treated isolates assembled at 31 days of age began to fight immediately and nearly completed their peck order (42 out of 45) by the end of a week, or at about the time that pen-reared normals began to organize. The other two groups of isolates introduced sexual behavior as a complication. Androgen induces sexual behavior as well as aggressiveness and in the absence of penmates no conditioning could be made in relation to either type of behavior. Those assembled at 41 days displayed predominately sexual behavior and established only 20 out of 45 peck rights. Chicks assembled at 51 days apparently made some adjustment to the conflicting tendencies for either attack or for sexual approaches. None of the treated birds in flocks showed similar sexual behavior. These results may offer some evidence for the need for spacing, in time, the emergence of behavior patterns during development to permit social adjustments or learning. Treatment with androgen tended to reduce this interval and resulted in a conflict of sexual and dominance drives.

In another experiment of this series, an attempt was made to determine how early in the life of normal chicks that agonistic behavior might occur if there were no social inertia. Four sets of ten male chicks each were used, with one set in isolation and the other three in pens. Every other day one chick from each of two groups was interchanged. Similarly, a chick from isolation was placed into the third penned group as one was removed and placed into isolation. These two patterns of rotation in the membership of the groups were continued for 8 weeks from hatching. The data in Figure 3 are from daily observations of 15 min. per pen. All forms of agonistic behavior were not only at a higher frequency in the

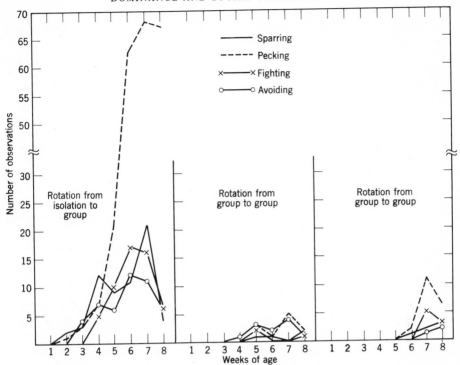

FIGURE 3. Differences in the frequency of some behavior patterns of male chicks rotated between isolation and a group and those rotated from group to group. (The latter show social inertia. See Guhl, 1958.)

pen containing chicks from isolation, but also started at an earlier age. Chicks introduced from isolation were a stronger stimulus for attack than those rotated between groups, and differences in their postures were apparent. The results showed that social adjustments may be made early in the life of chicks, even without exogenous androgen.

SOCIAL STATUS AND SEXUAL BEHAVIOR

A number of early studies on the reproductive behavior among chickens have shown marked differences between individual males and females in the number of matings which occurred. Some of these suggested a relation to social rank as one possible cause for variability. Since males do not peck the hens, and therefore have a peck order apart from that of the females, the sexes will be considered separately.

SOCIAL STATUS OF MALES. To determine the effects of social rank among males on their sexual activity they were tested singly, and in rotation, with a flock of hens to obtain an estimate of their sexual drive (Guhl, Collias, and Allee, 1945). All of the males in each of two flocks were then permitted to remain together with the same

hens used for testing. A summary of some of the data is given in Table 1. Although differences in sexual drive between some of the males are suggested in the summary chart, there is also an indication that social rank influenced performance when in competition for mates. Of particular interest at this point is the case of Male III who was psychologically castrated in the presence of his social superiors. He also failed to tread the hens when all three of his ranking males were removed temporarily, but did so readily when placed into a pen with strange hens. It should be added that the sex ratio was one male to two females, whereas a ratio of 1–12 is less competitive among Leghorns. However, when a more reasonable sex ratio was used (Guhl and Warren, 1946) similar results were obtained. This extension of the study showed that the highest ranking male may also sire the most chicks in this polygamous species.

It is of further interest that the passive dominance of males over females is associated with normal mating. Young cockerels will fail to mate with older hens and are dominated by them. Since estrogen may induce sexual behavior in capons without any increased aggressiveness over the castrate condition, such males were used to

determine the importance of heterosexual dominance (Guhl, 1949). Treated capons mated quite
normally with females over which they were able
to rank but were driven by those that held higher
status in the heterosexual peck order formed prior
to treatment. Thus social status may facilitate or
inhibit sexual performance. Furthermore, it has
been shown that the relative acquaintance between the sexes influences the frequency and type
of sexual behavior patterns displayed by both
sexes (Guhl, 1961).

TABLE 1. *Mean Rates per Hour of Sexual Activity
of Four Males in Each of Two Flocks Tested Singly
With the Hens and When All Four Males Were in
the Pen*

Peck Order	Courting		Treading	
	Singly	4 Males	Singly	4 Males
I	8.2	8.8	3.1	3.2
II	8.2	7.1	4.6	1.3
IV	7.2	6.4	2.5	2.5
III	8.7	2.5	3.3	0.0
V	24.3	11.3	9.5	1.6
VI	42.1	9.6	8.4	1.4
VII	21.1	3.2	7.1	0.2
VIII	41.3	10.0	9.2	1.1

Note.—From Guhl, Collias, and Allee (1945).

SOCIAL STATUS OF FEMALES. Negative correlations have been found between the social rank of
hens and the frequencies of submitting to a male
and of being mated (Guhl, Collias, and Allee,
1945). Hens may crouch in a submissive posture
when strongly dominated by certain of their
superior penmates, and pseudomatings between
hens have been reported under such circumstances
(Guhl, 1948). The sexual crouch displayed to a
male is essentially similar but may be of a higher
intensity. From such observations one might
assume hens that rank low and are in the habit of
submitting may display the receptive crouch more
readily than high ranking females that are more
inclined to domination. Such a hypothesis can be
tested by altering the relative intensities of these
habits.

The peck orders of three different flocks of
approximately 30 hens each were determined
(Guhl, 1950). Three or more males were caged
within the pen and released singly for short daily
periods. The number of crouchings observed for
the hens in the top, middle, and bottom thirds of
the social order was recorded. After several weeks
of observation each flock was divided (sub-
flocked) according to three levels in the peck

order. This procedure did not alter the peck rights
between the hens remaining in contact. Levels of
domination and subordination were altered
because the top third (as a group) now had two-
thirds fewer birds to dominate, the bottom birds
had two-thirds fewer dominating them, whereas
the habits of the middle third were altered about
equally between dominating and being dominated.
The same males were then rotated through each
of these respective subflocks. Subflocking occurred when the laying cycle was at its peak as
an attempt to balance the endocrine relationships.

The results for one flock are indicated in
Figure 4. Under the conditions of subflocking a
total reduction in rates of crouching was expected
because only one male was used per small flock
per day, whereas a sequence of three or more
males was used in the large flock. In this manner
the sex ratio remained essentially the same.
However, in each test, the top third hens showed
an increase in crouching response during subflocking, while the receptivity in the middle and
bottom thirds decreased. No estimates were
made of the sex drive of individual hens. Not all
of the comparisons between thirds (on a week
by week basis) within flocks were significant.
Nevertheless, it may be concluded that the habit
of dominating in females interferes with mating.

SOCIAL STRESS AND THE ENDOCRINE SYSTEM

Although the peck order is an adjustment to

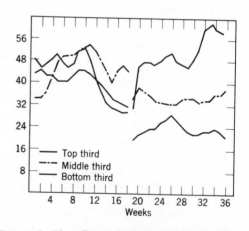

FIGURE 4. The effects of habits of domination and
submission on sexual receptivity as shown by subflocking that changed the intensity of domination in
highest ranking hens and of submission by hens in
lowest social level. (After subflocking by three levels in
the peck order the dominating females from the top
third became more receptive. See Guhl, 1950.)

the presence of penmates, individuals at the lowest levels of the social order may be under stressors related to their rank, especially if situations become highly competitive. Since these birds are usually confined and under high population density for economic reasons, the effects of social stress are now receiving some attention. According to Selye (1947) animals may adapt to various forms of stress by the increased activity of the pituitary-adrenal cortex axis, which becomes evident by the hypertrophy of the adrenals.

It has been shown that crowding, that is, the number of square feet per bird, results in an increase in the adrenal weight in hens (Siegel, 1959). However, there was no evidence of an acute hypertrophy of the cells or lipoidal depletion and the reaction was well within the birds' adaptive ability. Similar results were obtained with males (Siegel, 1960). When social stress was introduced experimentally among males (Siegel and Siegel, 1961) all individuals responded with increased adrenal weight when compared with controls in individual cages. The left adrenal, which is more sensitive than the right in chickens, was heavier in cockerels which ranked low in the social order. Within the limits of the stress to which they were subjected, the males remained in the adaptive phase. It was postulated that these birds may be less liable to psychological stimulation than mammals.

REFERENCES

ALLEE, W. C. and COLLIAS, N. E. The influence of estradiol on the social organization of flocks of hens. *Endocrinology*, 1940, **27**, 87–94.

ALLEE, W. C., COLLIAS, N. E. and LUTHERMAN, C. Z. Modification of the social order in flocks of hens by the injection of testosterone propionate. *Physiol. Zool.*, 1939, **12**, 412–440.

ALLEE, W. C. and FOREMAN, D. Effects of an androgen on dominance and subordinance in six common breeds of *Gallus gallus*. *Physiol. Zool.*, 1955, **28**, 89–115.

BEACH, F. A. Evolutionary aspects of psychoendocrinology. In A. Roe and G. G. Simpson (Eds.), *Behavior and evolution*. New Haven: Yale Univer. Press, 1958. Pp. 81–102.

CARPENTER, C. R. Characteristics of social behavior in non-human primates. *Trans. N.Y. Acad. Sci. Ser. II*, 1942, **4**, 248–258.

GUHL, A. M. Social discrimination in small flocks of the common domestic fowl. *J. comp. Psychol.*, 1942, **34**, 127–148.

GUHL, A. M. Unisexual mating in a flock of White Leghorn hens. *Trans. Kan. Acad. Sci.*, 1948, **51**, 107–111.

GUHL, A. M. Heterosexual dominance and mating behavior in chickens. *Behaviour*, 1949, **2**, 102–120.

GUHL, A. M. Social dominance and receptivity in the domestic fowl. *Physiol. Zool.*, 1950, **23**, 361–366.

GUHL, A. M. The development of social organization in the domestic chick. *Anim. Behav.*, 1958, **6**, 92–111.

GUHL, A. M. The effects of acquaintance between the sexes on sexual behavior in White Leghorns. *Poult. Sci.*, 1961, **40**, 10–21.

GUHL, A. M. The behaviour of chickens. In E. S. E. Hafez (Ed.), *The behaviour of domestic animals*. Baltimore, Md.: Williams & Wilkins, 1962. Ch. 17. (a).

GUHL, A. M. The social environment and behaviour. In E. S. E. Hafez (Ed.), *The behaviour of domestic animals*. Baltimore, Md.: Williams & Wilkins, 1962. Ch. 5. (b).

GUHL, A. M. and ALLEE, W. C. Some measurable effects of social organization in flocks of hens. *Physiol. Zool.*, 1944, **17**, 320–347.

GUHL, A. M., COLLIAS, N. E. and ALLEE, W. C. Mating behavior and the social hierarchy in small flocks of White Leghorns. *Physiol. Zool.*, 1945, **18**, 365–390.

GUHL, A. M., CRAIG, J. V. and MUELLER, C. D. Selective breeding for aggressiveness in chickens. *Poult. Sci.*, 1960, **39**, 970–980.

GUHL, A. M. and WARREN, D. C. Number of offspring sired by cockerels related to social dominance in chickens. *Poult. Sci.*, 1946, **25**, 460–472.

MASLOW, A. H. Dominance quality and social behavior in infra-human primates. *J. soc. Psychol.*, 1934, **11**, 313–324. (a).

MASLOW, A. H. Dominance and social behavior in monkeys. *Psychol. Bull.*, 1934, **31**, 688. (b).

SELYE, H. The general-adaptive-syndrome and the diseases of adaptation. In, *Testbook of endocrinology*. Montreal: Montreal University, 1947. Pp. 837–866.

SIEGEL, H. S. The relation between crowding and weight of adrenal glands in chickens. *Ecology*, 1959, **40**, 495–498.

SIEGEL, H. S. Effect of population density on the pituitary-adrenal cortical axis of cockerels. *Poult. Sci.*, 1960, **39**, 500–510.

SIEGEL, H. S. and SIEGEL, P. B. The relationship of social competition with endocrine weights and activity in male chickens. *Anim. Behav.*, 1961, **9**, 151–158.

SIEGEL, P. B. and HURST, D. C. Social interactions among females in dubbed and undubbed flocks. *Poult. Sci.*, 1962, **41**, 141–145.

WILLIAMS, C. and McGIBBON, W. H. An analysis of the peck-order of the female domestic fowl, *Gallus domesticus*. *Poult. Sci.*, 1956, **35**, 969–976.

GROUP SOCIAL PATTERNS AS INFLUENCED BY REMOVAL AND LATER REINTRODUCTION OF THE DOMINANT MALE RHESUS*

Irwin S. Bernstein

CARPENTER (1942) reported that, when the dominant male of the supreme rhesus group on Cayo Santiago was removed, the group lost status relative to the other groups on the island. Three weeks later the male was returned, was immediately accepted as the dominant male in his original troop, and his troop immediately regained its former status. Both processes seem to indicate recognition of the individual since no fighting was required to reestablish the previous order. Kawai (1960), moreover, has reported the reorganizations which occurred in the status hierarchy in troops of Japanese macaques during the period following removal of leader males and indicates a complicated pattern of readjustment, which must be negated to reestablish the former order.

The present experiment was concerned with the changes in activity patterns among the remaining males after the removal of the dominant male from a rhesus group and with the effects subsequent to the reintroduction of this male.

METHOD

SUBJECTS. Nine rhesus monkeys were used in this experiment. The group consisted of the dominant male (26 lb), a second male (20 lb), an adolescent male (15 lb), a juvenile male (9 lb), four adult females (12–15 lb) and a juvenile female (6 lb). The group had been established several months prior to the start of the present experiment. The juvenile female was removed for health reasons prior to the return of the dominant male.

APPARATUS. A 24-ft. by 48-ft. by 8-ft. compound constructed of 2-in. by 4-in. welded wire over a wood frame served to house the group. A system of platforms and runways 6 ft. above the floor extended throughout the compound. An observation post was used for all recording sessions. Details of the apparatus appear in Bernstein and Mason (1963).

PROCEDURE

Two observers working together used a standardized social activity vocabulary to record the behavior of the dominant male for 2 hours in the morning and 3 hours in the afternoon on 17 days between October 22 and November 30. During 5 days of the following week the social activities of the other three males in the group were recorded for 30 min. in the morning and 30 min. in the afternoon.

The dominant male was removed on December 10 and reintroduced on January 8. During this period the social activities of the remaining three males were scored on 17 days. After the reintroduction of the dominant male, data were recorded on the social activities of each male for 30 min. each day for 5 days.

A standard vocabulary was used and the time of onset and cessation of all activities was recorded. A voicewriter was used to record during periods of rapid interaction.

RESULTS

The hourly rates of selected activities for the males appears in Table 1. Specific responses have been combined into broader categories.

The most common social interaction was approaching and remaining in proximity (less than 3 ft.) with another animal. The next most

* From *Psychological Reports*, 1964, **14**, 3–10. Copyright 1964 by Southern University Press, Missoula, Montana. This research was supported by National Science Foundation Grant NSF G-22637, and in part by National Institutes of Health Grant H-5691.

common interaction was social grooming. Both of these responses lasted extended but variable periods of time and so duration scores were used in making comparisons between animals. Other social activities except for huddling were far less frequent and, when combined, totaled less than 5 per cent of the duration of proximity plus grooming scores. Most of the low frequency interactions, moreover, lasted only a few seconds per episode and so only frequency counts were used in comparing animals on these scores. Huddling [definitions from Bernstein and Draper (in press)], on the other hand, lasted for some time once initiated, but both frequency and duration was variable over days (apparently related to weather). Frequency of initiation of huddling was therefore used in comparing animals on this category.

REMOVAL OF THE DOMINANT MALE. The major effect of removal of the dominant male was to increase the frequencies of most social interactions in the remaining three males. The relative frequencies for activities of the three males remained the same with the juvenile male most active and the second adult least active.

The over-all response pattern of the second adult male did not change after the removal of the dominant male. The second male continued a general avoidance of all animals although it was less pronounced than when the dominant male was present. His status in the group remained low even though little aggression was directed toward him.

The juvenile male became even more active socially after the dominant male was removed, but also showed little change in response patterns or status.

The adolescent male, on the other hand, tended to modify his behavior toward the pattern shown by the dominant male. He was only moderately successful in this inasmuch as although his status rose rapidly, the responses of other animals toward him were not what they had been to the dominant male.

Proximity. The dominant male had approached others more often than he was approached by them with the exception of one adult female. The adolescent male showed the same pattern after the removal of the dominant male. The adolescent male also spent less time in proximity with other animals after the removal of the dominant male, but continued to be with others more frequently than had the dominant male. Further, whereas the dominant male seldom remained with other males, the adolescent male continued to spend considerable time with the other males.

The juvenile male, on the other hand, increased

TABLE 1. *Mean Frequencies per Hour for Group Males Before, During, and After Removal of the Dominant Male*

Categories	Dominant Male		Adolescent Male			Juvenile Male			Second Adult Male		
	Before Removal	After Return	Before Removal	During Removal	After Return	Before Removal	During Removal	After Return	Before Removal	During Removal	After Return
Total activity rate	9.4	36.5	11.8	15.9	5.2	18.3	29.3	4.0	2.2	4.2	0.8
Proximity*	1469	1246	5599	3905	686	1909	4298	1519	990	939	599
Grooms*	46	8	116	24	170	419	301	318	42	119	25
Is groomed*	230	460	299	215	7	179	163	56	27	78	95
Solicits grooming*	0.1	0.4	0	0.3	0	0	0.1	0	0	0	0
Masculine sex responses	2.8	17.8	2.4	5.3	0	10.4	20.3	0	0	0.6	0.4
Feminine sex responses	0.3	0	0.2	0.4	0	0.8	0.9	0.4	0	0.3	0
Aggressive role	3.3	6.0	2.6	5.0	1.6	1.8	0.5	0	0.2	0.8	0
Submissive role	0	0	0.6	0.3	1.2	1.0	1.3	0.4	0.6	1.1	0.2
Huddle and nonspecific contact	1.0	0	2.2	1.7	0.8	3.2	3.6	1.6	0.4	0.8	0

* Scores show time in seconds per hour for hourly rate. For proximity, each animal in proximity was scored separately, therefore, scores are "animal seconds" and may exceed 3600 per hour.

his proximity scores and approached and remained with all animals in the group for extended periods. The second adult male, however, maintained the lowest proximity score in the group both before and after the removal of the dominant male. He seldom approached other animals and usually withdrew when others approached him.

Grooming. The dominant male had groomed others only 20 per cent as long as they had groomed him. Almost all of his grooming partners had been adult females. After the removal of the dominant male the adolescent male decreased the frequency of his grooming and showed a pattern more similar to that of the dominant male. The adolescent male, however, included the other two males as grooming partners and was not groomed by the highest ranking female although he did groom her. This female was often in proximity with the adolescent male and it is difficult to say whether one was dominant over the other as they seldom showed agonistic behavior to one another, although when the dominant male had been present, the adolescent male had been subordinate to all adult females.

The juvenile male maintained his grooming pattern after the removal of the dominant male, grooming all animals frequently and persistently. The second adult male on the other hand, groomed with the two other males at times but seldom with any females and in general maintained a low grooming score.

Approaching another and then presenting for grooming was seldom seen. No animal ever did so to the dominant or adolescent males. The adolescent and juvenile males never showed the response until after the removal of the dominant male.

Agonistic Behavior. The status hierarchy was reflected in the direction of agonistic responses. After removal of the dominant male, the adolescent male supersede the adult females, except possibly one, in status.

The most striking change was the increase in frequency and intensity of aggressive responses by the adolescent male after the removal of the dominant male. The other males did not change appreciably.

The juvenile male was threatened more often than he threatened and the second adult male continued to be threatened by others.

Every animal in the group had fled before aggressive displays by the dominant male. After he was removed, every animal except the first female fled before aggressive displays by the adolescent male. On the other hand, whereas most animals had fear grimaced to the dominant male,

few did so to the adolescent male. Females had also enlisted aid from the dominant male in threatening others whereas they only rarely enlisted aid from the adolescent male.

Sexual Activities. All three males hip touched, mounted, and pelvic thrusted more often after the dominant male was removed from the group. The second adult male, however, restricted his sexual activity to the juvenile male. The juvenile male was the most active of the three males, and had even exceeded the dominant male in frequency before he had been removed from the group. Sexual presentations, however, remained unchanged except that the adolescent male had shown this response frequently to the dominant male.

Other Social Interactions. Play was seen only by the juvenile male who usually approached the juvenile female but played with the second adult male at least once.

Lipsmacking was seldom seen when not associated with grooming.

Huddling and nonspecific contact had seldom involved the dominant male, but the adolescent male continued to participate in the interactions; the juvenile male maintained the highest score in this category.

RETURN OF THE DOMINANT MALE. Upon his return the dominant male assumed his previous role. The other males resumed their former roles and the only changes from the pre-removal period were the increased aggressive and sexual activity of the dominant male and the depressed activity scores of the other males.

Immediately upon return to the group the dominant male raised his tail in the question mark position, typical of strutting dominant males, and lipsmacked to all animals in the group, vocalizing as he walked. Within 45 sec. of reentry, an adult female approached the dominant male who grimaced to her and approached the juvenile male, who fled. Shortly thereafter the dominant male threatened the juvenile male. The same adult female again approached and sex presented to the dominant male. Three minutes after reentry, the dominant male lipsmacked to the adolescent male who fear grimaced and sex presented. The dominant male approached and mounted. Afterward the adolescent male repeatedly threatened one of the adult females while the dominant male returned to the first female and was approached by a second. The adolescent male sex presented to the dominant male again. One adult female enlisted aid from the dominant male who responded by mounting her. Afterwards there ensued a period of alternate grooming by the female and mounting by the dominant male.

The dominant male continued to be socially active for the next week. He was particularly active mounting females, but also energetically threatened and chased the juvenile and adolescent males. On several occasions the adolescent was caught and briefly bitten. The whole group fear grimaced to the dominant male more frequently than formerly and also enlisted his aid more frequently. The second male lipsmacked frequently to the dominant male but otherwise avoided him.

DISCUSSION

It thus appears that either the dominant male was remembered as an individual after a period of 1 month, or that his posture and size immediately earned the position for him upon reentry. Kawai (1960) in forming a Japanese macaque group has noted that a large male introduced to a group of smaller animals assumes the highest status position without resort to physical fighting. Bernstein and Mason (1963) also report the immediate recognition of status in rhesus monkeys between animals of different size and sex classifications on initial introduction. A period of 30 days, however, does not preclude individual recognition in rhesus monkeys.

During the period of absence of the dominant male, the remaining males became more active socially. This perhaps suggests that the presence of the dominant male had previously inhibited the expression of many social activities in these animals much as Bernstein and Draper (in press) have demonstrated that juveniles are inhibited by the presence of adults in a rhesus group. Further, the behavior of the juvenile male assumed much of the character of the behavior of juveniles in the complete absence of adults (Bernstein and Draper, in press). Those activities showing the greatest increase were the same which were markedly higher in the all-juvenile group. This suggests perhaps that it is the presence of a dominant adult male in particular that inhibits the activities of juvenile rhesus monkeys.

The adolescent male rather than the other adult male appeared to assume the role of dominant male after the removal of the original dominant male. The second adult male had behaved previously as a social isolate whereas the adolescent male had been socially active despite his low status. When the dominant male was removed, the second adult male remained essentially an isolate, although more frequently interacting with group members, whereas the adolescent male promptly raised his status. This

finding is in accord with that of Kawai (1960) who reported an immature male in the center group assuming the role as leader male after the death of the original leader despite the presence of fully adult males in the peripheral group.

The activities of the adolescent male, however, only approximated those of an adult dominant male. Many response patterns were incomplete and contained elements characteristic of juvenile animals. Further, other group members did not always respond as they had to the dominant male, e.g., the adult females did not fear grimace, enlist aid in fighting, or sex present to the adolescent male although they did flee when he was aggressive. In these respects, the behavior of the adolescent male was similar to the pattern described by Bernstein and Draper (in press) for the highest status animal in an all-juvenile group.

The immediate resumption of the former order upon the reintroduction of the dominant male does not mean that no effect of removal could be seen. The dominant male appeared more active and the other males less active than formerly, probably indicating an increased tension or excitement. Further, the dominant male mounted the adolescent male shortly after entry, which Kawai (1960) interprets as an affirmation of status in Japanese macaques. It is less clear in meaning for rhesus monkeys in that reciprocal mounting between animals of clearly different status occurs (Bernstein and Sharpe, unpublished manuscript). In this case, however, the dominant male afterwards was repeatedly aggressive to the adolescent male, and, to a lesser extent, also to the juvenile male. This aggression may be responsible for the inhibition of social interactions on the part of these males when the dominant male was present in the group again.

Overview. The dominant male in a rhesus monkey group was removed from the group and returned after a period of 1 month. Prior to removal, during the period of removal, and after return, data were collected on the social activities of all males in the group.

Upon the removal of the dominant male, the social activities of the remaining males all increased. A socially inactive adult male, however, remained essentially peripheral to the group whereas a socially active adolescent male achieved highest status in the group. The activities of the latter male, however, although similar in many respects to the pattern previously shown by the dominant male, lacked certain elements considered typical in the behavior of dominant males. Other animals in the group did not respond to the activities of the adolescent male in the same fashion in which they had responded to the dominant male displaying the same behavior. A juvenile male became the most socially active animal, showing the same increases in activities which were characteristic of juveniles in the absence of adults (Bernstein and Draper, in press). Upon his return, the dominant male immediately assumed his previous role although he was

markedly more active socially on reintroduction than he had been prior to removal. Increased aggression was noted toward the adolescent and juvenile males both of which showed submissive behavior in each instance. The activities of all three subordinate males were markedly depressed during the time of increased social activity on the part of the dominant male.

REFERENCES

BERNSTEIN, I. S. and DRAPER, W. A. The behavior of juvenile rhesus monkeys in groups. *Animal Behav.*, in press.

BERNSTEIN, I. S. and MASON, W. A. Activity patterns of rhesus monkeys in a social group. *Animal Behav.*, 1963, **11**, 455–460.

CARPENTER, C. R. Societies of monkeys and apes. *Biol. Symposia*, 1942, **8**, 177–204.

KAWAI, M. A field experiment on the process of group formation in the Japanese monkey [*Macaca fuscata*] and the releasing of the group at Ohiryama. *Primates*, 1960, **2**(2), 181–253.

ROLE OF THE DOMINANT MALE RHESUS MONKEY IN RESPONSE TO EXTERNAL CHALLENGES TO THE GROUP*

Irwin S. Bernstein

DOMINANCE among macaques and baboons has been defined by priority to incentives (Carpenter, 1954; Maslow, 1936). It has become evident, however, that the dominant males in natural troops show consistent behavior patterns which go beyond simple preferential access to incentives. Studies of Japanese macaques have revealed a complex relationship between "leader" males and the other troop members (Imanishi, 1960). Baboon dominant males have been shown to protect the troop (Hall, 1962; Washburn and DeVore, 1961).

Inasmuch as these studies suggested important characteristics for dominant males in addition to preferential access to incentives, the present experiment was designed to investigate the responses of a dominant male rhesus in a series of laboratory tests designed to challenge the group, or individuals, in specified ways.

METHOD

Subjects

Seven feral rhesus monkeys were used in the present experiment. All animals had been living as a group for several months prior to participation in the present experiment. The *S*s were: the dominant male (25+ lb), three adult females (12–15 lb), a second adult male (20+ lb), an adolescent male (15 lb), and a juvenile male (10 lb).

Apparatus

The apparatus consisted of a 24 ft. by 48 ft. by 8 ft. compound constructed of 2 in. by 4 in. welded wire over a wooden framework. A network of runways 6 ft. above the ground extended throughout the compound. The two rear corners

* From the *Journal of Comparative and Physiological Psychology*, 1964, **57**, 404–406. Copyright 1964 by the American Psychological Association. This

research was supported by Grant NSF-6-22637 from the National Science Foundation.

of the compound were partially isolated from the rest of the area by welded wire partitions and could be entered only through a single doorway. From an observation post in front of the compound a 3-ft. cube cage, also constructed of welded wire over a wood frame, could be rolled 3 ft. into the compound, or withdrawn. A more detailed description of the compound appears in Bernstein and Mason (1963).

Procedure

TEST 1. The dominant male was removed from the group and placed in the 3-ft. cube cage. The cage was rolled into the compound and the responses of all Ss and the caged male were recorded for 500 sec. Then E entered the compound with a large net and walked a fixed pattern threatening the group. After 100 sec. E withdrew and remained behind a screen for the next 100 sec. This sequence of alternate entry and withdrawal was repeated 10 times. Two Os recorded all responses from the observation post during the first five sequences and from a position opposite the observation post for the last five sequences.

At the end of the series the dominant male was released into the group and 1 week later the test series was repeated.

TEST 2. The procedure of alternate entry and withdrawal by E used in Test 1 was repeated ten times with the full group. Due to the activities of the dominant male, it was not possible to walk the fixed pattern during all trials.

TEST 3. Each of the group in turn except for the dominant male was individually placed in the 3-ft. cube cage and exposed to the remainder of the group for 100 sec. during which time E threatened the caged animal with a stick. Males and females were used alternately and the responses of all Ss in the compound were recorded.

TEST 4. All of the group except the dominant male were removed from the compound. Males and females were alternately captured from a small cage and held near the wire of the compound by E for 100 sec. each. The responses of the dominant male were recorded.

TEST 5. With the dominant male alone in the compound E first threatened with the large net from outside the compound for 500 sec. and then entered the compound ten times, each trial lasting 100 sec.

TEST 6. Finally, a white rat was placed within the cube cage as a novel stimulus for the group. Approaches by all members of the group were

recorded and each 30 sec. for 30 min. the animal closest to the cage was indicated.

RESULTS

TEST 1. During the trials when the dominant male was caged and E was out of sight, the juvenile male approached the cage and was threatened by the dominant male. Two females approached the cage three times and one of them was also threatened. In contrast, when E was in the compound every S approached the caged dominant male repeatedly ($p < 0.001$, binomial probability). There were 47 episodes when Ss stood directly on the cage in addition to 34 others when Ss only approached the cage. Only once did the dominant male threaten any animal on these approaches whereas he threatened E repeatedly. (An accurate count was not possible owing to the position of the recording E.)

Further, when E was out of sight, the juvenile and adult males fear-grimaced to the dominant male and one female twice attempted to enlist the aid of the dominant male in threatening other Ss. Nothing of this nature occurred with E in the compound, but one female repeatedly reached into the cage and the dominant male reached out to two other Ss. Moreover, with E in the compound two females and the juvenile male crouched next to the cage 19 times whereas with E out of view only eight such episodes were recorded ($p < 0.03$, binomial probability). The dominant male also shook the cage ten times with E out of sight and two times when E was present ($p < 0.02$, binomial probability).

TEST 2. When E attempted to enter the compound with the full group present, the dominant male charged on eight trials forcing E from the compound, and threatened seven times. A female threatened five times and the second adult male threatened two times. The group usually remained close together with the dominant male between them and E. Only twice did the dominant male retreat from E whereas on three other occasions he neither charged nor retreated.

TEST 3. When each of the group members in turn was placed in the 3-ft. cube cage and harassed with a stick, the dominant male was the only S to approach the cage, although one female threatened the Os once. The dominant male threatened E 12 times when females were in the cage and once when the juvenile male was in the cage. In addition the dominant male charged the observation post reaching through the wire six times when females were in the cage. During one trial, he fear grimaced to the female in the cage. After each *S* was returned to the group the dominant male vocalized repeatedly and lip-smacked to the returned *S*.

TEST 4. When *S*s were held by *E*, the dominant male charged *E* threatening vigorously 20 times and threatened 26 other times. In addition, when the females and the juvenile male was being held, he reached through the wire and pulled at and otherwise attacked *E*; these attacks were sustained for 20 sec. or more, or repeated in rapid sequence. Except for the vocalizations of the second male, vocalization by the restrained animal produced strong aggressive behavior to *E* on the part of the dominant male.

TEST 5. When alone in the compound and threatened by *E*, the dominant male made seven charges but none of them carried him to within 10 ft. of *E*, and most carried less than 10 ft. Twenty-two threats were recorded but there were 13 times, including those following a charge, that the dominant male retreated to the farthest corner of the compound. One episode of teeth-grinding and nine yawns were recorded in addition to five vocalizations.

TEST 6. The dominant male was found to be closest to the white rat in 74 per cent of time units ($p < 0.001$, binomial probability). The juvenile male was closest during 24 per cent of time counts whereas other group members avoided the area or approached only with caution.

DISCUSSION

1. The dominant male in at least some rhesus monkey groups actively protects the group from external challenges by threatening or attacking the source of disturbance, or by maintaining a position between the source of disturbance and the group.

2. The dominant male may come to the aid of a group member in distress and his responses become more intense as the distress of the harassed member is increased. Vocalization by the distressed animal also appears to increase the intensity of response by the dominant male.

3. When the group is challenged and the dominant male is restricted in space, group members will approach and remain near the dominant male whereas ordinarily few animals approach his immediate vicinity.

4. The role of the dominant male, in at least some cases, clearly exceeds definitions relying on preferential access to incentives.

REFERENCES

BERNSTEIN, I. S. and MASON, W. A. Activity patterns of rhesus monkeys in a social group. *Anim. Behav.*, 1963, **11**, 455–460.

CARPENTER, C. R. Tentative generalizations on the grouping behavior of non-human primates. *Hum. Biol.*, 1954, **26**, 269–276.

HALL, K. R. L. The sexual agonistic and derived social behavior patterns of the wild chacma baboon (*Papio ursinus*). *Proc. Zool. Soc., Lond.*, 1962, **139**, 283–327.

IMANISHI, K. Social organization of subhuman primates in their native habitat. *Curr. Anthropol.*, 1960, **1**, 393–409.

MASLOW, A. H. The role of dominance in the social and sexual behavior of infra-human primates: Observations at Vilas Park Zoo. *J. genet Psychol.*, 1936, **48**, 261–279.

WASHBURN, S. L. and DEVORE, I. Social behavior of baboons and early man. In S. L. Washburn (Ed.), *Social life of early man*. Chicago: Aldine 1961. Pp. 91–105.

AUTHOR INDEX

SUBJECT INDEX